This volume provides a case study of a marine pollutant of exceptional potency: tributyltin (TBT). TBT compounds are extremely poisonous, and have been widely utilised as active ingredients in marine anti-fouling paint formulations to obtain increased fuel efficiencies during ship operations, and long lifetimes between repainting for maritime vessels and structures. However, its extreme toxicity has resulted in numerous adverse biological effects on non-target organisms. The environmental persistence of TBT ensures that such problems are likely to continue for some time to come.

This authoritative synthesis reviews the environmental chemistry and toxicological effects of TBT and its degradation products, and outlines the international response to control TBT. A wide variety of disciplines are brought together to illustrate the general principles, pathways and problems involved in identifying and quantifying an environmental toxin, elucidating deleterious biological consequences, and the legal framework that can invoke mitigation via regulation.

This text serves a dual purpose. Firstly, from a research perspective, it provides a benchmark for assessing environmental recovery and therefore has wide appeal for undergraduate courses in environmental science, chemistry, ecology and marine biology. Secondly, by depicting the evolution of environmental legislation, it forms a valuable sourcebook for environmental planners and serves as a 'successful' case study for undergraduate courses in environmental law, planning and science.

TRIBUTYLTIN: CASE STUDY OF AN ENVIRONMENTAL CONTAMINANT

CAMBRIDGE ENVIRONMENTAL CHEMISTRY SERIES

Tributyltin: case study of an environmental contaminant

Edited by

STEPHEN J. DE MORA

Département d'océanographie, Université du Québec à Rimouski

CAMBRIDGE
UNIVERSITY PRESS

CAMBRIDGE UNIVERSITY PRESS
Cambridge, New York, Melbourne, Madrid, Cape Town, Singapore, São Paulo, Delhi

Cambridge University Press
The Edinburgh Building, Cambridge CB2 8RU, UK

Published in the United States of America by Cambridge University Press, New York

www.cambridge.org
Information on this title: www.cambridge.org/9780521105125

First published 1996
This digitally printed version 2009

A catalogue record for this publication is available from the British Library

Library of Congress Cataloguing in Publication data

Tributyltin : case study of an environmental contaminant / edited by Stephen J. de Mora
    p. cm. – (Cambridge environmental chemistry series : 8)
  Includes index.
  ISBN 0 521 47046 3
  1. Tributyltin – Environmental aspects. 2. Tributyltin – Toxicology.
3. Water–Pollution – Law and legislation. I. De Mora, S. J. II. Series.
TD427.T73T75 1996
363.73´84 – dc20 96–1585 CIP

ISBN 978-0-521-47046-9 hardback
ISBN 978-0-521-10512-5 paperback

This book is dedicated to my dad.

Both his silent encouragement and gentle cajoling inspired me to set goals which seemingly daunting were nevertheless attained.

Contents

†Geoffrey Bryan died 17 September 1993.

Contributors

○ ○

Dr Claude Alzieu
IFREMER, Technopolis 40, 155 rue Jean-Jacques Rousseau, 92138 Issy les
Molineaux, France

Dr Graeme Batley
Centre for Advanced Analytical Chemistry, CSIRO Division of Coal and Energy
Technology, Private Mail Bag 7, Menai, NSW 2234, Australia

Mr R. F. Bennett
Denton Lodge, 11 Whitstable Road, Blean, Canterbury, Kent CT2 9EA

Dr Klaus Bosselmann
Faculty of Law, University of Auckland, Private Bag 92019, Auckland, New
Zealand

Dr Geoffrey W. Bryan†
NERC Plymouth Marine Laboratory, Citadel Hill, Plymouth, Devon PL1 2PB

Professor Stephen J. de Mora
Département d'océanographie, Université du Québec à Rimouski, 300 allée des
Ursulines, Rimouski, Québec, Canada G5L 3A1

Dr Peter E. Gibbs
NERC Plymouth Marine Laboratory, Citadel Hill, Plymouth, Devon PL1 2PB

Dr R. James Maguire
National Water Research Institute, Department of the Environment, Canada
Centre for Inland Waters, Burlington, Ontario, Canada L7R 4A6

Dr Ph. Quevauviller
European Commission, Measurements and Testing Programme (BCR), 200 Rue
de la Loi, B-1150 Brussels, Belgium

Dr Carol Stewart
Department of Chemistry, University of Auckland, Private Bag 92019, Auckland,
New Zealand

Preface

○ ○ ○ ○ ○ ○ ○ ○ ○ ○ ○ ○ ○ ○ ○ ○ ○ ○ ○

Tributyltin (TBT) compounds are extremely poisonous, an attribute that has seen them utilized as the active ingredient in marine anti-fouling paint formulations. The potency of TBT ensures good fuel efficiencies during ship operation and long lifetimes between repainting for both boats and structures used in mariculture. However, its extreme toxicity has resulted in numerous adverse biological effects on non-target organisms, of particular importance being the shell deformations observed in oysters and the masculinization of female marine snails. The environmental persistence of TBT ensures that such problems are likely to continue for some time to come. In the first instance, this book describes the manufacture and industrial applications of TBT compounds, reviews their distribution and behaviour in the environment, and summarises their deleterious effects on organisms and aquatic ecosystems.

The widespread use of these anti-fouling paints has meant that TBT has become a contaminant of global concern. Many countries have taken diverse steps to regulate the use of TBT-based products in order to protect coastal and fresh water ecosystems, together with their resources. The political response to this pollutant has been notably faster and more widespread globally than that shown towards other pollutants in the environment, such as lead, for example, despite the fact that public health was never threatened. The steps taken have not been without controversy, considering the economic advantages to shipping on the one hand and the ecological damage on the other hand. The diversity of opinion is reflected in the range of authors presenting material here. Hopefully this book portrays a balanced point of view reflecting the responsible actions that have been taken worldwide. The history of the international legislative response to this toxic contaminant is documented. Also included is a consideration of the efficacy, to date, of the recently imposed TBT controls. TBT usage is now restricted and as the flux of TBT into the environment has diminished lately, environmental concentrations are set to decrease, albeit slowly in some circumstances. Thus, the maximum extent of the distribution is likely to have been already achieved. This is probably also the case for biological damage, but population dynamics for some species may be such that recovery to 'pre-TBT' levels may be prolonged.

Thus, this book reviews the environmental chemistry and toxicological effects of TBT and its degradation products. Hand in hand with this, it outlines the international response to control TBT. These two facets serve a dual purpose. Firstly, from a research perspective, the text provides a benchmark for assessing environmental recovery and should have wide appeal for teaching in tertiary courses for environmental science and chemistry, as well as ecology and marine biology. Secondly, depicting an evolution of environmental legislation, the book is aimed at environmental planners in general but should be useful as a 'successful' case study for tertiary courses in environmental law, planning and science. The material is presented in such a way that this text will act as a general guide illustrating the principles, pathways and problems involved in identifying and quantifying an environmental toxin, elucidating deleterious biological consequences, and the process of mitigation and legislation to protect the environment.

Stephen de Mora
Rimouski

1

○ ○ ○ ○ ○ ○ ○ ○ ○ ○ ○ ○ ○ ○ ○ ○ ○ ○ ○

The tributyltin debate: ocean transportation versus seafood harvesting

Stephen J. de Mora

1.1   Introduction

Tributyltin is not a compound in its own right, but only a constituent part of molecules in this class of organotin substances. It comprises three *n*-butyl chains attached to a single tin atom via covalent Sn–C bonds. As the tin exists in the form Sn(IV), the moiety displays an univalent positive charge. Commercial products are typically available in the form of *bis*(tributyltin) oxide (known as TBTO), acetate (TBT–OAc), halides (TBTF, TBTCl) and as the co-polymer with methylmethacrylate (TBTM). Industrial applications of TBT compounds followed the recognition of its biocidal properties, first noted in the early 1950s (Bennett, 1983). Although present applications include molluscicides, stone preservation, and disinfectants, the most important usage remains in wood preservatives and anti-fouling paints (Blunden & Evans, 1990). The production and use of TBT compounds in these two contexts are reviewed here in Chapter 2. Several other organotin compounds are routinely utilised in agricultural biocides, catalysts and as heat and light stabilisers for PVC. Reviews of environmental behaviour of such substances can be found elsewhere (Blunden, Hobbs & Smith, 1984; Maguire, 1991).

Because TBT can act as a broad spectrum biocide, it is useful in the preservation of wood, which is subject to decomposition via bacteria, fungi and insects. Unprotected soft woods degrade relatively quickly giving useful lifetimes of only one to two years. TBT treatment can extend this time to decades. TBT formulations have also proved useful for the preservation of marine timbers (Hill & Killmeyer, 1988). Although there are isolated exceptions, such as the deliberate discharge of 500–800 l of solvent containing approximately 40% TBT (de Mora & Phillips, 1992), wood treatment processes are relatively unimportant as a means of entry into the environment.

For the aquatic environment, tin-based anti-fouling paint represents the most important source of TBT. For a given location, the highest concentrations of TBT are generally found close to drydock and shipyard facilities where TBT-based paints are stripped from ships and yachts (de Mora, Stewart & Phillips, 1995; Stewart & de Mora, 1992; Waldock & Thain, 1988). However, anti-foulant paints work by the slow release into the water of a toxin that forms a thin veneer around the hull, thereby repelling nuisance organisms. Thus, TBT continually leaches from vessels utilising this anti-fouling agent.

As depicted in Figure 1.1, there are three main types of TBT paints (Champ & Pugh, 1987): contact leaching (conventional or insoluble matrix), ablative (soluble matrix), and self-polishing co-polymer (SPC). In conventional paints, relatively elevated amounts of TBT are incorporated into an insoluble matrix. The initial rate of TBT release tends to be high, but diffusion controlled release through microchannels from deeper within the matrix causes an exponential decrease of TBT loss with time. Leaching is further attenuated as the pores become clogged with carbonate, limiting the effective lifetime to less than two years and generally leaving a significant amount of TBT-based paint that is ineffective and must be removed prior to repainting. Ablative paints hold the TBT in a sparingly soluble matrix and TBT is again lost via diffusion. Periodic sloughing of the TBT-depleted matrix gives rise to erratic efficacy but extended lifetime, typically up to two years, with respect to anti-fouling performance.

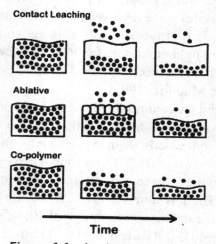

Figure 1.1 A schematic diagram of the three types of TBT-based anti-fouling paints highlighting the ageing characteristics and the mode of action.

Both of these types of paint are known as 'free association' paints as they incorporate relatively high concentrations of TBT compounds in dispersion and leaching is controlled via diffusion. In contrast, modern anti-fouling paints utilise TBT co-polymerised with methylmethacrylate (TBTM). A relatively slow and constant release rate of TBT throughout the lifetime of the paint is achieved by the gradual alkaline hydrolysis of the surface. The term self-polishing is applied as the hydrolysis continually erodes the surface exposing fresh TBTM. SPC paints contain lower concentrations of TBT than found in free association paints and the effective lifetime of the anti-foulant can be extended to five years.

First gaining EPA registration in 1978, the use of TBT-based SPC paints has increased dramatically. In 1991, 80% of ships greater than 4000 DWT utilised this anti-foulant. For pleasure craft, an added attraction of colourless TBT was that pigments could be incorporated into the paint giving rise to various colours not previously possible with copper-based coatings. The economic benefits of such paints, particularly for large oceangoing vessels should not be understated. Financial savings through enhanced fuel efficiency, reduced maintenance requirements and extended drydocking intervals amount to nearly $US3 billion annually for the world marine fleet (Anon., 1992).

Thus, there are several advantages to TBT substances that have consequently seen widespread use as the active ingredient in anti-fouling paints. This serves as the prime mode of entry into the aquatic environment. However, TBT compounds have marked biocidal properties towards many organisms. Global attention was focussed on TBT when it was suggested that non-target organisms, particularly marine snails (Smith, 1981) and oysters (Alzieu *et al.*, 1981), were extremely sensitive and adversely affected by this anti-fouling agent. The conflicting interests of ocean transportation and sea harvesting has brought the utilisation of TBT-based products under close scrutiny.

1.2   Environmental chemistry

The speciation of organotin in the natural environment is complicated owing to the large number of possible species. Considering simply the butylated forms, $(C_4H_9)_x Sn$, all species for $X = 1$ to 4 have been found in the environment (Kuballa *et al.*, 1995). Methylation leads to the formation of mixed $Bu_m SnMe_n^{(4-m-n)+}$ species (Maguire, 1984; Rapsomanikis & Harrison, 1988), but such species are apparently of little environmental significance. Although (hydroxybutyl)tin compounds are formed via biodegradation reactions (Lee, 1985; Lee, Valkirs & Seligman, 1989), their presence in the environment has yet to be ascertained. Free TBT ions do not exist in solution to any significant extent, but rather TBT exists in solution as the *bis*(tributyltin) oxide, the chloride or the carbonate, depending upon the dissolved CO_2 concentration, pH and salinity of the waters (Laughlin, Guard & Coleman, 1986). The solubility, estimated to fall in the range of $1-10 \, \text{mg} \, l^{-1}$ as TBTO in sea water, is not likely to be exceeded (Blunden *et al.*, 1984).

The fate of tributyltin species in the water column is determined by a range of biogeochemical processes. The principal abiotic removal mechanisms are volatilisation, photolysis and adsorption onto suspended particles, in this case followed by sedimentation. Biotic pathways include biosorption, uptake and biodegradation. Degradative processes have been the subject of several reviews (Blunden & Chapman, 1982; Cooney, 1988; Maguire, 1987; Stewart & de Mora, 1990).

Volatilisation is not likely to be an important loss mechanism. The vapour pressure of TBTO is quite low, with estimates ranging from 6.4×10^{-7} to 1.2×10^{-4} mm Hg at 20 °C (Blunden *et al.*, 1984). Several laboratory-based experiments have failed to observe losses from fresh waters by this means over several months (Maguire, Carey & Hale, 1983; Maguire & Tkacz, 1985). Results from a marine mesocosm experiment suggested that volatilisation of TBT may have been a significant sink, but the influence of photolysis of TBT in the microlayer could not be ruled out (Adelman, Hinga & Pilson, 1990). As documented in Chapter 5, TBT does seem to accumulate at the sea–air interface. This behaviour is greatly facilitated in incubation experiments by the presence of phytoplankton cultures (St-Louis, 1995). This study also demonstrated that bubble formation could enhance the transfer of TBT to the atmosphere.

Photolysis can certainly be an important removal process for butyltin compounds (Blunden & Chapman, 1982). Typical Sn–C bond cleavage energies are about $210 \, \text{kJ} \, \text{mol}^{-1}$; c.f. the energy of blue light of 400 nm is approximately $300 \, \text{kJ} \, \text{einstein}^{-1}$. All experiments of the degradation of TBT in water–sediment systems have demonstrated faster rates in the light than in the dark (Maguire *et al.*, 1983; Maguire & Tkacz, 1985; Seligman, Valkirs & Lee, 1986; Watanabe, Sakai & Takatsuki, 1992). However, the photolytic mechanism may be biologically mediated as the degradative products under light conditions include (hydroxybutyl)tins in addition to DBT, observed as the sole yield in the dark (Lee *et al.*, 1989). The Devonport Drydock facility operated by the New Zealand Royal Navy has made use of TBT photochemistry. The TBT-contaminated sludge accumulated during the removal of paint from ships cannot be released directly in Waitemata Harbour near Auckland. Instead, H_2O_2 is added, and the material is recirculated in a tank exposed to sunlight until TBT levels have declined to acceptable levels.

Several studies have demonstrated the importance of adsorption of tributyltin onto suspended material. Adsorption is due both to the hydrophobic character of the moiety and, to a lesser extent, its univalent charge. The bulky nature of the species would also provide a contribution from the entropy effect of displacing several water molecules. Laboratory-based adsorption experiments have shown that 72–100% of the TBT burden is adsorbed (Randall & Weber, 1986). TBT adsorption onto inorganic materials (bentonite, humic acid, ferric hydroxide and 'natural' suspended solids) followed a Langmuir adsorption isotherm (Leguille *et al.*, 1993). Perhaps surprisingly a second order rate law was obeyed. The activation energy was low, about $42\,kJ\,mol^{-1}$, and consistent with weak physical adsorption. A Freundlich adsorption model has also been used to explain the behaviour of TBT (Dowson, Bubb & Lester, 1993b). The adsorption of TBT is enhanced with a decrease in pH or with a decrease in salinity (Dowson *et al.*, 1993b; Randall & Weber, 1986). TBT adsorption was found to be reversible, which has implications for the mobility and partitioning in boundary zones, such as estuarine and interstitial waters (Unger, MacIntyre & Huggett, 1988).

The role of adsorption is important in the aquatic environment and serves to remove TBT from the water column relatively quickly. Such rapid kinetics can also have useful applications. About 81–92% of the TBT present in untreated waster water was associated with suspended solids (Fent & Muller, 1991). During treatment, adsorption onto sludge was found to be the most important elimination process for butylated tin compounds. Compared to the water column, degradation rates in the sediments are much slower, with half-lives on the order of years (de Mora, King & Miller, 1989; de Mora *et al.*, 1995; Dowson, Bubb & Lester, 1993a; Kilby & Batley, 1992). Sediments thus act as a relatively long term repository for TBT, and help maintain its persistence in the environment. The accumulation of TBT onto suspended material cannot be construed solely as a sink process because the TBT remains bioavailable, particularly for benthic organisms (Maguire & Tkacz, 1985) and filter feeders (Langston & Burt, 1991). Because the adsorption is relatively weak (Leguille *et al.*, 1993) and reversible (Dowson *et al.*, 1993b; Maguire & Tkacz, 1985), the sediments can act as a source of TBT should they be remobilised by storms or dredging. Some environmental evidence exists for localised contamination via desorption following dredging activities (Dowson, Bubb & Lester, 1992). Dredge spoil dump sites can certainly contribute TBT to an otherwise unaffected area (de Mora *et al.*, 1995).

Living and detrital cells can also act as a substrate for TBT adsorption. Microbial biofilms have been shown to accumulate TBT, without subsequent degradation (Blair *et al.*, 1988). This behaviour is not surprising given that the water–octanol partition coefficient for TBT is about 5000 (Laughlin *et al.*, 1986). However, a free ion activity model has been utilised to describe the uptake of TBT by the marine microflagellate *Pavlova lutheri* (St-Louis, 1995). Surface adsorption was essentially an ionic process, well described by a Freundlich isotherm. Accumulation was the result of subsequent lipophilic intracellular absorption. Exposure of starfish *Leptasterias polaris* to dissolved TBT promoted accumulation on the external tissues and subsequent translocation to internal organs was retarded (Rouleau, Pelletier & Tjälve, 1995). Passive uptake also seems to be important for fresh water fish (Wong *et al.*, 1994). Pike exhibited TBT concentrations that varied seasonally, with highest values in the spring. Concentrations were independent of fish size, thereby suggesting that a food chain accumulation mechanism was unlikely.

The biodegradation of TBT is well documented and it can proceed under a wide range of conditions. Because of the importance of TBT as a biocide, early studies focussed on its degradation in soils. Loss was accelerated in unsterilised soils relative to sterilised ones (Barug & Vonk, 1980) and biodegradation during bacterial monoculture incubation was soon demonstrated (Barug, 1981). Fungal cultures were also shown to be capable of degrading TBT, a finding of importance due to the use of TBT as a wood preservative (Orsler & Holland, 1982). The fresh water green alga *Ankistrodesmus falcatus* was able to accumulate and debutylate TBT to produce DBT predominantly, but with some MBT and inorganic Sn (Maguire, Wong & Rhamey, 1984). With respect to the marine environment, diatoms and dinoflagellates were observed to debutylate TBT under dark conditions (Lee *et al.*, 1989). Alternatively, DBT and (hydroxybutyl)tins were produced in the light. The marine diatom *Skeletonema costatum* was found to be effective for debutylation even at 4 °C (Reader & Pelletier, 1992). In eelgrass *Zostera marina*, TBT was decomposed via a consecutive debutylation pathway with first order kinetics (François, Short & Weber, 1989). Crabs, oysters and fish have demonstrated the ability to metabolise tributyltin, administered in both *in vitro* and *in vivo* experiments as *bis*(tributyltin) oxide (TBTO), forming a range of hydroxylated products (Lee, 1985). Such products had previously been shown to result from the reaction of TBT derivatives with the cytochrome *P*-450 monooxygenase enzyme system (Fish, Kimmel & Casida, 1976).

Considering sediments, the mechanism of microbial degradation under both aerobic and anaerobic conditions was observed to occur via debutylation (Shizhong, Chau & Liu, 1989). From the interpretation of profiles of butylated tin compounds in marine sediments, it has been suggested that the degradation is via methylation or debutylation, the specific pathway being dependent upon the dominant microbial activities in the sediment (Yonezawa *et al.*, 1994). The role of non-biological processes that might occur within sediments remains under dispute. The addition of cyanide was able to suppress degradation in water–sediment studies (Maguire & Tkacz, 1985). However, as noted above, biodegradation was not totally suppressed in sterilised soils (Barug & Vonk, 1980). Moreover, a rapid abiotic degradation has also been observed under experimental conditions (Stang, Lee & Seligman, 1992).

For a given environment, it is very difficult to estimate the relative importance of the various loss processes. However, their role can be inferred from the pattern of butyltin species preserved in the sediments. Considering in particular the sediments of Arcachon Bay (Sarradin *et al.*, 1991), sediments within the harbour exhibited high levels of TBT and the relative importance of MBT increased with distance from the TBT source. This behaviour was explained by differential rates of debutylation in waters versus sediments. Thus, much of the TBT initially released from small boats accumulated quickly via adsorption onto and deposition of particulate material near the source. As degradation rates within the sediments are slow, TBT was essentially preserved. In contrast, the TBT that was advected from the harbour into the adjacent bay, as either particulate or dissolved forms, underwent accelerated photolytic and microbial degradation while in the water column. As a result, comparatively more DBT and MBT was deposited in the bay than in the harbour.

1.3 Biological effects

The biological response of organisms to TBT varies enormously. TBT provokes a wide range of harmful effects (sublethal to mortal) to numerous organisms (bacteria to fish) at greatly differing scales (RNA damage to local extinctions). In this book, Chapters 4 and 6 survey such deleterious effects towards fresh water and marine organisms. Chapter 7 is devoted to the description of TBT-induced responses in marine snails. Hence, the intent here is to highlight briefly only two aspects of TBT pollution, namely the plight of oyster shellfisheries and the loss of some species of snails from coastal ecosystems, in order to unfold the case study of TBT as a contaminant of global concern.

In terms of public and political impact, the seminal TBT publication was that of Alzieu *et al.* (1981) ascribing shell deformation of Pacific oysters *Crassostrea gigas* to tributyltin derived from the anti-foulant in marine paints. As shown in Figure 1.2, the shells of contaminated oysters are severely thickened or 'balled' and contain multiple cavities or chambers, the number being a function of the TBT burden of the individual (King, Miller & de Mora, 1989). Despite some initial scepticism, the role of TBT in oyster shell deformation became well established following several experiments and field observations in both France and the United Kingdom (Alzieu *et al.*, 1986; Chagot *et al.*, 1990; Paul & Davies, 1986; Waldock & Thain, 1983). Shell calcification has also been observed for *Crassostrea gigas* in Australia (Batley *et al.*, 1989) and New Zealand (King *et al.*, 1989). Considering shell deformities in other oyster species, thickening was observed in the native rock oyster of New Zealand *Saccostrea glomerata* (Smith & Curtin, 1986) and shell curl has been

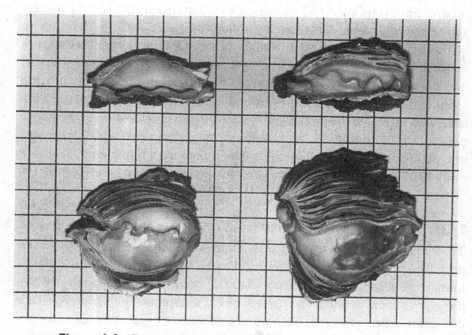

Figure 1.2 Examples of TBT-induced shell thickening in Pacific oysters *Crassostrea gigas* from New Zealand (de Mora *et al.*, 1989).

reported for the Sydney rock oysters *Saccostrea commercialis* (Batley *et al.*, 1989). As an obvious consequence of shell thickening, the volume of the body cavity is relatively small, resulting in unacceptable body weights for commercial exploitation. This feature led to a decline in the shellfisheries of oysters at sites in France and, in turn, prompted the first political action in the form of legislation controlling the use of TBT on small boats. As with the effect on oyster shells, this response was also global in character.

Imposex, a term coined by Smith (1971), in marine gastropods refers to the masculinization of females manifested by the formation of a penis and vas deferens. This mutation was subsequently shown to result from TBT (Smith, 1981). Marine snails are so sensitive to TBT that a no observed effect limit (NOEL) has yet to be confirmed. Certainly, the level is less than typical detection limits for present analytical techniques. Furthermore, it is of some concern that imposex has been observed in whelks *Buccinum undatum* from the open North Sea (Ten Hallers-Tjabbes, Kemp & Boon, 1994). In the extreme case, it is possible that the females are transformed into functioning males (Stewart *et al.*, 1992). At a community level, the reproductive failure prevents local recruitment to the site (Bryan *et al.*, 1986; Gibbs & Bryan 1986). This can cause populations to comprise ageing males and masculinised females. For neogastropods that lack a motile larval stage, this ultimately leads to local extinctions of populations as found for *Nucella lapillus* along parts of the English coastline (Gibbs, Bryan & Pascoe, 1991). Imposex has been observed in 45 species of neogastropods from locations throughout the world (Ellis & Pattisina, 1990), thereby providing a compelling example of the global nature of TBT as a pollutant.

Given that TBT has been extensively used because of its biocidal properties, it is of interest to note that increasingly there are studies characterising TBT-resistant bacteria. Such microorganisms have been isolated from sea water (Suzuki, Fukagawa & Takama, 1992) and from sediments underlying both fish and estuarine waters (Wuertz *et al.*, 1991a). As described previously, several marine phytoplankton demonstrate the ability to degrade TBT to DBT and (hydroxybutyl)tin species without apparent harm to themselves (Lee *et al.*, 1989; Reader & Pelletier, 1992). Similarly, higher organisms can metabolise TBT (Lee, 1985). One can speculate that such biological processes will play an important role in the long term decontamination of TBT polluted environments.

The toxic effects of TBT to humans have not been well documented. Occupational exposure in various incidents has led to dermatitis, and the irritation of eyes and the respiratory tract (Blunden & Evans, 1990).

However, worker protection during spray paint applications can be ensured by using suitable disposable clothing and respirators. Non-occupational exposure to TBT is slight, and possible pathways are via skin absorption or food intake. A wide range of household commodities was analysed in order to evaluate exposure via skin contact, and TBT was undetected in all 95 items (Yamada *et al.*, 1993). Considering foodstuffs, a recent survey of wines in Canada found TBT to be below detection (i.e. $<0.1 \, \mu g \, l^{-1}$) in most wines, but a maximum concentration of $18.7 \, \mu g \, l^{-1}$ was recorded (Forsyth, Weber & Dalglish, 1992). Higher concentrations were observed for DBT, which was believed to have been leached from PVC. TBT content in commercially available salmon ranged from 0.081 to $0.20 \, \mu g \, g^{-1}$ (Short & Thrower, 1986). Sea urchins (*Tripneustes gratilla*, local name *Cowaki*) and marine bivalves (*Anadara scapha*, local name *Kaikoso*), purchased at the Suva markets in Fiji, have also been analysed for TBT content (Stewart & de Mora, 1992). The respective concentrations were <15 and $90 \, ng \, TBT–Sn \, g^{-1}$ (dry weight). Hence, such levels in seafood pose no threat to public health given that an acceptable daily intake (ADI) for TBT has been suggested to be $6 \, \mu g \, kg^{-1} \, d^{-1}$ (O'Grady & Lawrence, 1988).

1.4   Legislative response

The dilemma of tributyltin was succinctly enunciated by Goldberg (1986) who noted that two superlatives apply. Firstly, TBT is an extremely effective anti-fouling agent that saves the US navy, among others, an estimated $US150 million annually. Secondly, TBT is a potent toxin only of anthropogenic origin that threatens seafood resources in coastal environments. The political response to the observation of TBT-induced damage to oysters in France was swift, but cautious. Swift because controls in France were established in 1982, based on the circumstantial evidence of Alzieu *et al.* (1981). Cautious because initially only a temporary ban was imposed on the use of anti-fouling paints containing more than 3% organotin on non-aluminium boats less than 25 m in length in ports along the French Atlantic coast. This was subsequently extended to become a permanent ban for all organotin-based paints at all coastal sites in France.

The database on which the first restrictions were imposed was perhaps limited (Huggett *et al.*, 1992; Stebbing 1985). Nevertheless, the United Kingdom also introduced regulations in 1985, setting a water quality target of $20 \, ng \, TBT–Sn \, l^{-1}$ and prohibiting the sale of paint containing more than 7.5% total tin in co-polymer paints or more than 2.5% total tin in free association paints, effectively removing the later from the market. This initial British approach was more conservative than the French strategy and failed to achieve the desired aim of protecting marine ecosystems. The water quality target was modified to only 2 ng TBT–Sn l^{-1} in 1987 and a French-style ban was enforced on the use of TBT-based anti-foulants for small craft. On a state by state basis, controls based on the French model were imposed in the USA starting from 1987. Thereafter, several countries worldwide enacted legislation to regulate the utilisation of TBT in anti-fouling paints (i.e. the continued application of TBT in wood preservatives has generally been permitted). Various control strategies have been utilised to restrict the introduction of TBT into the aquatic environment:

- registration of TBT as a pesticide
- limiting the TBT content in paint
- regulating the sale of TBT-based paint
- prohibiting the use of TBT on small vessels (partial ban)
- completely banning the use of organotin compounds as anti-foulants (total ban).

Serendipity plays a role in environmental protection. This can be eloquently illustrated with a couple of examples from the South Pacific. Firstly, New Zealand acts as the principal trading partner of the island nations in the south west Pacific Ocean and, hence, is the main supplier for most commodities. Once New Zealand restricted the use of TBT-based anti-foulant paints in 1989, manufacturers quickly reverted to the production of copper-based marine paints (marketed under the 'environmentally friendly' slogan on account of being tin-free). As a result and without national legislative controls, they became the only paints readily available in countries such as the Cooks Islands. Secondly, copper is known to be a potent toxin to marine organisms (Lewis & Cave, 1982) and Goldberg (1992) has recently cited copper fouling of oysters in Taiwan as representative of recent marine pollution. The change from TBT-based paints back to copper-based paints has often raised fears about the danger posed by this alternative biocide. Thus, it is with considerable interest to learn that, owing to the synergistic effect of TBT on the bioaccumulation of copper, the change to copper-based anti-foulants has led to a decrease in the copper content of Sydney rock oysters *Saccostrea commercialis* (Batley, Scammell & Brockbank, 1992).

The extreme position of banning the use of TBT-based anti-foulants altogether has been limited to landlocked countries (Austria and Switzerland) and New Zealand, a maritime nation but without an important role in shipping, shipbuilding or drydocking. The question remains as to whether or not this will be the ultimate fate of TBT-based anti-foulants globally. From a legislative perspective, future prohibitions hinge on the demonstrated success of less severe restrictions. From an economic point of view, marine fleet operators could be persuaded to change from TBT-based products given an equally cost effective alternative.

1.5   The efficacy of controls

Initial restrictions on the use of TBT had a clearly defined goal, namely the restoration of oyster shellfisheries at coastal sites. The apparently rapid recovery of oysters in Arcachon Bay on the French Atlantic coast (Alzieu et al., 1986), manifested by lower TBT body burdens and decreased incidence of shell deformation, did much to promote the French model of TBT legislation. Similar improvements have since been noted for oysters and mussels in British estuaries (Waite et al., 1991) and in the USA (Uhler et al., 1993). Moreover, dogwhelk populations (Nucella lapillus) have shown signs of improved reproductive health along the Northumbrian coast in the UK (Evans et al., 1994). There have also been several sedimentary studies demonstrating that the TBT flux to the environment has diminished. This can be evident as a significant decrease in the TBT concentration of surficial sediments (Wuertz et al., 1991b). Alternatively, sedimentary profiles at some sites now display a subsurface maximum in the TBT content (de Mora et al., 1995; Dowson et al., 1993a; Kilby & Batley, 1992).

However, ecosystem restoration has not been universal. Environmental quality targets were not met at 7 out of 17 locations along the east coast of the United Kingdom, in part ascribed to dredging activity (Dowson et al., 1992). Similarly, elevated TBT concentrations in sea water at sites in the Netherlands have recurred despite legislation (Ritsema & Laane, 1991). In this case, the continued but now illegal use of TBT has been suggested as the cause. Sedimentary TBT concentrations in a number of British estuaries exhibited no consistent trends in the first few years following controls (Waite et al., 1991). Shell thickening continues to be a problem for Pacific oysters in Poole Harbour, UK (Dyrynda, 1992). As described in Chapter 7, local extinctions of dogwhelks persist in the UK and recolonisation at the most severely TBT polluted localities is likely to take several years.

Ocean going ships continue to act as a source to TBT to the marine environment. This is especially true for freshly painted vessels (Waldock & Thain, 1988). Elevated TBT concentrations have been found in sediments near commercial wharves at several locations in the Mediterranean Sea (Gabrielides *et al.*, 1990), Suva Harbour in Fiji (Stewart & de Mora, 1992), Boston Harbour in the USA (Makkar, Kronick & Cooney, 1989) and Auckland, New Zealand (de Mora *et al.*, 1995). Taking into account the larger surface area but slower release rates, the TBT leached from a ship approximates that from 100 yachts. Given that the use of TBT-based paints on yachts has been declining dramatically, the relative importance of ships as the major contributor of TBT to the environment is increasing.

Assessing the environmental harm from TBT anti-fouled ships is rather difficult. Imposex has been observed in dogwhelks around an oil terminal in the Shetland Islands (Bailey & Davies, 1988). The only significant exposure for these organisms was the large tankers frequenting the harbour. Furthermore, the waters generally conformed to the water quality target of $2\,ng\,TBT-Sn\,l^{-1}$. More disturbing is the report of imposex in whelks from the open North Sea, at sites where this phenomenon was not observed 20 years ago (Ten Hallers-Tjabbes *et al.*, 1994). The widespread distribution of TBT and its breakdown products is becoming increasingly appreciated. Butyltin species have been detected in the blubber of several marine whales and seals (Iwata *et al.*, 1994) and have also been measured in sediments collected at a depth of 377 m in the Bellenas Basin of British Columbia, Canada (Stewart & Thompson, 1994). Whether such instances comprise sufficient threat to marine resources and ecosystems to justify further restrictions on the use of TBT remains debatable. There always remains the danger of administrative complacency as a consequence of the introduction of controls (Ellis, 1991).

1.6   The end of the beginning

According to Goldberg (1992), pollution and contamination both refer to a degradation of the environment. They are distinguished by their severity in that pollution induces the loss of resources, such as seafood. Furthermore, for a substance to be classified as a pollutant towards a particular organism, a cause–effect relationship must be clearly established. In this context, tributyltin definitely constitutes a contaminant in the marine environment. Moreover, its deleterious effects on marine molluscs verifies that it is also a pollutant in many instances. The potency of this toxin and the widespread use of tin-based anti-fouling paints have ensured that TBT-induced damage to aquatic ecosystems is a matter of global concern.

The TBT issue has largely escaped the emotive arguments that have plagued other environmental debates. The most devastating biological effects of TBT on non-target organisms seem to have been unleashed on marine snails. Clearly, they lacked the public appeal of, for instance, seals and whales. The outcry for their plight has been muted and generally limited to marine ecologists. On the other hand, oysters are popular and the economic decline of shellfisheries provoked prompt mitigation. Perhaps it is not surprising that legislative control of TBT-based anti-foulant paints was initiated in France, but several countries worldwide quickly followed suit. Such action is specially noteworthy in that firstly, the incriminating evidence was largely circumstantial and, secondly, public health was never threatened. However, on account of the importance of TBT to the shipping industry, the onus was on the scientific community to prove a cause–effect relationship. Initially, the legislative response was measured. Small craft were identified as the most likely source of TBT affecting oyster mariculture. Accordingly, TBT was banned from use only on boats less than 25 m in length. Regulations evolved with the scientific verification of TBT as the causitive agent for much biological damage to non-target organisms. Ecosystem recovery has since been observed, much to the benefit of both oysters (Alzieu *et al.*, 1989; Batley *et al.*, 1992; Waite *et al.*, 1991) and neogastropods (Evans *et al.*, 1991, 1994).

That lessons can be learned from the case study of TBT biocides is a recurring theme in the literature. Proving TBT induction of oyster shell calcification was a challenge that exposed weaknesses in marine ecotoxicology and prompted the plea for developing better simulation techniques (Stebbing, 1985). In contrasting the inexpensive ease of detecting imposex in marine snails to the costly difficulty of measuring butyltin concentrations, Ellis (1991) has presented a case for the use of bioindicator organisms for the rapid assessment of ecosystem health. Given their extreme sensitivity towards TBT, they also serve as an excellent gauge for environmental remediation. The lack of a predictive capability regarding the behaviour and possible ecosystem damage of substances newly introduced into the environment is an ongoing concern. However, TBT constitutes a classic example of a marine pollution episode (an organometallic constituent adversely affecting mariculture in the coastal environment) that can be used as a guiding principle to anticipate potential problems in the future (Goldberg, 1992).

As a case study, TBT exemplifies a responsible approach taken in the identification and regulation of an environmental pollutant. This was a multi-faced problem. From an ecological perspective, the suspicion of adverse biological effects stimulated toxicological investigations of contaminants at concentrations perceived to be environmentally significant. Several organisms were examined for a wide range of responses, in some cases being sublethal for individuals but devastating at the community level. This in turn prompted technique development for the analysis of organotin speciation at ultratrace concentrations in numerous matrices, a problem exacerbated by the absence of suitable certified reference materials for technique validation. Only then could a reliable database of the environmental distribution be constructed and the behaviour of TBT be established. The legal task was to evolve controls for a substance of enormous economic benefit to shipping that could ruin maricultural activities and wreak havoc on ecosystems. Rather than imposing a total ban, legislation was designed to decrease the flux of TBT into the aquatic environment, particularly with respect to sensitive coastal zones used for seafood production. Such avenues are explored in the subsequent chapters.

The unrestricted use of TBT has ended, but challenges remain. TBT continues to be introduced into the environment from vessels not subject to restrictions. Furthermore, TBT in sediments poses a long term threat as it persists for decades and can be readily remobilised. Ongoing investigations are required to quantify the current flux, to monitor ecosystem recovery and to verify the effectiveness of the limited controls. Finally, the case study of TBT contamination serves to illustrate that the cornerstone of global stewardship is vigilance.

1.7 References

Adelman, D., Hinga, K. R. & Pilson, M. E. Q. 1990. Biogeochemistry of butyltins in an enclosed marine ecosystem. *Environmental Science and Technology*, **24**, 1027–32.

Alzieu, C., Héral, M., Thibaud, Y., Dardignac, M. J. & Feuillet, M. 1981. Influence des peintures antisalissures a base d'organostanniques sur la calcification de l'huitre- *Crassostrea gigas*. *Revue des Travaux des Pêches maritimes*, **45**, 100–16.

Alzieu, C., Sanjuan, J., Deltriel, J. P. & Borel, M. 1986. Tin contamination in Arcachon Bay: effects on oyster shell anomalies. *Marine Pollution Bulletin*, **17**, 494–8.

Alzieu, C., Sanjuan, J., Michel, P., Borel, M. & Dreno, J. P. 1989. Monitoring and assessment of butyltins in Atlantic coastal waters. *Marine Pollution Bulletin*, **20**, 22–6.

Anon. 1992. *TBT copolymer anti-fouling paints: the facts.* The ORTEP Association, Vlissingen, Netherlands, p. 16.

Bailey, S. K. & Davies, I. M. 1988. Tributyltin contamination around an oil terminal in Sullam Voe (Scotland). *Environmental Pollution*, **55**, 161–72.

Barug, D. 1981. Microbial degradation of *bis*(tributyltin) oxide. *Chemosphere*, **10**, 1145–54.

Barug, D. & Vonk, J. W. 1980. Studies on the degradation of *bis*(tributyltin) oxide in soil. *Pesticide Science*, **11**, 77–82.

Batley, G. E., Fuhua, C., Brockbank, C. I. & Flegg, K. J. 1989. Accumulation of tributyltin by the Sydney Rock oyster, *Saccostrea commercialis*. *Australian Journal of Marine and Freshwater Research*, **40**, 49–54.

Batley, G. E., Scammell, M. S. & Brockbank, C. I. 1992. The impact of the banning of tributyltin-based antifouling paints on the Sydney rock oyster, *Saccostrea commercialis*. *The Science of the Total Environment*, **122**, 301–14.

Bennett, R. F. 1983. Industrial development of organotin compounds. *Industrial Chemistry Bulletin*, **2**, 171–6.

Blair, W. R., Olson, G. J., Trout, T. K., Jewett, K. L. & Brinckman, F. E. 1988. Accumulation and fate of tributyltin species in microbial biofilms. *NISTIR Report* **88/3852**, 1668–72.

Blunden, S. J. & Chapman, A. H. 1982. The environmental degradation of organotin compounds – a review. *Environmental Technology Letters*, **3**, 267–72.

Blunden, S. J. & Evans, C. J. 1990. Organotin Compounds. In Hutzinger, O. (Ed) *The Handbook of Environmental Chemistry*. Springer-Verlag, Berlin, pp. 1–44.

Blunden, S. J. Hobbs, L. A. & Smith, P. J. 1984. The environmental chemistry of organotin compounds. In Bowem, H. S. M. (Ed) *Environmental Chemistry*. The Royal Society of Chemistry, London, pp. 49–77.

Bryan, G. W., Gibbs, P. W., Hummerstone, L. G. & Burt, G. R. 1986. The decline of the gastropod Nucella lapillus around the southwest coast of England: evidence for the effect of tributyltin from antifouling paints. *Journal of the Marine Biological Association of the United Kingdom*, **66**, 611–40.

Chagot, D., Alzieu, C., Sanjuan, J. & Grizel, H. 1990. Sublethal and histopathological effects of trace levels of tributyltin fluoride on adult oysters *Crassostrea gigas*. *Aquatic Living Resources*, **3**, 121–30.

Champ, M. A. & Pugh, W. L. 1987. Tributyltin antifouling paints: Introduction and overview. *Oceans '87 International Organotin Symposium*, **4**, 1296–308.

Cooney, J. J. 1988. Microbial transformations of tin and tin compounds. *Journal of Industrial Microbiology*, **3**, 195–204.

de Mora, S. J., King, N. G. & Miller, M. C. 1989. Tributyltin and total tin in marine sediments: profiles and the apparent rate of TBT degradation. *Environmental Technology Letters*, **10**, 901–8.

de Mora, S. J. & Phillips, D. R. 1992. *Report on the concentration of tri(n-butyl)tin in sediments from the Waikumete Stream and Henderson Creek following an accidental spill from a timber treatment works in west Auckland*. Auckland Uniservices Limited. Report no. 4374, p. 12.

de Mora, S. J., Stewart, C. & Phillips, D. 1995. Sources and rates of degradation of tri(*n*-butyl)tin in marine sediments near Auckland, New Zealand. *Marine Pollution Bulletin*, **30**, 50–7.

Dowson, P. H., Bubb, J. M. & Lester, J. N. 1992. Organotin distribution in sediments and waters of selected East Coast estuaries in the UK. *Marine Pollution Bulletin*, **24**, 492–8.

Dowson, P. H., Bubb, J. M. & Lester, J. N. 1993a. Depositional profiles and relationships between organotin compounds in freshwater and estuarine sediment cores. *Environmental Monitoring and Assessment*, **28**, 145–60.

Dowson, P. H., Bubb, J. M. & Lester, J. N. 1993b. A study of the partitioning and sorptive behaviour of butyltins in the aquatic environment. *Applied Organometallic Chemistry*, **7**, 623–33.

Dyrynda, E. A. 1992. Incidence of abnormal shell thickening in the Pacific oyster *Crassostrea gigas* in Poole Harbour (UK) subsequent to the 1987 TBT restrictions. *Marine Pollution Bulletin*, **24**, 156–63.

Ellis, D. V. 1991. New dangerous chemicals in the environment: lessons from TBT. *Marine Pollution Bulletin*, **22**, 8–10.

Ellis, D. V. & Pattisina, L. A. 1990. Widespread neogastropod imposex: a biological indicator of global TBT contamination? *Marine Pollution Bulletin*, **21**, 248–53.

Evans, S. M., Hawkins, S. T., Porter, J. & Samosir, A. M. 1994. Recovery of dogwhelk populations on the Isle of Cumbrae, Scotland, following legislation limiting the use of TBT as an antifoulant. *Marine Pollution Bulletin*, **28**, 15–17.

Evans, S. M., Hutton, A., Kendall, M. A. & Samosir, A. M. 1991. Recovery in populations of dogwhelks *Nucella lapillus*, suffering from imposex. *Marine Pollution Bulletin*, **22**, 331–3.

Fent, K. & Muller, M. D. 1991. Occurrence of organotins in municipal wastewater and sewage sludge and behaviour in a treatment plant. *Environmental Science and Technology*, **25**, 489–93.

Fish, B. H., Kimmel, E. C. & Casida, J. E. 1976. Organotin chemistry: reactions of tributyltin derivatives with a cytochrome P-450 dependent monooxygenase enzyme system. *Journal of Organometallic Chemistry*, **118**, 41–54.

Forsyth, D. S., Weber, D. & Dalglish, K. 1992. Survey of butyltin, cyclohexyltin, and phenyltin compounds in Canadian wines. *Journal of the Association of Official Analytical Chemists International*, **75**, 964–73.

François, R., Short, F. T. & Weber, J. H. 1989. Accumulation and persistence of tributyltin in eelgrass. *Environmental Science and Technology*, **23**, 191–6.

Gabrielides, G. P., Alzieu, C., Readman, J. W., Bacci, E., Aboul Dahab, O. & Salihoglu, I. 1990. MED POL survey of organotins in the Mediterranean. *Marine Pollution Bulletin*, **21**, 233–7.

Gibbs, P. E. & Bryan, G. W. 1986. Reproductive failure in populations of the dog whelk caused by imposex induced by tributyltin from antifouling paints. *Journal of the Marine Biological Association of the United Kingdom*, **66**, 767–77.

Gibbs, P. E., Bryan, G. W. & Pascoe, P. L. 1991. TBT-induced imposex in the dogwhelk, *Nucella lapillus*: geographic uniformity of the response and effects. *Marine Environmental Research*, **32**, 79–87.

Goldberg, E. D. 1986. TBT: an environmental dilemma. *Environment*, **28**, 17–44.

Goldberg, E. D. 1992. Marine metal pollutants: a small set. *Marine Pollution Bulletin*, **25**, 45–7.

Hill, R. & Killmeyer, A. J. 1988. Chemical and biological investigations of organotin compounds as wood preservatives. *Proceedings of the American Wood-Preservers' Association*, **84**, 1–16.

Huggett, R. J., Unger, M. A., Seligman, P. F. & Valkirs, A. O. 1992. The marine biocide tributyltin: assessing and managing the environmental risks. *Environmental Science and Technology*, **26**, 232–7.

Iwata, H., Tanabe, S., Miyazaki, N. & Tatsukawa, R. 1994. Detection of butyltin compound residues in the blubber of marine mammals. *Marine Pollution Bulletin*, **28**, 607–13.

Kilby, G. W. & Batley, G. E. 1992. Chemical indicators of sediment chronology. *Australian Journal of Marine and Freshwater Research*, **44**, 635–47.

King, N., Miller, M. C. & de Mora, S. J. 1989. Tributyltin levels for seawater, sediment and selected marine species in coastal Northland and Auckland, New Zealand. *New Zealand Journal of Marine and Freshwater Research*, **23**, 287–94.

Kuballa, J., Wilken, R. D., Jantzen, E., Kwan, K. K. & Chau, Y. K. 1995. Speciation and genotoxicty of butyltin compounds. *Analyst*, **120**, 667–73.

Langston, W. J. & Burt, G. R. 1991. Bioavailability and effects of sediment-bound TBT in deposit-feeding clams, *Scrobicularia plana*. *Marine Environmental Research*, **32**, 61–77.

Laughlin, R. B., Guard, H. E. & Coleman, W. M. 1986. Tributyltin in sea water: speciation and octanol-water partition coefficient. *Environmental Science and Technology*, **20**, 201–4.

Lee, R. F. 1985. Metabolism of tributyltin oxide by crabs, oysters and fish. *Marine Environmental Research*, **17**, 145–8.

Lee, R. F., Valkirs, A. O. & Seligman, P. F. 1989. Importance of microalgae in the biodegradation of tributyltin in estuarine waters. *Environmental Science and Technology*, **23**, 1515–18.

Leguille, F., Castetbon, A., Astruc, M. & Pinel, R. 1993. Study of the tributyltin water–solid partitioning. *Environmental Technology*, **14**, 949–55.

Lewis, A. G. & Cave, W. R. 1982. The biological importance of copper in oceans and estuaries. *Oceanography and Marine Biology Annual Reviews*, **20**, 471–695.

Maguire, R. J. 1984. Butyltin compounds and inorganic tin in sediments in Ontario. *Environmental Science and Technology*, **18**, 291–4.

Maguire, R. J. 1987. Environmental aspects of tributyltin. *Applied Organometallic Chemistry*, **1**, 475–98.

Maguire, R. J. 1991. Aquatic environmental aspects of non-pesticidal organotin compounds. *Water Pollution Research Journal of Canada*, **26**, 243–360.

Maguire, R. J., Carey, J. H. & Hale, E. J. 1983. Degradation of tri-n-butyltin species in water. *Journal of Agricultural and Food Chemistry*, **31**, 1060–5.

Maguire, R. J. & Tkacz, R. J. 1985. Degradation of the tri-n-butyltin species in water and sediment from Toronto Harbor. *Journal of Agricultural and Food Chemistry*, **33**, 947–53.

Maguire, R. J., Wong, P. T. S. & Rhamey, J. S. 1984. Accumulation and metabolism of tri-*n*-butyltin cation by a green alga, *Ankistrodesmus falcatus*. *Canadian Journal of Fisheries and Aquatic Science*, **41**, 537–40.

Makkar, N. S., Kronick, A. T. & Cooney, J. J. 1989. Butyltins in sediments from Boston Harbour, USA. *Chemosphere*, **18**, 2043–50.

O'Grady, J. & Lawrence, J. 1988. *Report of the working party reviewing the use of antifoulants containing organotins in New Zealand*. Ministry for the Environment (New Zealand), Wellington, p. 112.

Orsler, R. J. & Holland, G. E. 1982. Degradation of tributyltin by fungal culture filtrates. *International Biodeterioration Bulletin*, **18**, 95–8.

Paul, J. D. & Davies, I. M. 1986. Effects of copper- and tin-based anti-fouling compounds on the growth of scallops (*Pectin maximus*) and oysters (*Crassostrea gigas*). *Aquaculture*, **54**, 191–203.

Randall, L. & Weber, J. H. 1986. Adsorptive behaviour of butyltin compounds under simulated estuarine conditions. *The Science of the Total Environment*, **57**, 191–203.

Rapsomanikis, S. & Harrison, R. M. 1988. Speciation of butyltin compounds in oyster samples. *Applied Organometallic Chemistry*, **2**, 151–7.

Reader, S. & Pelletier, E. 1992. Biosorption and degradation of butyltin compounds by the marine diatom *Skeletonema costatum* and the associated bacterial community at low temperature. *Bulletin of Environmental Contamination and Toxicology*, **48**, 599–607.

Ritsema, R. & Laane, R. W. P. M. 1991. Dissolved butyltins in fresh and marine waters of the Netherlands in 1989. *The Science of the Total Environment*, **105**, 149–56.

Rouleau, C., Pelletier, E. & Tjälve, H. 1995. Short-term bioconcentration and distribution of methylmercury, tributyltin and corresponding inorganic species in the starfish *Leptasterias polaris*. *Applied Organometallic Chemistry*, **9**, 327–34.

Sarradin, P. M., Astruc, A., Desauziers, V., Pinel, R. & Astruc, M. 1991. Butyltin pollution in surface sediments of Arcachon Bay after ten years of restricted use of TBT-based paints. *Environmental Technology*, **12**, 537–43.

Seligman, P. F., Valkirs, A. O. & Lee, R. F. 1986. Degradation of tributyltin in San Diego Bay, California, waters. *Environmental Science and Technology*, **20**, 1229–35.

Shizhong, T., Chau, Y. K. & Liu, D. 1989. Biodegradation of bis(tri-*n*-butyltin) oxide. *Applied Organometallic Chemistry*, **3**, 249–55.

Short, J. W. & Thrower, F. P. 1986. Accumulation of butyltins in muscle tissue of the chinook salmon reared in sea pens treated with tri-*n*-butyltin. *Marine Pollution Bulletin*, **17**, 542–5.

Smith, B. S. 1971. Sexuality in the American mud-snail, *Nassarius obsoletus* Say. *Proceedings of the Malacological Society of London*, **39**, 377–8.

Smith, B. S. 1981. Tributyltin compounds induce male characteristics on female mud snails *Nassarius obsoletus = Ilyanassa obsoleta*. *Journal of Applied Toxicology*, **1**, 141–4.

Smith, P. & Curtin, L. 1986. TBT paints and the health of the marine environment. *Shellfisheries Newsletter*, **32**, 10–11.

St-Louis, R. 1995, *Rôle du phytoplancton dans le devenir des organoétains à l'interface air–mer*. PhD, Université du Québec à Rimouski, p. 118.

Stang, P. M., Lee, R. F. & Seligman, P. F. 1992. Evidence for rapid, nonbiological degradation of tributyltin compounds in autoclaved and heat-treated fine-grained sediments. *Environmental Science and Technology*, **26**, 1382–7.

Stebbing, A. R. D. 1985. Organotins and water quality – some lessons to be learned. *Marine Pollution Bulletin*, **16**, 383–90.

Stewart, C. & de Mora, S. J. 1990. A review of the degradation of tri(*n*-butyl)tin in the marine environment. *Environmental Technology*, **11**, 565–70.

Stewart, C. & de Mora, S. J. 1992. Elevated tri(*n*-butyl)tin concentrations in shellfish and sediments from Suva Harbour, Fiji. *Applied Organometallic Chemistry*, **6**, 507–12.

Stewart, C., de Mora, S. J., Jones, M. R. L. & Miller, M. C. 1992. Imposex in New Zealand neogastropods. *Marine Pollution Bulletin*, **24**, 2204–9.

Stewart, C. & Thompson, J. A. J. 1994. Extensive butyltin contamination in southwestern coastal British Columbia, Canada. *Marine Pollution Bulletin*, **28**, 601–6.

Suzuki, S., Fukagawa, T. & Takama, K. 1992. Occurrence of tributyltin-tolerant bacteria in tributyltin- or cadmium-containing seawater. *Applied and Environmental Microbiology*, **58**, 3410–12.

Ten Hallers-Tjabbes, C. C., Kemp, J. F. & Boon, J. P. 1994. Imposex in whelks (*Buccinum undatum*) from the open North Sea: Relation to shipping traffic intensities. *Marine Pollution Bulletin*, **28**, 311–13.

Uhler, A. D., Durell, G. S., Steinhauer, W. G. & Spellacy, A. M. 1993. Tributyltin levels in bivalve mollusks from the east and west coasts of the United States: Results from the 1988–1990 national status and trends mussel watch project.

Environmental Toxicology and Chemistry, **12**, 139–53.

Unger, M. A., MacIntyre, W. G. & Huggett, R. J. 1988. Sorption behaviour of tributyltin on estuarine and freshwater sediments. *Environmental Toxicology and Chemistry*, **7**, 907–15.

Waite, M. E., Waldock, M. J., Thain, J. E., Smith, D. J. & Milton, S. M. 1991. Reductions in TBT concentrations in UK estuaries following legislation in 1986 and 1987. *Marine Environmental Research*, **32**, 89–111.

Waldock, M. J. & Thain, J. E. 1983. Shell thickening in *Crassostrea gigas*: organotin antifouling or sediment induced. *Marine Pollution Bulletin*, **14**, 411–15.

Waldock, M. J. & Thain, J. E. 1988. Inputs of TBT to the marine environment from shipping activity in the U.K. *Environmental Technology Letters*, **9**, 999–1010.

Watanabe, N., Sakai, S. & Takatsuki, H. 1992. Examination for degradation paths of butyltin compounds in natural waters. *Water Science Technology*, **25**, 117–24.

Wong, P. T. S., Chau, Y. K., Brown, M. & Whittle, D. M. 1994. Butyltin compounds in Severn Sound, Lake Huron, Canada. *Applied Organometallic Chemistry*, **8**, 385–91.

Wuertz, S., Miller, C. E., Pfister, R. M. & Cooney, J. J. 1991a. Tributyltin-resistant bacteria from estuarine and freshwater sediments. *Applied and Environmental Microbiology*, **57**, 2783–9.

Wuertz, S., Miller, C. E., Doolittle, M. M., Brennan, J. F. & Cooney, J. J. 1991b. Butyltins in estuarine sediments two years after tributyltin use was restricted. *Chemosphere*, **22**, 1113–20.

Yamada, S., Fujii, Y., Mikami, E., Kawamura, N., Hayakawa, J., Aoki, K., Fukaya, M. & Terao, C. 1993. Small-scale survey of organotin compounds in household commodities. *Journal of the Association of Official Analytical Chemists International*, **76**, 436–41.

Yonezawa, Y., Fukui, M., Yoshida, T., Ochi, A., Tanaka, T., Noguti, Y., Kowata, T., Sato, Y., Masunaga, S. & Urushigawa, Y. 1994. Degradation of tri-*n*-butyltin in Ise Bay sediment. *Chemosphere*, **29**, 1349–56.

2

○ ○ ○ ○ ○ ○ ○ ○ ○ ○ ○ ○ ○ ○ ○ ○ ○ ○ ○

Industrial manufacture and applications of tributyltin compounds

R. F. Bennett

2.1 Organotin compounds

Although organotin compounds were first identified in the mid nineteenth century, industrial development of these chemicals did not occur until nearly a century later. Over the past 40 years world production and usage of organotin compounds has grown to a level that is currently estimated to be at around 50 000 tonnes per annum. The major industrial application of organotin compounds is with the mono- and di- derivatives, which are used as heat and light stabilizers in PVC processing. Mono/dibutyltin derivatives find widespread application in pipe, sheeting and other rigid PVC manufacture. Octyltin thioglycolate derivatives are approved worldwide as stabilizers for PVC used in food packaging materials and in bottles used for potable liquids. Thio-organotin stabilizers have the unique advantage of producing crystal clear PVC.

Triorganotin compounds are biologically active. They represent a smaller tonnage market – between 15 and 20% of total organotin world manufacture. Triorganotin compounds have an important place in the agrochemical field as fungicides and miticides, with the production and use of about 5000 tonnes per annum. Triphenyltin compounds (the acetate and hydroxide) are of particular importance in potato crop blight control. There was usage of triphenyltin derivatives in anti-fouling paint manufacture during the period 1960–85. Such usage has virtually disappeared from all world markets. Trineophyltin and tricyclohexyltin derivatives are used as miticides/acaricides. Tributyltin compounds were first examined for their biological properties in the early 1950s at the Institute for Organic Chemistry TNO, Utrecht, Holland (van der Kerk & Luijten, 1954). They were first used commercially as replacements for biocides such as DDT, organomercurials and arsenical compounds. Tributyltin compounds are now predominantly used as active ingredients in the formulation of anti-fouling coatings for commercial shipping and to a smaller extent in solvent based, industrial wood preservation processes. World production of primary tributyltin compounds is estimated to be several thousand tonnes per annum. In commercial use, the primary tributyltin compounds are formulated in anti-fouling paints and wood preservative preparations at relatively low percentage levels.

2.2 Tributyltin compounds

The term 'tributyltin' and its acronym 'TBT' has in recent years quite incorrectly become accepted to represent a chemical compound in its own right. 'Tributyltin' is a chemical moiety which is a part of a tributyltin compound. *Bis*-tributyltin oxide, generally known as 'TBTO', is a colourless liquid of molecular weight 596. Tributyltin fluoride is a white crystalline solid with a molecular weight of 309. Tributyltinmethacrylate–methylmethacrylate copolymer is a solid resinous material with a molecular weight in excess of 40 000. These products are examples of tributyltin compounds each exhibiting different physical and chemical properties. The feature which is common to all such compounds is the tributyltin moiety, 'TBT'. In aqueous media, the 'TBT' moiety can be released by a hydrolysis reaction in which other tributyltin speciations are formed in a complex system of equilibrium reactions; in these systems, the tributyltin moiety 'TBT', as such, is not freely available as a cationic species.

Tributyltin compounds are organic derivatives of tetravalent tin, characterized by three covalent tin–carbon bonds. They form part of the large family of organotin compounds which encompass mono-, di-, tri- and tetraorganotin derivatives. Chemically the compounds are represented by the structures $RSnX_3$, R_2SnX_2, R_3SnX and R_4Sn where R is any alkyl or aryl group and X represents any organic or inorganic derivative, for example a carboxylate or a halide.

The first recorded synthesis of organotin compounds was by Löwig in 1852. In the following year Frankland completed a study of a series of organotin compounds. It was not until the late 1940s that commercial development and use of organotin compounds began in the USA. A detailed review of the history of the subsequent worldwide development has been published by the Royal Society of Chemistry (Bennett, 1983).

2.3 Synthesis of tributyltin compounds

It is normal industrial practice to combine manufacture of tributyltin compounds with tetrabutyltin, monobutyltin and diorganotin compounds in the same production facility. Organotin compounds can be synthesized by several methods – Grignard route, Würtz route, alkyl aluminium route and by direct synthesis.

2.3.1 The Grignard route

A Grignard reagent is formed by the interaction of magnesium with an alkyl halide in the presence of a complexing agent such as ether

$$Mg + RX \rightarrow RMgX$$

Tin tetrachloride can be reacted with the Grignard reagent to form the tetraalkyltin compound

$$4RMgX + SnCl_4 \rightarrow R_4Sn + 4MgXCl$$

Alkylation of tin tetrachloride with a Grignard reagent is difficult to control cleanly at a partial stage in order to produce either the trialkyltin, di- or monoalkyltin derivatives. Accordingly in commercial operations, the tetraalkyltin compound is produced first. The reaction is carried out at about 80 °C in a suitable solvent such as toluene. Tetraalkyltin compounds are used as the source material for the production of the other derivatives. The tri-, di- and monoalkyltin compounds are produced by a Kocheshkov disproportionation reaction. Tributyltin compounds that are manufactured by the Grignard route use butyl chloride as the source material from which a butyl Grignard compound is derived

$$Mg + BuCl \rightarrow BuMgCl$$

(for simplicity the butyl radical, C_4H_9, is shown as 'Bu'). Reaction with tin tetrachloride yields tetrabutyltin and magnesium chloride as a by-product

$$4BuMgCl + SnCl_4 \rightarrow Bu_4Sn + 4MgCl_2$$

Tributyltin chloride is produced by a Kocheshkov disproportionation reaction

$$3Bu_4Sn + SnCl_4 \rightarrow 4Bu_3SnCl$$

Tributyltin chloride is used as the starting material for production of various derivatives where other groups can be introduced by nucleophilic substitution of the chloride. The oxide, TBTO, is the major derivative produced commercially

$$2Bu_3SnCl + 2NaOH \rightarrow Bu_3Sn–O–SnBu_3 + 2NaCl + H_2O$$

Bis-tributyltin oxide, widely known by the acronym 'TBTO', is itself used by the chemical and paint industries for the production of other TBT derivatives.

Commercial production of organotin compounds by the Grignard route began in the USA at the Metal & Thermit Corporation's plant in Rahway, New Jersey, in the late 1940s. Industrial production was soon to follow in Europe and Japan. The Grignard route is additionally used for the commercial production of phenyltin compounds.

2.3.2 The Würtz route
Alkyl halides can be reacted with sodium and tin tetrachloride to form tetraalkyltin compounds. The reaction probably proceeds via the formation of an intermediate alkyl sodium compound which reacts *in situ* with tin tetrachloride. The reaction requires to be carried out in a large volume of solvent in order to suppress the conversion of the alkyl halide to the corresponding alkyl hydrocarbon. Tetrabutyltin has been manufactured commercially by the Würtz route according to the reaction

$$4BuCl + 8Na + SnCl_4 \rightarrow Bu_4Sn + 8NaCl$$

Disproportionation by the Kocheshkov method is required to yield the tributyltin and mono/dibutyltin chlorides.

Various modifications of the Würtz route have been reported in order to yield mixtures of tetra- and tributyltin compounds. The disadvantage of using very large solvent volumes, sodium metal, together with uncompetitive economics, have rendered the process obsolescent.

2.3.3 Alkyl aluminium route
Following the development of alkyl aluminium compounds by Ziegler in 1955, examination of their use in organotin synthesis demonstrated a suitable route for the manufacture of tetraalkyltin compounds

$$4R_3Al + 3SnCl_4 \rightarrow 3R_4Sn + 4AlCl_3$$

The aluminium chloride that is produced during the reaction can form complexes with trialkyltin chloride that is formed, thereby inhibiting further alkylation. However, in the presence of a complexing agent such as an ether or an amine, alkylation proceeds to completion. The production and handling of alkyl aluminium compounds is technically a highly sophisticated process. Large industrial facilities which have been installed combine the manufacture of alkyl aluminium compounds for use as Ziegler–Natta catalysts with their use as starting materials for organotin production.

The alkyl aluminium route for synthesizing organotin compounds has a number of significant advantages. Alkenes are used instead of alkyl halides as the basic starting material. Thus, butene is used in the production of butyltin compounds instead of butyl chloride which is required for the Grignard process. The butyl aluminium compound is prepared from aluminium, hydrogen and butene according to the reaction

$$Al + 1.5H_2 + 3C_4H_8 \rightarrow Al(C_4H_9)_3$$

The alkylation of tin tetrachloride with tributyl aluminium can be operated in a relatively small volume reaction vessel. The reaction can be operated as a continuous process and carried out without solvents.

A further significant advantage is that the alkyl aluminium reaction can be controlled to produce tributyltin compounds directly

$$Al(C_4H_9)_3 + SnCl_4 \rightarrow (C_4H_9)_3SnCl + AlCl_3$$

The capability to form tributyltin chloride by direct reaction avoids the necessity of using the Kocheshkov disproportionation. After hydrolysis of the tributyltin chloride–aluminium chloride complex, the chloride, TBTCl, can be converted by alkaline hydrolysis to form the oxide, TBTO

$$2Bu_3SnCl + 2NaOH(aq) \rightarrow Bu_3Sn-O-SnBu_3 + 2NaCl + H_2O$$

Production of organotin compounds by the alkyl aluminium route started in Germany at Schering AG, Industrial Chemicals Division, Bergkamen, in 1962. The Industrial Chemicals Division has since been acquired by the USA Witco Corporation in 1992. Production in the USA by the alkyl aluminium route commenced in 1981 at the Axis, Alabama, plant of M&T Chemicals. This company is now part of Elf Atochem, a French group of chemical companies.

2.3.4 Direct synthesis

Organotin halides can be synthesized directly from tin metal, tin alloys, or tin halides. Such a method was the basis of Frankland's preparation of diethyltin diiodide in 1853. Direct synthesis routes using alkyl iodides and bromides were developed in Japan in the early 1950s. In 1954, Sankyo Organic Chemicals Co. Ltd developed a sodium–tin alloy process reacting butyl bromide to produce a mixture of dibutyltin- and tributyltin bromide

$$BuBr + (Sn-Na) \rightarrow Bu_2SnBr_2 + Bu_3SnBr + NaBr$$

The order of reactivity of alkyl halides with tin by direct synthesis declines from the iodide to the chloride

$$RI > RBr > RCl$$

The iodide process is operated commercially in Japan.

For a given halogen, reactivity of the alkyl halide with tin declines with the increasing size of the alkyl group

Me > Et > Pr > Bu

Methyltin stabilizers are manufactured by direct synthesis in the USA. In direct synthesis reactions a suitable catalyst is required, typically a quaternary halide such as a phosphonium or ammonium compound.

A development of the direct synthesis route for the production of triorganotin compounds has been described by Holland (1987). An alkyl halide is reacted with tin metal dispersed in a molten quaternary halide to yield the trialkyltin halide and a by-product containing the quaternary salt as a tin halide complex. The quaternary halide, tin and halide ions are recovered from the by-product by electrolysis and reused. Synthesis of tributyltin chloride is described using butyl chloride as the starting material.

2.4   Industrial manufacture of tributyltin compounds

The manufacture of primary organotin compounds (characterized by the synthesis of the tin–carbon covalent bond) is undertaken in Europe (nine producers), Japan (nine producers), USA (six producers), and in Brazil, Korea and China. Additionally there are a number of secondary producers worldwide who purchase primary organotin intermediates such as dibutyltin oxide or dioctyltin oxide in order to produce their own in-house speciality PVC stabilizers. Production of tributyltin compounds is included within the organotin product ranges of several chemical companies in Europe, Japan, USA and Korea. Major international producers of TBT compounds are Acima Chemical Industries Ltd Inc, Switzerland; ElfAtochem North America Inc; ElfAtochem Vlissingen BV, Netherlands; Witco GmbH, Germany; Nitto Kasei Co. Ltd, Japan; and Song Woun Industrial Co. Ltd, Korea. Additionally there are a number of companies such as marine paint manufacturers, who purchase TBTO for the production of their own in-house speciality copolymer derivatives. A detailed list of individual products and producers is given in the *World Directory of Tin Chemical Suppliers* published by the International Tin Research Institute in 1994.

Primary tributyltin compounds are synthesized commercially via the Grignard or the alkyl aluminium route to yield tributyltin chloride. Since other functional groups can be introduced by nucleophilic substitution of the chloride, TBTCl is mainly used in industry as the starting material for the manufacture of the oxide, TBTO, which itself is used as the precursor for the manufacture of other TBT compounds.

2.4.1 Tributyltin oxide

Tributyltin oxide is produced by aqueous saponification of the chloride

$$2Bu_3SnCl + 2NaOH \rightarrow Bu_3Sn\text{–}O\text{–}SnBu_3 + 2NaCl + H_2O$$

Sodium chloride is removed from the process in the aqueous layer separation. *Bis*-tributyltin oxide is for simplicity usually referred to as 'tributyltin oxide' or 'TBTO'. The term 'TBT' which has been widely used in the media (particularly with reference to the environmental effects of TBT compounds) is ambiguous, since it does not characterize a specific TBT compound.

Decomposition of TBTO can occur quite rapidly either by thermal induction and/or photochemical reaction to yield a white precipitate of dibutyltin oxide, 'DBTO'

$$Bu_3Sn\text{–}O\text{–}SnBu_3 \rightarrow Bu_4Sn + Bu_2SnO$$

A glass container of TBTO that is left on a laboratory shelf and exposed to occasional sunlight will show white flecks of DBTO precipitating within days. In the aquatic environment TBT compounds are degraded primarily by micro-organisms and also by sunlight. Half-lives reported in the more recent literature are generally between a few days and a few weeks. The rate of degradation is temperature related.

Commercial grades of TBTO are produced to a 95/97% specification. The impurities present are tetrabutyltin, added stabilizer and dibutyltin derivatives. Higher grade purities are produced by vacuum rectification. The tin metal content of TBTO is just under 40%. TBTO is a colourless liquid with a faint typical odour. Its solubility in water is low, varying between 1 and $10\,\mathrm{mg\,l}^{-1}$ according to pH, temperature and salinity. TBTO is soluble in most organic solvents.

2.4.2 Tributyltinmethacrylate methylmethacrylate copolymers

The principal tributyltin compounds which are manufactured commercially are tributyltin methacrylate/methylmethacrylate copolymers – 'TBTM/MMA' compounds. Synthesis of the polymers is a complex process requiring sophisticated technological control. Although most TBTM/MMA copolymers which are commercially available are similar, they may vary with regard to tin content and with molecular weight. This will relate to the functional performance that is required of the polymer when used in specific anti-fouling paint coatings. Anti-fouling performance requirement varies according to ship size and design, and the climatic conditions of the oceans in which the ship is to operate.

Methacrylic acid, TBTO and methylmethacrylate are used as the starting materials for the production of TBTM/MMA copolymers. Methacrylic acid is a highly reactive compound, capable of esterification via the carboxylate group and polymerization via the carbon double bond

$$
\begin{array}{ccc}
H & & CH_3 \\
\diagdown & & \diagup \\
& C = C & \\
\diagup & & \diagdown \\
H & & C = O \\
& & \diagdown \\
& & OH
\end{array}
$$

In the esterification reaction TBTO combines with the carboxylate group to yield tributyltin methacrylate monomer 'TBTM' with water as a by-product

$$
\begin{array}{ccc}
H & & CH_3 \\
\diagdown & & \diagup \\
& C = C & \\
\diagup & & \diagdown \\
H & & C = O \\
& & \diagdown \\
& & O - Sn - Bu_3 + H_2O
\end{array}
$$

Methylmethacrylate monomer 'MMA' is highly reactive via the carbon double bond

$$
\begin{array}{ccc}
H & & CH_3 \\
\diagdown & & \diagup \\
& C = C & \\
\diagup & & \diagdown \\
H & & C = O \\
& & \diagdown \\
& & O - CH_3
\end{array}
$$

Monomeric TBTM and monomeric MMA can be reacted together to yield a high molecular weight TBTM/MMA copolymer

$$\left[\begin{array}{c} CH_3 \\ | \\ -CH_2-C- \\ | \\ C=O \\ | \\ O-Sn \\ {\diagup}{|}{\diagdown} \\ Bu\ \ Bu\ \ Bu \end{array} \right]_x \quad --- \quad \left[\begin{array}{c} CH_3 \\ | \\ -CH_2-C- \\ | \\ C=O \\ | \\ O-CH_3 \end{array} \right]_y \right]_n$$

The copolymerization reaction requires an initiator to introduce free radicals which attack the carbon double bonds. The activated double bonds in turn generate further free radicals in a chain reaction, to produce a continuous process of polymerization. The reaction is highly exothermic and reaction rates need to be carefully controlled. The reaction requires to be carried out in a suitable solvent medium. TBTM/MMA copolymer products are produced commercially in high aromatic organic solvents. The copolymer products usually contain about 50% solids. Molecular weights are in excess of 40 000.

2.4.3 Tributyltin naphthenate

Tributyltin naphthenate 'TBTN' is a yellowish brown liquid which is soluble in most organic solvents but only slightly soluble in water at about $1.5\,\text{mg}\,\text{l}^{-1}$. The product can be synthesized by reaction of sodium naphthenate and TBTCl yielding sodium chloride as a by-product. Alternatively naphthenic acids can be reacted directly with TBTO. Tributyltin naphthenate possesses a very low vapour pressure, $9 \times 10^{-5}\,\text{Pa}$ ($20\,^{\circ}\text{C}$), lower than that of TBTO, $1 \times 10^{-3}\,\text{Pa}$ ($20\,^{\circ}\text{C}$). This property has led to the development of the usage of the naphthenate in wood preservation applications.

2.4.4 Tributyltin fluoride

Tributyltin fluoride has been produced commercially from TBTCl

$$TBTCl + NaF \rightarrow TBTF + NaCl$$

The product is a white solid, sparingly soluble in water. It was formerly used as a biocide in the manufacture of 'free association' anti-fouling paints. Manufacture of TBTF is now virtually obsolete, associated with the decline in usage of free-association paint systems which incorporate TBT derivatives.

2.4.5   Other tributyltin compounds

Industrial manufacture of other TBT compounds relates to relatively small tonnages of speciality products. Manufacture is undertaken to meet specific customer requirements in anti-fouling compositions and biocides. An indication of the range of some of the TBT compounds that have been produced is shown in a classification listing of TBT products, issued by the Japanese Environmental Agency in September 1990. As well as TBTO, some other 13 TBT compounds are listed – TBT-methacrylate, fumarate, fluoride, 2,3-dibromosuccinate, acetate, laurate, phthalate, sulphamate, maleate, chloride, naphthenate, rosinate, and copolymers of alkylacrylate–methyl methacrylate–TBTmethacrylate. Additional products listed in other international publications include benzoate, linoleate and salicylate. By 1994 manufacture of many of the specialist derivatives had ceased with production concentrating on TBTcopolymer compounds.

Over the past 40 years a very large number of TBT derivatives have been synthesized and examined for their biological activity. In 1984, Ameron Inc, California, developed a number of TBT–polysiloxane resins. By hydrolysis of the tin–oxygen–silicon bonds, these compounds release the tributyltin moiety at what is claimed to be a controlled rate. A number of the TBT siloxane anti-fouling coatings were examined but they do not appear to have been developed commercially.

In 1990, the US Naval Oceans Systems Center, San Diego, announced the synthesis of tributenyltin compounds – tributyltin analogues containing double bonds at C-1 or C-3 or combinations thereof. These compounds were examined for use as anti-fouling toxicants in marine anti-fouling coatings and other biocidal applications. The compounds are suggested to be environmentally acceptable as they have increased degradation rates.

2.5   TBT industrial waste and by-product disposal

Solid and liquid organotin waste material that arises during industrial manufacturing processing is incinerated in specially constructed incinerators. Temperatures in excess of 850 °C are achieved. The products of incineration are stannic oxide (SnO_2), water and carbon dioxide.

Waste water arises from process waters in the aqueous saponification of TBTCl, from the hydrolysis of $TBTCl_xAlCl_3$ and from $MgCl_2$ separation. Purification of the waste water can effectively be achieved by a number of methods. Tributyltin compounds can be absorbed or decomposed by chemical or microbiological action. Adsorption is achieved via activated carbon filters. Chemical precipitation can be produced at specific pH values. Other methods that can be used industrially include decomposition by UV radiation or oxidation with $KMnO_4$ or ozone. Different manufacturing companies use in-house technology that has been developed to meet the specific requirements of their own production facilities.

2.6   Industrial development of TBT compounds

When TBT compounds were first shown to exhibit biologically active properties during the early 1950s, industrial biocides that were in use at that time included DDT, organomercurials and arsenical compounds. During the following decade, TBT derivatives were examined as potential replacements for such biocides because they offered effective bacterio-static and fungicidal activity with a lower mammalian toxicity. TBT compounds were found to be exceptionally active against Gram-positive bacteria. Arnold & Clarke (1956) examined the fungicidal properties of TBTO comparing its activity with organomercurials, PCP, and *p*-toluene sulphonamide in emulsion paints. An abstract of their test results comparing mould growth is shown in Table 2.1.

During the period up to the early 1980s, TBTO or its derivatives were used in the manufacture of fungicidal paints to inhibit the growth of moulds. Aqueous formulations of TBTO with quaternary ammonium compounds found application in masonry cleaning and surface preservation to prevent the growth of lichens and mosses. Over the past ten years these

Table 2.1   *Abstracts of Arnold & Clarke trials: mould growth after 126 + 12 days*

Compound	Concentration: one part in				
	250	1000	4000	16 000	64 000
Phenylmercury acetate	**	***	***	***	****
Phenyl 8-mercury oxyquinoline	**	**	***	***	****
Pentachlorphenol	**	****	****	****	****
p-Toluene sulphonamide	***	***	***	***	***
Tri-*n*-butyltin oxide	—	—	—	*	**
Controls (5)	****	***	***	****	****

Notes: — no growth; * trace of growth; ** appreciable growth; *** substantial growth; **** luxuriant growth.

applications have become obsolete with the introduction of newer biocides such as organobiborates and dodecylamine lactate. Other applications for tributyltin compounds which formerly existed include cooling water biocides, anti-slime agents in paper making, disinfectants and textile rot-proofing. Today these applications have been replaced by other biocides with the major usage of tributyltin compounds remaining in anti-fouling paint manufacture and to a smaller extent in timber preservation.

2.7   The application of tributyltin compounds in wood preservation

2.7.1   Decay by insect, fungal and bacterial activity

The biological decay of wood is a natural process which occurs worldwide. Wood is attacked by insects, fungi and bacteria resulting in the breakdown of the cellulose and other complex substances from which it is composed. Carbon dioxide and water are produced together with various organic compounds which become available as nutrients, thereby assisting regenerative growth of new trees and other plant life. Slow growing hard woods, such as teak and oak, possess a natural resistance to decay because of the presence of tannins and other preservatives which are synthesized in the heart wood during growth. Teak possesses a potential immunity to decay of about 25 years, oak between 15 and 20 years. By comparison soft woods, in an untreated state, can show decay relatively quickly in a period of between one to five years (Figure 2.1). Natural decay is an unwelcome phenomenon in wood and timbers which are used in industrial or house construction.

Chemical preservation treatments provide an extended life span for soft woods such that resistance to fungal and insect attack can be effective for many decades. The preservation of soft woods, thereby extending their natural durability for use in building and industrial applications, is of environmental importance, conserving the consumption of increasingly scarce and valuable hard wood resources. Exterior softwood joinery (doors and window frames), if untreated, can exhibit decay within a few years of installation (Figure 2.2).

2.7.2   Wood preservation

The preservation of wood has a long history. Ancient civilizations attempted to limit the decay of timber by the application of animal,

Figure 2.1 An example of soft wood decay caused by condensation on an untreated flat roof joint.

Figure 2.2 Typical decay as shown in an untreated window frame. The corner sill joint is severely degraded. Joint areas are particularly vulnerable to the commencement of fungal and insect attack.

vegetable and mineral oils. The age of the industrial revolution saw the introduction of preservation with coal derived products such as creosote. By the mid-nineteenth century, preservation with copper, zinc and arsenic compounds had been demonstrated. In 1933, a process for timber preservation with an aqueous copper chrome arsenate ('CCA') formulation was patented. During the past 60 years the worldwide growth in consumption of preserved timber for building and industrial applications has grown dramatically. Connell & Nicholson (1990) estimated that some 15 million cubic metres of timber are treated annually with copper chrome arsenate formulations.

2.7.3   Tributyltin compounds in wood preservation

By the late 1950s, light organic solvent preservatives containing penta-chlorophenol ('PCP') and organochlorine insecticides had come into widespread usage. During this period, the Osmose Wood Preserving Company of America, at Buffalo, New York, began active investigation into wood preservation with organotins. In 1960 they marketed a new wood preservative 'OZ' which included a tributyltin compound. Formulated in mineral oil, the preservative was sold for application by dipping, spray or brush. The name 'OZ' was derived by analogy to the Tin Man in 'the Wizard of Oz'.

During the following 30 years the usage of solvent based wood preservatives containing TBTO developed worldwide. Tributyltin com-pounds are very effective fungicides and bacteriocides. They do not show good activity against Gram-negative bacteria but are extremely active against Gram-positive organisms. Formulations which combine TBTO with an insecticide have been widely used in industrial applications and for domestic remedial treatments. By the mid-1980s, over 600 'TBT' containing products had received pesticide registration in the UK for use as wood preservatives or surface coating biocides. Usage of the naphthenate, TBTN, was introduced to the wood preservative market in the mid-1980s.

The introduction of new alternative biocides for wood preservation, particularly for domestic applications, began to take place during the 1980s. Manufacturers promoted new products which were considered to be effective but potentially of less environmental concern. Sale and usage of TBTO based preservative products for amateur or domestic application declined and by 1990 many countries had introduced legislation not to renew registrations for domestic product formulations and most professional applications.

Approvals for TBTO and TBTN based formulations which are used in industrial and for a limited number of professional applications have remained in existence. Products used for professional applications are supplied in paste form in the UK. *Pesticides 1993*, a UK publication, lists approvals covering 50 TBTO and 10 TBTN wood preservative formulations. The products listed are supplied by Becker, Fosroc, Hickson, Protim Solignum and Rentokil. Aqueous preservation of timber for joinery products is not generally favoured because of the potential distortion which can result if the wood swells and because of the long drying times involved. Since the 1960s, TBTO, and TBTN have become the major fungicides, formulated in light organic solvents, used in joinery preservation. The level of TBTO that is used in treatment compositions is about 1%, a level which has been found to give an equivalent performance to 5% PCP. Formulations, which may include an insecticide product, are prepared in either kerosene or a white spirit type solvent and may contain waxes, resins or other additives.

Originally, industrial applications were made by dipping. The wood was immersed in an open vessel for a few minutes, removed and allowed to drain and dry. End results were difficult to control, often with uneven distribution of the preservative. Such processes are becoming obsolete and light organic solvent preservatives ('LOSPs') are applied in closed systems with solvent and preservative recovery employed.

2.7.4 Double vacuum processing – the VAC-VAC process

The first attempts at improving the immersion system with 'LOSPs' for joinery treatment were made in the UK in 1959 by Hickson Timber Products Ltd. A cage containing fabricated joinery frames was lowered into a tank, a lid placed over and a vacuum applied, resulting in a solution uptake of 0.25 gallon per cubic foot. The process was subsequently improved and developments led to the installation of the first double vacuum type plant in 1964 (Figure 2.3). The history and development of the double vacuum system is recorded in a paper entitled *Twenty five years of joinery treatment* by Lewis & Aston, presented to the British Wood Preservers Association Convention in 1989.

The process operates two distinct vacuum cycles – giving rise to the name of the 'VAC-VAC' system and 'VAC-VAC' plant. Whereas the first vessels to be used were cylindrical, all plants now are rectangular in shape due to the low pressure requirements of the VAC-VAC plant. The rectangular pressure vessel is evacuated to a pre-determined vacuum and then flooded with preservative, some preservative being immediately taken up by the evacuated timber cells. The vessel is ventilated so that atmospheric pressure forces more preservative into the wood over a period of a few minutes. The vessel is then emptied of preservative fluid and the second stage vacuum applied for about 20 minutes. The difference in pressure between the interior of the wood and the space around it causes air to escape from the wood, expelling preservative from the wood cells, to leave them coated but not flooded. Recovered preservative solution is pumped back to storage. The pressure vessel is finally vented to the atmosphere and treated timber is removed from the cylinder in a touch dry state.

The original preservative solutions used in the VAC-VAC process were based on TBTO and organo-chlorine insecticides. The efficiency of the treatment for joinery has been outstanding with remarkably few recorded failures. The level of confidence in the performance of TBTO treated window frames is such that many suppliers have underwritten performance guarantees for 25–30 years. The Western European Institute for Wood Preservation published data in 1989, placing the annual volume of timber treated with Light Organic Solvent Preservatives at 1.2 million cubic metres. The double vacuum process is also used for the treatment of less durable hardwood species, where it has been recognized that preservative treatment is necessary. Increasingly, tributyltin compounds are finding use for the protection of tropical timbers, such as Rubberwood, from fungal attack in SE Asia (Smith & Kumar Das, 1993).

Figure 2.3 One of the first VAC-VAC plants installed in the UK in the early 1960s. Photo by courtesy of Hickson Timber Products Ltd.

The use of organo-chlorine insecticides in timber preservation has in recent years been progressively replaced by the use of the synthetic pyrethroids, cypermethrin and, in particular, permethrin. These products are incompatible with TBTO. Tributyltin naphthenate, TBTN, however, is compatible and light organic solvent preservatives have been developed which formulate TBTN in conjunction with pyrethroids.

Usage of a double vacuum process has developed in many parts of the world, particularly Europe, the Middle East, SE Asia, Australia and New Zealand. The outstanding success of the use of TBT solvent-based formulations when applied to wood by vacuum technique, directly relates to the long term resistance to decay which results. The process ensures that high risk joints and end grain zones are well treated. Wood dimensions are unchanged and the wood can be overpainted within a minimum time of treatment. Organic solvent emissions to the atmosphere are reduced and site hygiene improved. Industrial experience, over a long number of years, has shown that the process operates in a safe and environmentally acceptable manner.

2.7.5 Aqueous application of TBT compounds in wood preservation

The use of TBTO and its derivatives has been examined for use in timber preservation by pressure impregnation with aqueous dispersions. In 1972, work at the Penarth Research Centre, UK, sponsored by Albright & Wilson Ltd, led to the development of the use of aqueous TBTO/quaternary ammonium compound formulations. The TBTO formulations were applied by pressure and a programme of soil burial stake tests undertaken. Initial results of the work are described by Richardson & Cox (1974). The grant of a number of patents was obtained and the industrial application of aqueous TBT/quaternary ammonium compounds developed. By comparison with CCA applications, however, the usage of aqueous TBTO systems has been very small. An aqueous formulation was formerly used in the remedial treatment of dry rot. TBTN with co-biocides in aqueous detergent formulations, has been developed for pressure application.

A recent development of interest is described in the *Annual Report for 1988* of the International Tin Research Institute. The compounds tributyltin methanesulphonate (TBTMS) and tributyltin ethanesulphonate (TBTES) are reported to exhibit fungicidal properties which are of potential interest in the preservation of timber used in ground contact applications. The products TBTMS and TBTES are water soluble (solutions of 2–3% w/v can be prepared). Pressure impregnation of timber with aqueous formulations of these compounds has been undertaken. Stake trials are in process at a number of wood preservative companies.

2.7.6   Health & safety and environmental considerations

The safe application of tributyltin compounds in the preservation of joinery by solvent/double vacuum application over the past 30 years has been well demonstrated. Processing is undertaken by professionally trained industrial applicators. The long service life of joinery products that is achieved has reduced the requirement for domestic remedial treatments. With pre-treated wood, the final consumer, the home owner, does not come into contact with the application of preservative chemicals.

Health & safety aspects relating to the use of tributyltin compounds in wood preservation were reviewed by Schweinfurth at the British Wood Preservers Association (BWPA) Convention, Cambridge, in 1987. Schweinfurth illustrates the relative toxicity of TBTO and TBTN by showing that both have a slightly higher toxicity than copper sulphate pentahydrate and range in the neighbourhood of the naturally occurring alkaloids caffeine and nicotine (Table 2.2).

Exposure to tributyltin compounds at high concentrations may cause skin irritation, but no damage to red blood cells has been observed in organotin production workers exposed to TBTO. Tributyltin compounds have not been demonstrated to be neurotoxic or to represent mutagenic, teratogenic or carcinogenic hazards to humans.

Although TBT compounds have been in use for more than 30 years and there is extensive experience from the medical surveillance of employees in organotin production, there are no reports on cases of acute systemic poisoning or of long term adverse effects in humans.

Environmental issues and regulations in Europe affecting the pre-treatment preservation of wood were reviewed by Anderson, Connell & Waldie at the BWPA Convention in 1987. A revised code of practice for

Table 2.2   *Relative toxicity of TBTO and TBTN*

Compound	LD_{50} (mg/kg)	Species
Copper(II) sulphate $5H_2O$	300	rat
TBTN	224	rat
Caffeine	192	rat
TBTO	127	rat
Nicotine	50	rat

the *Safe Design and Operation of Timber Treatment Plants* was published in 1991 by the British Wood Preserving and Damp-proofing Association. This code was developed in consultation with Her Majesty's Inspectorate of Pollution, the National Rivers Authority and the Health & Safety Executive. The code gives guidance on the safety of plant equipment, its operation and the treatment site, and makes special reference to measures to be taken to ensure environmental protection.

Potential environmental risk from the use of TBT light solvent preservative formulations primarily relates to spillage. Bund capacities are recommended for new installations to be a minimum of 110% of the total preservative storage on site. Bulk storage tanks are recommended to be fitted with high level alarm indicators to prevent overfill. TBT evaporative losses to the atmosphere from pre-treatment plants are unlikely to pose risk to the environment due to the low vapour pressure of the compounds. TBT compound leaching from treated wood that has been applied by double vacuum treatment is considered to be negligible, because of the strong adsorption to wood which occurs.

2.8 The use of tributyltin compounds in anti-fouling coatings

2.8.1 Fouling and anti-fouling agents

Fouling, the growth of shell, weed and algae on immersed surfaces, is a natural phenomenon which occurs continuously in the marine environment. Fouling occurs on the hulls of ships and boats, oil rig support structures, buoys and fish nets. The fouling that is encountered on ships relates to a wide variety of marine organisms. Several hundred species of organisms have been identified on the fouled hulls of ships – these include diatoms, algae (red, brown and green weeds) as well as a wide variety of marine animals such as barnacles, hydroids, tube worms and bryozoa. The intensity and distribution of fouling organisms varies widely between different world locations and with seasonal climate. An example of such variation is shown in the yearly fouling spectrum at Miami, USA, as compared with that at Plymouth, England, in data prepared in 1952 by Woods Hole Oceanographic Institution, for the US Navy Bureau of Ships (Figure 2.4).

Fouling organisms have the ability to attach themselves quickly and firmly to a ship's hull with the capacity for very rapid initial growth and a vast reproductive potential. Growth and reproduction can, under favourable climatic conditions, be such that on a very large tanker some hundreds of tonnes of fouling can accumulate in a period of less than one year (Figure 2.5).

Fouling of a ship's hull leads to an increase in friction between the hull and sea water, causing 'hull roughness'. This, together with the added weight of the fouling, leads to a considerable increase in fuel consumption, loss of speed and manoeuvrability.

Fouling has caused considerable problems to shipping throughout history. Marine borers such as the *Teredo* have been responsible for the destruction of many wooden sailing ships down the ages. A wide variety of agents were used in classical times to combat fouling – pitch, lead, tin and

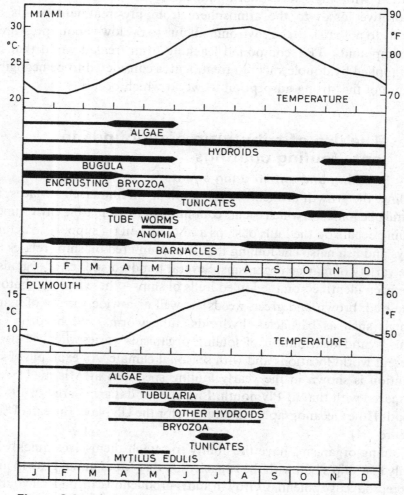

Figure 2.4 Comparison of fouling seasons at Miami and Plymouth.

copper sheeting. By the eighteenth century, tar and bitumen were in widespread use. Copper–rosin based anti-foulants were introduced nearly a century ago and have remained in use since that time. During the 1950s, in order to improve the effectiveness and life expectancy of copper based anti-foulants, newly available toxicants were added as 'boosters'. Such products as DDT, organomercurial, organolead and arsenical compounds were employed. By 1960 triorganotin compounds were used as 'boosters' in copper anti-fouling compositions. Development of their use as the principal toxicant in vinyl and acrylic substrate systems soon followed. Phillip (1973) in a review of developments of anti-fouling coatings, summarizes comparative toxicity data of various biocides which were in use at that time (Table 2.3).

Due to relatively high toxicity and adverse environmental effects, organolead and organomercurial compounds were withdrawn from use. With the introduction of spray painting techniques, arsenical additives were withdrawn from use in the 1970s.

2.8.2 Early use of triorganotin compounds in anti-fouling paints
The effectiveness of triethyltin and triphenyltin derivatives for use in anti-fouling formulations was described by Tisdale (1943). During the

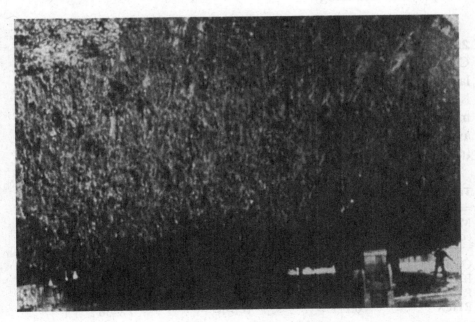

Figure 2.5 An example of the heavy degree of fouling that can occur as a consequence of inadequate anti-fouling protection.

early 1960s M & T Chemicals Inc. USA, investigated the performance of a number of triorganotin compounds in anti-fouling paint formulations. The history and development of these systems is recorded by Bennett & Zedler (1966). The original formulations that were developed used tributyltin oxide and tributyltin sulphide in acrylic, vinyl and alkyl substrates. Anti-fouling performance testing was undertaken along the East Coast of America and along the Gulf of Mexico. The success of the introduction of TBT based paints related not only to their effective anti-fouling performance but also to their ability to be presented in any colour. Yachting marinas were soon to change from vessels with hulls painted with copper based red/browns, to those which exhibited hulls painted with a range of the brightest colours. The usage of triorganotin based anti-fouling paints for commercial vessels was soon to follow.

In 1970, International Red Hand Marine Coatings had established 'wide spectrum' triorganotin anti-fouling compositions as a leading brand applied to some 200 vessels, representing 5 million gross tonnes of shipping. Development of different TBT systems during this period occurred world wide. The use of triphenyltin derivatives as anti-foulants became established, particularly in Japan. The major development with TBT anti-fouling coatings, however, occurred in 1974, with the introduction of copolymer technology.

2.8.3   Anti-fouling paint systems

Conventional anti-fouling paints, which have been in existence since the last century, are based on a soluble matrix. This type of anti-fouling coating is sometimes referred to as 'free association'. The biocide is physically mixed into a rosin matrix which dissolves slowly in sea water. As it dissolves away, the biocide is released. Rosin is a natural product, obtained from trees, mainly consisting of a mixture of acids. Product quality is variable so

Table 2.3   *Comparative toxicity data*

Metal compound	Algae (ppm)	Barnacles (ppm)	Rats (oral) (mg/kg)
Cu_2O	1–50	1–10	470[a]
R_3SnX	0.01–1	0.1–1	130–600
R_3PbX	0.1–1	0.1–1	30–200
RHgX	0.1–1	0.1–1	8–36

[a]US Nat. Int. Occup. Saf. 1987. R = alkyl group.

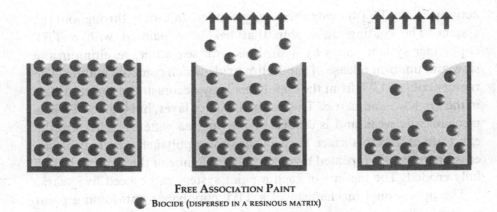

FREE ASSOCIATION PAINT
BIOCIDE (DISPERSED IN A RESINOUS MATRIX)

Figure 2.6 The biocide leaches freely from the resinous matrix. The initial release is rapid and uncontrolled. Subsequent release declines steadily from the matrix such that the anti-fouling performance of the paint diminishes with time.

that the lifetime of a paint is accordingly unpredictable. The products have poor mechanical strength which limits the thickness of application as paint. Biocide release rates are initially very high, declining to exhaustion over a relatively short period of time, of one to two years. The operational mechanism of a 'free association' paint system is illustrated in Figure 2.6.

Insoluble matrix paints based on chlorinated rubber and vinyl resins were developed during the late 1940s. Their higher mechanical strength allows the application of thicker coatings and effective anti-fouling performance can be maintained for a two year period. Insoluble matrix systems exhibit a high initial biocide leaching rate and an exponentially declining rate of exhaustion. The operational mechanisms of insoluble matrix paints is similar to that of the 'free association' paint illustrated in Figure 2.6.

2.8.4   TBT copolymer systems

Development of TBT copolymer systems relates back to early patents held by M & T Chemicals Inc., USA, which describe the chemical incorporation of a biocide such as tributyltin acrylate in a methylmethacrylate polymer. International Paint extended this technology to develop anti-fouling paint systems based on TBT *copolymers*. The International Paint patents which were granted describe the invention of the 'SPC' – 'self-polishing copolymer' – system. The polymer is physically robust and enables coatings of $2 \times 150\,\mu m$ to be applied, a thickness about twice that which was used in other systems. The copolymer resins are biologically active with the TBT biocide chemically bonded uniformly throughout the system. The coating on a ship that has been painted with a TBT copolymer system reacts by hydrolysis with sea water, resulting in the slow and uniform release of tributyltin oxide which combats fouling. The removal of the TBT from the copolymer only occurs at the surface layer, in the top few nanometres. The residual surface layer, having lost TBT, is mechanically weak and is eroded by moving sea water, resulting in the exposure of a fresh surface layer of organotin polymer. The hydrolysis/erosion process is repeated throughout the lifetime of the paint until it is fully eroded. The lifetime of such a paint system can exceed five years.

The operational mechanism of a TBT copolymer anti-fouling paint system is illustrated in diagrammatic form in Figure 2.7.

The hydrolysis/erosion process which operates has several unique and beneficial properties.

i. The system is able to provide controlled biocide release at a constant and minimal rate.

ii. The life of the coating is proportional to its thickness and is accurately predictable.

iii. The erosion process results in a smoother, 'polished surface', which reduces hull roughness. This has a significant effect on the reduction of frictional drag, leading to conservation in fuel consumption.

iv. Hulls can be repainted directly without the need to remove any remaining old copolymer coating.

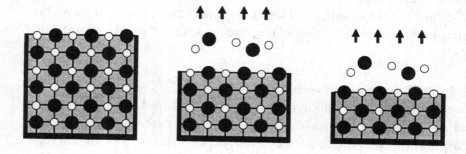

SELF POLISHING COPOLYMER SYSTEM
○ TBT ● COPOLYMER RESIN

Figure 2.7 At the paint surface, sea water hydrolyses the TBT copolymer bond and the TBT biocide and copolymer resin is slowly released at a controlled rate. A uniform anti-fouling performance is achieved throughout the life of the paint.

The 'polishing' effect that occurs gave rise to the nomenclature of 'SPC' – 'self-polishing copolymer'. A detailed review of the mechanism and performance of SPC paints is given in *Self-polishing antifoulings: a scientific perspective* (Anderson, 1993). The review presents interesting data on the measurement of the leached layer (also called the 'active zone') in TBT polishing systems. The leached layer is shown to stay very thin (in the 10–20 μm range) throughout the lifetime of the paint with a virtually linear relationship existing between the polishing rate (loss of micrometres of film thickness) and speed. The non-zero intercept shown on graphed data indicates that polishing can operate at zero speed, although this is dependent upon the TBT level in the copolymer. Underwater inspection of ships' hulls confirms that under static conditions fouling is resisted, even in a most severe environment.

The SPC system was introduced to major shipping lines in Europe and the Far East in the mid-1970s. In the USA, it received full EPA registration in 1978. In that year the passenger liner *QE*2 was first painted with an SPC system at Southampton. Fuel cost savings of 12% were subsequently reported to have been achieved over a two year period of operation. In the past 20 years, developments and improvements to TBT copolymer technology have been made. Fast polishing products have been designed to give maximum protection against exceptionally severe fouling in tropical waters. High performance products are available which can be directly applied over non-copolymer substrates.

Although 'TBT' is a broad spectrum biocide which is active against most fouling organisms, there are a number of slime forming diatoms which are 'TBT' resistant. TBT copolymer compositions normally include copper compounds and organic 'boosters' as co-biocides in order to provide effective performance against the whole fouling spectrum.

2.8.5 Anti-fouling performance of TBT copolymer systems

TBT copolymer anti-fouling systems have been mainly used for large commercial vessels where anti-fouling performance is monitored by vessel inspection on drydocking. In order to obtain accurate comparisons (regardless of the drydock location or the inspector), standard inspection aids are employed. Vessels are assigned a fouling rating (FR) where the total absence of fouling is rated 'zero FR' and a totally fouled vessel rated '100 FR'. An inspection rating of up to '10 FR' is accepted as 'satisfactory'. Monitoring of the effectiveness of anti-fouling coatings is undertaken by ship owners and anti-fouling paint manufacturers.

Results of surveys made between 1987 and 1992 are reported in a publication produced by the ORTEP Association and Marine Painting Forum (July 1992). The data were presented at the 33rd Session of the Marine Environment Protection Committee (International Maritime Organization), in October 1992. The results of surveys on 766 commercial vessels painted with TBT copolymer systems are shown in the histogram in Figure 2.8. The surveys record the fouling rating achieved after the vessels had been four years out of dock.

The results achieved on 20 very large vessels, 'VLCCs', painted with TBT copolymer systems after five years out of dock are shown in the histogram in Figure 2.9.

Results recorded with conventional 'free association' anti-fouling systems applied to 650 commercial vessels during the same time show a substantially inferior performance. After two years out of dock, 36% of the vessels were rated as satisfactory as compared with TBT copolymer figures of 90–92% as shown in Figures 2.8 and 2.9.

During the past decade, a wide variety of anti-fouling paint systems have been developed which use alternative biocides as replacements for 'TBT'. These anti-foulant systems have been introduced in response to legislation which prohibited the use of tributyltin based paints on small boats. It has been difficult to match the performance of 'TBT' as a biocide and to emulate the self polishing characteristics of the SPC system. Organotin-free coatings use rosin as the soluble medium in combination with an insoluble polymer to form a matrix. Toxicants used include

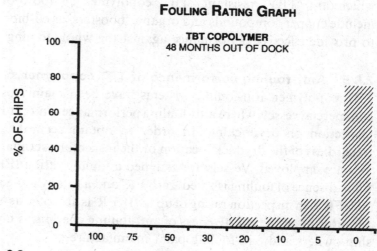

Figure 2.8

cuprous oxide with an organic biocide, such as amino, azine or thio-organic derivatives. The anti-fouling performance of the new 'tin free' products show a general improvement over the older, conventional, free association paints. For service lives up to three years, these products are considered by the shipping industry to be satisfactory except in the most challenging fouling conditions. Claims for service lives to achieve four years and over to match TBT copolymer product performance have yet to be substantiated by independent assessors.

2.8.6 Market usage of TBT copolymer coatings

Lloyd's Register identifies over 75 000 ships in the world commercial sea-going fleet. Estimates made by the marine coatings industry place the value of total world sales of anti-fouling paints in 1993 at about $US260 million. Commercial shipping usage accounts for about 90% of the total volume with the remaining 10% relating to the yacht and small pleasure craft market. During the period 1965–87, a large proportion of anti-fouling paint used on small pleasure craft was based on both free association and copolymer formulations containing TBTO, and to a lesser extent triphenyltin (TPT) compounds. Following the introduction of legislation in several countries during the late 1980s to ban the use of TBT containing paints on small boats (of length less than 25 metres), this market has been supplied with 'tin free' coatings.

Anderson (1993) reports that, excluding Japan, TBT self-polishing

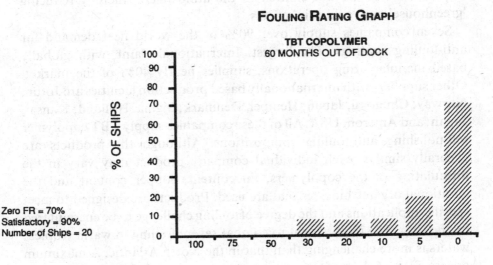

Figure 2.9

copolymer paints account for 69% of the total volume of world anti-fouling paints sales. Since 1977, over 13 000 applications of TBT copolymer coatings have been applied in the world fleet, representing 13 million DWT (dead weight tonnes). It is only in Japan that there has been any significant change to other anti-foulant systems which do not contain a triorganotin compound. Overall, however, the world volume of TBT copolymer products sold has been remarkably static, with the decline in volume in Japan being offset by a corresponding increase in usage in Singapore and South Korea. During the past 20 years major contracts for world shipbuilding and repair have increasingly been undertaken in these three Asian countries, particularly large cargo ships, product tankers, crude carriers and VLCCs (very large crude carriers). About 60% of the world fleet is represented by ships of gross tonnage greater than 20 000 tonnes (*Lloyd's Register's Statistical Tables 1992*). Crude carriers are of several hundred thousand gross tonnes weight. A VLCC when loaded is in excess of 500 000 DWT. Such vessels require to operate with good anti-fouling performance and minimal frictional resistance. The use of TBT self-polishing copolymer anti-fouling coatings provides a guaranteed life expectancy for these large vessels of at least five years out of dock before repainting is required. The conservation of fuel which is obtained, relating to the performance of the self-polishing system, is significant. Fuel conservation has a direct environmental benefit by the saving of natural resources. A further environmental benefit is gained as a consequence of not burning additional fuel – additional tonnages of carbon dioxide and sulphur dioxide are not emitted to the atmosphere, thereby reducing 'greenhouse' and 'acid rain' effects.

Seven companies supply over 90% of the world fleet demand for anti-fouling coatings. The largest, International Paint, with globally based manufacturing operations, supplies nearly 40% of the market. Other suppliers with internationally based production facilities are Jotun, Norway; Chugoku, Japan; Hempel, Denmark; Sigma, Holland; Kansai, Japan; and Ameron, USA. All of these companies supply TBT copolymer self-polishing anti-fouling compositions. Although the products are generally similar, each individual company product may vary in the formulation of the copolymers, tin content, copper content and the additional organic biocides that are used. Products are designed to meet specific applications and the degree of fouling challenge to be encountered by the ship. The requirement to combat severe fouling in warm tropical waters is more challenging than that in the North Atlantic. A maximum release rate of 4 µg of organotin per square centimetre per day was proposed by the US EPA in 1989 (and subsequently by the International Maritime Organization (IMO), United Nations, in 1990). At present no international standard method exists for the determination of leaching rates of biocides from anti-fouling paints. It will accordingly be difficult to establish an international target maximum leaching rate until such a method has been established. The IMO has proposed that this matter should be considered with the involvement of an international group such as the International Organization for Standardization, Geneva (ISO).

Reviews of the potential environmental effects of marine anti-fouling coatings have been made in recent years by the International Maritime Organization. At the 35th session of the Marine Environment Protection Committee (MEPC) in March 1994, a critical review of current and future marine anti-fouling coatings was presented by the UK delegation. The report, which was prepared for the UK by Lloyd's Register Engineering Services (Pidgeon, 1993), gives a useful performance and environmental review of major TBT and 'tin free' anti-fouling products. Twenty-eight TBT copolymer products that are in commercial use are listed.

Anderson (1993), in his 'scientific perspective' of self-polishing anti-foulings, compares the critical differences between TBT self-polishing paints with the more recently introduced 'TBT free' products. He concludes that 'tin free' anti-foulings are inferior to the TBT self-polishing copolymer products in several fundamental ways. The conclusions presented in his 'scientific perspective' are based on performance data of TBT copolymer anti-foulings from the past 20 years as compared with performance data gained from 'TBT free' polishing paints applied to over 2000 ships in the past four years. The comparison summary table is reproduced in Table 2.4.

TBT copolymer formulations were for some years used in the treatment of fish-farm nets to resist the growth of fouling. Treated nets were widely used, particularly in Scotland and Norway for salmon farming, and in Japan, with fish nets remaining free of fouling for periods of up to two years. Although the anti-fouling treatment was successful, concern over bioaccumulation of tributyltin in fish led to the termination of this application, with legislative bans being introduced.

2.8.7 Health & Safety and environmental considerations

TBT copolymer paints are applied to the hulls of ships by airless spray. Spray painting is universally employed for the painting of large vessels where speed and uniformity of application are essential and the paints are specifically formulated for this type of application. Roller and brush methods are only used for the repair and touch-up of small areas. Protective clothing (often disposable) and respirators are worn by operators. Over the past 20 years there have been no reports of occupational health problems arising from the application of TBT copolymer anti-fouling coatings.

Potential environmental problems which can arise from painting operations mainly relate to overspray drift. Operator training, level of skill and good supervision has much to do with the efficient use of paint. Wash-down waters used to clear overspray deposits will contain TBT and suitable disposal or treatment of the contaminated waste water is required. The control of VOC (volatile organic content) release from anti-fouling paint systems, as with all solvent based paints, is currently under review internationally, relating to proposed reductions in emission levels. Considerable research has been undertaken to develop aqueous based formulations but problems associated with sea water corrosion and film breakdown have proved insoluble to date.

Table 2.4 *Performance data of TBT copolymer anti-foulings compared with TBT free polishing paints*

TBT polishing anti-foulings	TBT-free polishing anti-foulings
Small active zone, with size independent of immersion time.	Large active zones possible, dependent on immersion time.
Active zone has good film integrity, and is easy to recoat.	Active zone can have poor film integrity, and must be removed before recoated.
Polishing rate is directly proportional to speed.	Polishing rate is not directly proportional to speed.
Proven in-service performance of >5 years, on all vessel types	Performance variable, with limited long-term ship results.
Good static performance, even after prolonged periods.	Performance deteriorates after prolonged static periods.
Media contributes to biocidal effectiveness.	Media is non-biocidal.
Chemical composition is precisely known.	Variable chemical composition possible, due to rosin variability.
Excellent weathering durability	Performance deteriorates after prolonged weathering.
Low copper content is possible.	High copper content is essential.

The surface cleaning of ships' hulls and removal of old paint before repainting gives rise to contaminated waste. High pressure hosing is applied to TBT copolymer painted vessels before repainting. The waste waters generated accordingly contain TBT and other biocides as contaminants. Recommended codes of practice have been developed for the application and removal of TBT anti-fouling coatings. The UK Ministry of Defence established guideline procedures in 1987 which highlighted the need to avoid drift and contamination during spraying and recommended the attempt of containment and subsequent disposal of waste arising from the removal of old coatings. The monitoring and control of dockyard activities which involve TBT-containing paints has been reviewed by Allison, Guest & Stott (1990). Methods that have been used (in combination) for the treatment of TBT contaminated dockyard waste water include flocculation and settling, filtration, activated carbon adsorption and UV light irradiation. Incineration or landfill to an authorized site has been used for the disposal of solid waste. Waste water arising from the above mentioned treatments has also been disposed of by landfill to authorized sites. The disposal of waste water from very large drydocks is achieved by discharge to the ocean with dilution, dispersal and subsequent biodegradation occurring.

Environmental concern about the effects of TBT in the aquatic environment has mainly related to the usage of TBT paints on small boats, particularly where the boats have been harboured in shallow, enclosed or semi-enclosed areas. Various studies have been made in locations in close proximity to centres of shellfish farming. TBT compounds, as with all anti-fouling agents, exhibit biocidal properties which resist fouling but they can also affect other aquatic organisms. As discussed in Chapters 6 and 7, the most prominent effects of TBT compounds against non-target organisms have been observed in oysters and in dogwhelks (marine snails). Reduced growth and shell thickening has been noted in certain species of oyster and an increased incidence of 'imposex' (the development of a penis) and sterility can be found in female dogwhelks. Following the introduction of a ban on the use of triorganotin containing paints on small boats in the late 1980s, there has, in the ensuing years, been a significant reduction in TBT levels in water and in the tissues of aquatic organisms. A review paper summarizing the results of TBT monitoring studies in Europe, the USA and Japan was presented by CEFIC, 1994, at MEPC 35. The CEFIC paper includes survey data from the UK reporting that offshore concentrations of TBT in water are low. Few samples taken outside estuaries have higher concentrations than

$2\,\mathrm{ng\,l^{-1}}$ and no coastal sediments have been found to have measurable contamination by TBT. Evans *et al.* (1994) report on the recovery of dogwhelk populations following legislation introduced in 1987. It is important to note that during this period, 1987–94, the usage of TBT copolymer coatings has remained unchanged on large ocean going vessels. The UK delegation in its critical review of anti-fouling coatings, presented at MEPC 35, concludes that the economic benefits of TBT copolymer coatings are greatest, and the environmental impact is least, for large vessels in global trade.

The environmental impact of the new 'tin free' anti-fouling paints, since their introduction in the late 1980s, has not yet been fully assessed. A toxicological risk assessment was undertaken by the Swedish National Chemicals Inspectorate, KEMI in 1994. Approvals have not been recommended for the use of 'tin free' anti-fouling paints on small pleasure craft operating in fresh water. Possible areas of concern relate to potential bioaccumulation and adsorption characteristics with sediment.

Substantial research has been devoted to the development of low surface energy anti-fouling coatings, based on fluorinated silicone polymers. Such products contain no biocides. Aquatic organisms have great difficulty to attach themselves to such polymer surfaces. Over the past ten years a number of products have been evaluated but none have proved satisfactory for use on commercial shipping. Although the coatings have good anti-fouling performance, problems relate to bonding the polymer to the substrate and its poor mechanical properties give rise to damage from abrasion.

The International Maritime Organization, in its report of the proceedings of MEPC 35, March 1994, recognizes that an extension of TBT restrictions to a total ban is not justified on cost/benefit grounds. The MEPC also acknowledge that alternative anti-fouling systems are not at present available. Views from industry indicate that effective systems which could replace the performance of TBT copolymer anti-fouling coatings, are some way ahead – possibly up to 10–20 years away.

2.9 The Organotin Environmental Programme Association

At the initiative of the International Tin Research Institute (ITRI), a special environmental project was set up at the Institute for Organic Chemistry, TNO, Utrecht, Holland, in 1972. A four year organotin environmental research programme (1973–7) was undertaken. At that time no suitable analytical methods were available for the detection of trace quantities of organotin compounds. Methods were developed at TNO which led to the analytical procedures that are in current usage. The programme included a study of the degradation of organotin compounds and identification of the breakdown products. The cost of the study was substantial – in excess of $US500 000. Half of this cost was borne by TNO and half by various industrial companies and ITRI. The project was named the 'Organotin Environmental Project' – known by the acronym 'ORTEP'. At the conclusion of the research, companies whose operations included the production of organotin compounds founded, in January 1978, an environmental programme association, the 'ORTEP' Association. Membership came from industrial companies in Europe, Japan and the USA.

The ORTEP Association was registered at the Chamber of Commerce in the Hague, the Netherlands and within its articles of constitution, the primary aim and objective recorded is

to promote and foster the dissemination of scientific and technical information on the environmental effects of organo-tin compounds.

The Association has established a data bank of technical and environmental information relating to organotin compounds. This is maintained by the ITRI, based at Brunel University Science Park, Uxbridge, England.

The data bank now contains in excess of 4000 indexed references. The ORTEP Association has established a number of working groups

A medical working group, reviewing medical aspects of working in the organotin industry and monitoring employees.
An analytical working group.
A labelling group, which has worked closely with Government departments, establishing internationally acceptable labelling.

The Association has sub-groups in Europe, Japan and the USA, all of whom have co-operated with Governments and international authorities on environmental issues, particularly with the submission of toxicological and analytical data. With regard to tributyltin compounds member companies within the ORTEP Association have sponsored a number of environmental projects, including TBT monitoring, dogwhelk imposex and occupational exposure studies.

2.10 Environmental risk–benefit considerations

2.10.1 Wood preservation

Environmental or human health risk from the industrial use of TBTO/ TBTN in wood preservation of joinery products (by vacuum application) is considered to be minimal. Accidental spillage incidents have been the only potential cause for concern. No environmental or health risk has been reported over the past 35 years from the in service use of TBT treated door and window frames. Benefits obtained relate to the absence of timber distortion during treatment and the exceptionally long service life of the treated joinery products. Assessments which have been made to examine the effect of a total ban on the use of TBT in timber treatment indicate this would be unlikely to provide any significant benefit to the aquatic, terrestrial or atmospheric environment, because of the exceptionally low level of release of 'TBT' from treated timber (UK Department of the Environment, 1992). Additionally, an economic penalty is involved with the use of alternative wood preservatives for joinery treatment, associated with a potentially shorter lifetime and the degree of replacement/remediation work required.

2.10.2 Anti-foulants – TBTO free association paints

Although prohibition of the use of triorganotin paints for small boats was introduced in the late 1980s, such anti-fouling paints continued in use in some countries on vessels over 25 metres in length (although at a diminishing rate). In the UK legislation to prohibit their use became fully effective in April 1995. In the USA, by comparison, TBT free association paints disappeared from the market place following the introduction of Federal law in 1989 where a requirement to comply with a leaching rate performance of less than '4 microgrammes of organotin per square centimetre per day' was introduced. This rate figure is widely exceeded by TBT free association paints and initial leaching rates can be more than tenfold greater. Product trade listings (e.g. *Pesticides 1995*) show TBT free association anti-fouling paints to have had registrations revoked. Following discussions within the ORTEP Association, organotin manufacturers recommended in Spring 1995 the general phasing out of TBTO for sale for use in the manufacture of free association paints. Some ships, however, which were painted with TBTO free associations systems immediately before 1995, will remain in service for a short period. Accordingly the full effect of the regulations may not become apparent for a further one to two years when these vessels drydock for repainting.

Marine survey data show that TBT free association paints have been the major cause of 'TBT' environmental concern. These paints were introduced around 1960 at the time when there was a world-wide exponential growth rate in the ownership of small boats and yachts. Between 1960 and 1970 TBTO replaced copper systems which were 'boosted' by organomercury, organolead and arsenicals (to provide a full spectrum anti-fouling protection). Although the effect of TBT free association paints has been environmentally undesirable, it is interesting to consider what may have resulted if the usage of other organometallic compounds had continued for a further two decades at a time when small boat growth was at its height.

2.10.3 Anti-foulants – TBT copolymer systems

The potential environmental risk from TBT copolymer anti-foulant systems is still under evaluation. Leaching from these anti-foulant coatings is uniform and of a very low emission rate. In many marine areas it has been difficult to differentiate TBT contamination from the now obsolescent free association paints and that arising from TBT copolymers. Survey data from harbours and oil terminals where no small boat activity has existed but which receive a heavy density of large tankers and cargo vessels, indicate that TBT contamination is localized to within a radius of a few kilometers (Pidgeon, 1993). Contamination in adjacent coastal areas to harbours appears to be absent.

Data presented by CEFIC to the IMO Marine Environment Protection Committee in 1992 (compiled by the ORTEP Association and the Marine Painting Forum) review economic and environmental benefits arising from the use of TBT copolymer paints (*TBT Copolymer Anti-Fouling Paints: The Facts*). Benefits are shown to exist in four main categories – direct fuel savings, extended drydocking intervals, reduced maintenance, and indirect savings. The overall economic benefit is estimated to be around $US3 billion per annum. The saving in fuel usage is of major environmental significance. Fuel represents about 50% of marine transport operating costs with consumption of fuel by the world fleet being around 180 million tonnes. Based on a knowledge of the size of the world fleet and current operational data, calculations show that the introduction of TBT copolymer self polishing anti-foulants has yielded annual fuel savings of

- 2% attributed to better anti-fouling performance
- 2% associated with lower frictional surface resistance.

The combined saving of 4% equates to 7.2 million tonnes. Apart from the importance of this major conservation of fossil fuels, the CEFIC data demonstrate that as a consequence of not burning this fuel

- 22 million tonnes of carbon dioxide are not emitted to the environment each year ('greenhouse gas')
- 0.6 million tonnes of sulphur dioxide pollution is prevented each year ('acid rain gas').

This has particular relevance towards the reduction of 'air pollution from ships' (UNCED Conference, Rio, 1992).

Other environmental issues currently being examined relate to the extended period between drydockings, which is possible by the use of TBT copolymers. Less frequent drydocking visits for the renewal of anti-foulant coatings result in substantial reductions in

- waste washdown water
- waste blasting grit
- waste paint cans
- volatile organic solvent emission ('VOCs').

Trade estimates, based on current dockyard practices, indicate the potential size of the environmental savings per annum

- 0.8 billion litres of contaminated waste water prevented
- 2.3 million tonnes of contaminated grit prevented
- 1.8 million waste paint cans are not produced
- 15 000 tonnes VOC solvent emission is prevented.

In the light of current environmental data, benefit analysis relating to the use of TBT copolymers has to be measured primarily against environmental risk penalties arising in, and closely around, ports, terminals and dockyards and any potential environmental risk arising from replacement products. Improvements of paint application and removal procedures as recommended by the IMO (MEPC 30, 1990; MEPC 35, 1994) are supportive control measures that will further minimize the environmental impact of TBT. The introduction of such measures in the UK, involving dockyard codes of practice and the control of waste washdown water disposal, has demonstrated their effectiveness (HMIP-IPR, 1994).

2.10.4 Alternative anti-foulant systems
All anti-fouling products by definition exhibit biocidal properties in order to resist fouling. Ideally, products should not affect non-target organisms, and any residues leached into the aquatic environment should degrade

rapidly to harmless by-products. All tin-free anti-fouling products which are in current usage contain high amounts of copper and an organic biocide. Following the small boat TBT ban, and the increased usage of tin-free systems, current monitoring data identify a significant increase of copper in some marine coastal waters (Alzieu, 1995). Contamination of coastal waters from triazine derivatives, used in anti-fouling paints, is also reported (Readman and Tolosa, 1995). The environmental significance of such data is as yet unknown; however, a number of regulatory authorities have concluded that the potential environmental impact of organic boosters used in anti-fouling paints should be reviewed.

Following the ban in the late 1980s of the use of TBT containing paints on small boats (under 25 metres length) shipping lines began to evaluate, on an experimental basis, new 'tin-free' systems, which were considered to be 'more environmentally' friendly. Initial results obtained during the early 1990s, after two to three years in service, appeared promising. Based on what were effectively short term performance results, confidence in the new 'tin-free' coatings grew. In 1993, Pidgeon, in his *Critical Review of Current & Future Marine Anti-fouling Coatings*, reported that the performance gap between the new 'tin-free' systems and TBT copolymers appeared to be closing. Unhappily the confidence was unfounded. By the following year serious performance failures began to be observed with vessels showing premature fouling and poor structural integrity of the anti-fouling paint film. Cracking and actual detachment of the paint was observed. Fuel consumption records demonstrated a significant increase in usage.

In June 1995 the trade journal *MER* (*Marine Engineers Report*) published an article 'Tin-free paints not up to the job, says major European shipowner'. Significant fouling is described to have occurred on tin-free coated vessels within two years of application, necessitating premature drydocking. The vessels are reported to have returned to tin-based copolymer paint systems. Similar changes of policy are now reported by other international shipping organizations and Defence Ministries, for naval vessels.

The search to develop other alternative anti-fouling coatings which completely satisfy technical and environmental requirements has proved to be a complex problem. Today such research for new products continues world-wide. However, the time scale to develop a completely new technology to replace TBTM-MMA copolymers is now estimated to be in the range of 10–20 years before product registration and satisfactory in service usage can be established.

2.10.5 Legislation and risk–benefit considerations

The usage of TBTO in free association paints provides an example where risk–benefit analysis has actively been involved in global decision making. The environmental risk from TBT free association paints has been considered to outweigh any benefits offered, leading to a recommended ban on their usage (MEPC 35, March 1994). The effect of legislation has been positive with monitoring data showing a continued reduction of TBT environmental contamination and a progressive return to the normal situation for oyster farming (Waldock, 1995) and a recovery in dogwhelk populations (Evans, 1994).

IMO (and member world governments) have recognized 'that an extension of present TBT restrictions to a total ban is not justified at present on cost/benefit grounds' (MEPC 35, March 1994). The justification for the continued usage of TBTM-MMA copolymer anti-fouling coatings relates to a least environmental risk, compared with environmental benefits which the systems provide and the non-availability of any satisfactory alternatives.

Acknowledgements

The author wishes to express his thanks and acknowledgement for information provided and helpful comments made, by the ORTEP Association, in particular: Drs J. A. Jonker, Elf-Atochem Vlissingen BV, The Netherlands; Dr D. B. Russell, Elf-Atochem North America Inc., Philadelphia, USA; Dr U. Stewen, Witco GmbH, Bergkamen, Germany; and Dr A. Suzuki, Hokko Chemical Industry Co. Ltd, Tokyo, Japan; and Dr T. Tsutsui, Nitto Kasei Co. Ltd, Osaka, Japan.

Additionally, individual thanks are expressed to Dr C. D. Anderson and Dr J. Hunter, International Paint Ltd, UK; Mr M. Connell, Hickson Timber Products Ltd, UK; and Dr P. J. Smith, International Tin Research Institute, UK.

2.11 References

Allison, D*., Guest, N. & Stott, J. F. D. 1990. Monitoring and control of dockyard activities involving TBT-containing paints, presented at *3rd International Organotin Symposium, Monaco*, April 1990, 18 pages. (*UK Ministry of Defence, Procurement Executive, Foxhill, Bath BA1 5AB, UK.)

Alzieu, C*. Environmental Impact of TBT – The French Experience, presented at *Costings vs Benefits of TBT-based & Alternative Antifoulants*, International Conference, Malta, December 1995. (*IFREMER 155 Rue Jean-Jacques Rouseau, 92318 Issy-les-Molineaux, France.)

Anderson, C. D*. 1993. Self-polishing antifoulings: a scientific perspective, presented at the *Ship Maintenance Conference*, Olympia, London, November 1993. 18 pages. (*International Paints Ltd, Felling, Gateshead, Tyne & Wear NE10 0JY, UK.)

Anderson, D. G*., Connell, M. & Waldie, P. 1987. Environmental issues in Europe: a review of current EEC regulations affecting the pre-treatment preservation of wood. *BWPA Convention, UK.* 28 pages. (*Hickson Timber Products Ltd, Castleford, West Yorkshire WF10 2JT, UK.)

Arnold, M. H. M. & Clarke, H. J. 1956. The evaluation of some fungistats for paint. *Journal of the Oil and Colour Chemists' Association*, London, UK, **39**, 900–8.

Bennett, R. F. 1983. Industrial development of organotin chemicals. *Royal Society of Chemistry, Industrial Chemistry Bulletin*, **2**(6), 171–6.

Bennett, R. F. & Zedler, R. J. 1966. Biologically active organotin compounds in paint manufacture. *Journal of the Oil and Colour Chemists' Association*, London, UK, **49**, 928–53.

Connell, M. & Nicholson, J. 1990. *A review of fifty-five years' usage of copper-chrome-arsenate (CCA) and CCA-treated timber*, 14 pages. Hickson Timber Products Ltd (Castleford, West Yorkshire WF10 2JT, UK.)

Evans, S. M., Hawkins, S. T., Porter, J. & Samosir, A. M. 1994. Recovery of dogwhelk populations on the Isle of Cumbrae, Scotland, following legislation limiting the use of TBT as an antifoulant. *Marine Pollution Bulletin*, **28**, 15–17.

Frankland, E. 1853. Ueber eine neue Reihe organischer Körper, welche Metalle enthalen, *Annalen der Chemie und Pharmacie**, **85**, 329–73 (*from 1873 *Justus Liebigs Annalen der Chemie*).

HMIP*-IPR *The Application or Removal of Tributyltin or Triphenyltin Coatings*, Chief Inspector's Guidance to Inspectors, HMIP-IPR 6/1, 1994. (*Her Majesty's Inspectorate of Pollution, Government Buildings, Burghill Road, Westbury-on-Trym, Bristol BS10 6EZ, UK.)

Holland, F. S. 1987. The direct synthesis of triorganotin compounds. *Applied Organometallic Chemistry*, **1**, 185–7.

International Tin Research Institute*. 1988. Organotin chemistry – organotin wood preservatives, *Annual Report for 1988*, p. 16. (*Kingston Lane, Uxbridge, Middx. UB8 3PJ, UK.)

KEMI Report No. 2/93, *Antifouling products – pleasure boats, commercial vessels, fish cages and other underwater equipment*, 58 pages. The Swedish National Chemicals Inspectorate. (P.O. Box 1384, S-171 27 Solna, Sweden.)

Lewis, D. A.* & Aston, D. 1989. *Twenty five years of joinery treatments*, BWPA Convention UK, 9 pages. (*Hickson Timber Products Ltd, Castleford, West Yorkshire WF10 2JT, UK.)

Löwig, C. 1852. Ueber Zinnäthyle, *Annalen der Chemie und Pharmacie** **84**, 308–19 (*from 1873 *Justus Liebigs Annalen der Chemie*).

Pesticides 1993. Pesticides approved under the control of Pesticides Regulations 1986. HMSO, London, UK, 1993.

Phillip, A. T.* 1973. Marine science aids the development of antifouling coatings. *Oil and Colour Chemists' Association of Australia, Proceedings & News*, July, 17–22. (*Defence Standards Laboratories, Department of Supply, Maribyrnong, Victoria.)

Pidgeon, J. D.* 1993. *Critical Review of Current & Future Marine Anti-fouling Coatings*, 1993, research carried out for the U.K. Department of Transport and presented at *The IMO, Marine Environment Protection Committee, 35th Session*, London, 19 pages. (*Lloyd's Register Engineering Services, 29 Wellesley Road, Croydon

CR0 2AJ, UK.)

Readman, J. W.* & Tolosa, I. Contamination of Mediterranean coastal waters from antifouling paints, presented at *Costings vs Benefits of TBT-based & Alternative Antifoulants*, International Conference, Malta, December 1995. (*IAEA Marine Environment Laboratory, 19 Av.des Castellans, B.P.800, MC 98012, Monaco.)

Richardson, B. A. & Cox, T. R. G. 1974. Aqueous organotin wood preservatives. *Tin & Its Uses*, **102**, 6–10. (International Tin Research Institute, Kingston Lane, Uxbridge, Middx. UB8 3PJ, UK.)

Schweinfurth, H.* 1987. Tributyltin compounds in wood preservatives: an update of health and safety aspects. *BWPA Convention, UK*. 18 pages. (*Schering AG, Postfach 650311, D1000, Berlin 65, Germany.)

Smith, P. J.* & Kumar Das, V. G. 1993. Tin in relation to toxicity: myths and facts. *2nd ANAIC International Conference on Materials Science & Environmental Chemistry*, Kuala Lumpur, Malaysia, 1993. 12 pages. (*International Tin Research Institute, Kingston Lane, Uxbridge, Middx. UB3 3PJ, UK.)

TBT copolymer anti-fouling paints: the facts, ORTEP Association and Marine Painting Forum, July 1992, 16 pages. (Organotin Environmental Programme Association, PO Box 70, 4380 AB Vlissingen, The Netherlands. Marine Painting Forum, BMT SeaTech Ltd, Northumbria House, Davy Bank, Wallsend, Tyne & Weir NE28 6UY, UK.)

Tisdale, W. H. 1943. *British Patent* 578, 312.

van der Kerk, G. M. J. & Luijten, J. G. A. 1954. The biocidal properties of organotin compounds. *Journal of Applied Chemistry*, **4**, 314–19.

Waldock, M.* *Recovery of Oyster Fisheries following TBT regulation*, presented at the Malacological Society of London, Queen Mary & Westfield College, London, November 1995. (*UK Ministry of Agriculture Fisheries & Food, Remembrance Avenue, Burnham on Crouch, Essex CM0 8HA, UK.)

Woods Hole Oceanographic Institution. 1952. *Marine fouling and its prevention*, prepared for Bureau of Ships, Navy Department US Naval Institute, Annapolis, Maryland, pp. 59, 65.

Ziegler, K., Gellert, H. G., Zosel K., Lahmkuhl H., Pfohl W. & Zosel, K. 1955; 1960. Herstellung von Aluminiumalkylen und Dialkylaluminium-hydriden. *Angewandte Chemie*, **67**, 424; Aluminiumtrialkyle und Dialkyl-Aluminiumhydride aus Olefinen, Wasserstoff und Aluminium. *Justus Liebigs Annalen der Chemie*, **629**, 1–13.

Acronyms

The following acronyms are found in this chapter.

BWPA: British Wood Preserving Association. Now known as –

BWPDA: British Wood Preserving and Damp-proofing Association (P.O. Box 894, London E15 4ED, UK).

CEFIC: Conseil Européen des Fédérations de L'Industrie Chimique (European Council of Chemical Manufacturers' Federations).

DWT: dead weight tonne.

IMO: International Maritime Organization (United Nations) (4 Albert Embankment, London SE1 7SR, UK).

ISO: International Organization for Standardization, Geneva.
ITRI: International Tin Research Institute.
LOSP: light organic solvent preservative.
MEPC: Marine Environment Protection Committee (IMO).
ORTEPA: Organotin Environmental Programme Association.
VLCC: very large crude carrier.

3

○ ○ ○ ○ ○ ○ ○ ○ ○ ○ ○ ○ ○ ○ ○ ○ ○ ○ ○ ○

The analysis of butylated tin compounds in the environment and in biological materials

Ph. Quevauviller

The origin and fate of butyltin compounds in the environment have been extensively studied in the past few years [1–6]. As described in Chapters 6 and 7 of this book, the toxicological impact of tributyltin (TBT) on marine organisms has been the major reason for a multitude of studies elucidating the biogeochemical pathway of this contaminant, together with its degradation products mono- and dibutyltin (MBT and DBT), in the environment. This interest and concern for the determination of butyltin compounds led to the development in the early 1980s of a wide variety of techniques which involve, in most cases, several different analytical steps: extraction, clean-up, derivatization, separation and final detection [7]. It is clear that accurate results for butyltin determinations is a prerequisite for good comparability of data and related studies on the toxicity and biogeochemical processes. However, the proliferation of multi-step techniques, although very sensitive, was not accompanied with the necessary awareness for the quality control of analysis. Only in the past five years has the availability of certified reference materials (CRMs), certified for their butyltin concentrations, allowed an improvement in the quality assurance of butyltin determinations in marine sediments. The overall quality of butyltin determinations is, however, far from being under control. Many techniques still need to be validated, some analytical steps (e.g. hydride generation) are often badly understood, and the number of CRMs is not sufficient to ensure a wide coverage of environmental and biological matrices. This chapter critically discusses the analytical techniques used in butyltin determinations following a 'step-by-step' approach, i.e. from sampling to final determination.

3.1   Sampling and storage

There are no particular strategies established for the monitoring of butylated compounds in environmental and biological matrices. However, many authors have reported critical points to be considered with respect to sample collection and storage to ensure a good representativity of samples with regard to butyltin species. The following remarks mainly deal with the risks of error that can be encountered.

3.1.1   Sample collection

The sources of error related to sample collection are numerous at the level of butyltin concentrations usually found in the environment ($ng l^{-1}$ or $ng g^{-1}$ levels). Risks of contamination may arise from PVC materials because monobutyltin and dibutyltin may be leached, even at neutral pH [8]. PTFE and polythene have been successfully used for water samplers and sieves for sediment treatment. Glass bottles have also been shown to be adequate, i.e. no strong butyltin adsorption was observed during water collection [9]. For large volumes of water, PTFE pumping systems were found to be suitable in butyltin surveys [10]. It was recommended to rinse the collection devices with water samples to avoid risks of adsorption. Nalgene sterile filtration units were used for the filtration of water samples for butyltin determinations and no adsorption nor contamination were observed both with bulk and centrifuged samples [11]. For sediment samples, PTFE spatulas and other plastic devices were used to scrape surface layers without any contamination.

3.1.2   Sample pre-treatment

Sample pre-treatment may be another source of possible discrepancies. For water samples, many authors monitor butyltin cations in filtered waters as the toxicological impacts are usually assessed with dissolved concentrations. However, some compounds are mostly adsorbed onto particulate material as is the case for TBT [11]. Consequently, the contamination assessment may be biased if filtered waters only are analyzed. As an example, analyses of bulk water samples with low amounts of suspended matter from the Rhine Estuary did not reveal high butyltin contents (less than $2 ng l^{-1}$ as Sn) whereas TBT concentrations of up to $200 ng g^{-1}$ (as Sn) were detected in suspended particles collected by centrifugation [11]. This clearly poses the problem of reliability of data with regard to the sample pre-treatment if an accurate estimation of contamination is required.

For sediment samples, sieving is very often used both to homogenise the samples and to compare the data obtained on the same basis, e.g. contents in the less than 63 µm fraction. The choice of this fraction is justified by the fact that butyltin species, particularly TBT, are strongly bound to the fine-grained sediment fractions [12]. It has been recommended to wet-sieve the samples with overlaying water from the sediment sampler to avoid possible desorption of butyltin cations during sieving [11]. Finally, the presence of light detrital fragments (mostly algal and leaf debris) were shown to induce sources of high variability of butyltin concentrations in sands as these particles are an important sink for such compounds [12].

3.1.3 Sample storage

Sample storage is an additional cause for unacceptable losses or contamination, and to which insufficient care has been paid. For waters, chemical and physical changes may occur upon contact with air and container walls, as well as resulting from temperature and pressure variations [13]. Sediments may also be subject to physico-chemical alterations when in contact with air (e.g. oxidation, flocculation). Furthermore, ongoing microbial activity may change the butyltin speciation, especially due to degradation.

The choice of adequate containers for the sample storage is of paramount importance. It was shown that no considerable adsorption of TBT took place in non-acidified water stored at 4 °C in Pyrex glass bottles over five months [14,15]. Some TBT adsorption onto Pyrex glass has been observed from synthetic solutions stored at pH 5.3 which was limited by using pre-equilibrated bottles; pre-waxing or silanization was less effective [16].

The stability of butyltin species in water may be achieved by sample acidification. Synthetic solutions of butyltin cations in HCl used in interlaboratory studies were shown to remain stable over some months [17]. Storage in the dark of filtered natural water samples acidified at pH 2 with HCl was demonstrated to be suitable for achieving a good stability for TBT and MBT, both at 4 °C and ambient temperature (c. 20 °C) whereas the stability of DBT was more doubtful [18]. However, the stability of butyltin species appeared to be less easily achieved in samples containing high amounts of suspended matter due to possible interferences with particulates and/or microbial communities [18].

Systematic studies have shown that several different treatments for sediment storage (freezing, wet storage at 4 °C) and drying (air-, freeze-

and oven-drying) were all suitable to preserve the stability of TBT over at least four months. However, MBT and DBT were often subject to variations (mostly losses) in most of the experiments. Freezing followed by oven-drying (at 50 °C) was shown to be the most suitable treatment to avoid changes in concentrations for the three species, but this procedure should be applied to short term storage only as microbial activity may not be completely stopped at 50 °C. In case long term periods are needed, freeze-drying is a better compromise [18].

3.2 Analytical methods

The main analytical methods used for butyltin determinations have been described in extensive reviews [19–23]. Most of them are based on separation by gas chromatography (GC) and detection with the classical types of detectors. Some of the first methods appeared in the 1970s. Meinema and co-workers [24] presented a method for the determination of butyltin species by solvent extraction, Grignard derivatization prior to GC separation and detection by mass spectrometry. Since then, a wide variety of techniques has been developed. Almost all of them are based on at least four analytical steps: extraction/concentration; derivatization; separation; and detection. The different steps have to be adapted for the different hyphenated techniques and their multiplicity enhances the chances of error leading. The following section will discuss in detail the various procedures individually.

3.2.1 Pre-concentration

Sampling and pre-concentration can be performed simultaneously with the use of an *in situ* sampler. A time integrated, remotely moored, automatic sampling and concentration device allowed the direct concentration of butyltin species on an octadecyl (C-18) bonded reversed-phase solid sorbent [25]. This approach is promising but leads to risks of degradation of butyltin compounds. Moreover, the highly acidic SiOH groups found in the silicate sorbent phases may contribute to the misidentification of the butyltin species in water samples. Preconcentration of TBT from estuarine waters has also been achieved using C-18 bonded solid phase adsorbent prior to hydride generation [26].

3.2.2 Extraction

The extraction should be done in such a way that the butyltin compounds are separated from the matrix without loss or contamination [27], without changing the speciation, and with the minimum of interferences. In general, acid extraction procedures are carried out when AAS is used as the final measurement step, whereas extraction with organic solvents is performed for other hyphenated techniques such as GC/FPD or HPLC-coupled techniques.

A wide variety of acid extraction procedures have been used for sediment analyses, e.g. using leaching with HCl/CH_3OH [28] or $CaCl_2/HCl$ [29]. Glacial acetic acid, in preference to other acids, has been utilized to extract TBT efficiently from sediment and biological matrices when using AAS as final detection [30]. Other leaching procedures have been used for biological samples (algae and invertebrates) and vegetable tissue, in particular extraction with dilute acetic acid [31], hydrochloric acid [32] or sodium hydroxide [33]. Other tests have been conducted which demonstrated a higher extraction efficiency for acids in comparison to organic solvents [34].

Butyltin species have to be extracted into a non-polar solvent prior to derivatization by a Grignard reaction. Solvents commonly used are benzene, hexane and dichloromethane. Acidification of the samples is required and is usually achieved with HCl. The addition of HBr has been shown to enhance the recovery of butyltin species [24,35–37] probably by preventing adsorption on the walls of the containers. The first extraction scheme using a solvent is generally sufficient for TBT but mono- or dibutyltin require complexation with tropolone for an efficient extraction [35,37,38]. Extraction into pentane of diethyldithiocarbamate complexes was successfully applied after pentylation [39]. Toluene has also been used for the extraction of butyltin species from mussel tissues [40].

Simultaneous extraction into dichloromethane and hydride generation of butyltin species has demonstrated excellent recoveries of these compounds in natural water samples [41,42]. This method can process large volumes of water (0.8 to 1 l); however, errors are likely to occur due to a necessary concentration/evaporation step. This has been a problem for MBT, with observed losses up to 50%.

Solvent extractions are required for the HPLC/AAS technique. These are carried out for water, sediment and biological samples, with dichloromethane/chloroform [43], chloroform [44], or toluene [45,46] often with addition of tropolone. The best recoveries were obtained with toluene [47] or hexane [48]. The extraction of butyltins from water on a tropolone loaded reverse phase column has also been shown to be efficient [36]. However, tropolone forms stable complexes with MBT and DBT, but not with TBT, and extraction difficulties were noted for water containing suspended matter when using toluene/tropolone [49]. Solvents such as chloroform or hexane are suitable for strong cation exchange HPLC but tropolone has to be avoided due to its high complexing power [36]. The concurrent quantitative extraction of all butyltin species is thus not always possible. Simultaneous extraction of other organic compounds present in the sample and their subsequent injection in the HPLC column leads to a rapid degradation of the column. In other procedures, the solvent is evaporated and replaced by a more polar one, leaving an organic residue; this may protect the column but some losses of butyltin species in the residue may occur.

A new extraction procedure involving microwaves has recently been tested successfully for butyltin determinations [50]. Extraction could be performed in a few minutes using an energy-controlled microwave procedure which allowed the preservation of the butyltin species. Finally, supercritical fluid extraction is a very promising development for the butyltin determination in environmental matrices since this method seems to offer good possibilities for efficiently extracting the species without alteration [51].

A good assessment of quality assurance implies that the extraction recoveries are verified. This can be achieved by spiking a sample of similar composition as the sample analyzed with a known content of the analyte of concern, leaving them to equilibrate and subsequently determining the analyte after extraction. The major drawback is that the spike is not always bound the same way as the naturally occurring compounds. Alternatively, and only if the extraction procedure does not change the matrix composition, the recovery experiment may be carried out on the previously extracted real sample by spiking, equilibration and extraction. However, the recovery assessment can often be over-estimated and this risk should be faced. Certified reference materials (CRMs) may be a tool to ascertain accuracy; however, they are only useful when they contain incurred, and not spiked, species. The extraction recovery may vary from one butyltin species to another [52]. Consequently, the recovery should be assessed independently for each compound, as well as for the compounds together. It is also necessary to find a compromise between a good recovery (sufficiently strong attack) and the preservation of the sample integrity with respect to speciation.

An interlaboratory study on extraction efficiency has shown that the methods tested were, in general, correct for DBT and TBT, however, MBT was never recovered quantitatively [53]. Ten methods were tested: direct extraction with organic solvent with or without a chelating agent, acid leaching, alkali leaching, and combinations of these techniques. Table 3.1 gives some results of butyltin recoveries obtained in a certification campaign using a coastal sediment (BCR, CRM 462 [54]).

3.2.3   Clean-up

Some derivatization procedures, such as hydride generation, often require a clean-up step owing to the large amounts of lipids which are extracted from biological tissues with organic solvents. Direct analysis of the extract may be performed but the high lipid content may partly inhibit the derivatization reaction and deteriorate the GC column, thereby leading to poorer analytical performance. A clean-up procedure can remove more than 90% of the lipid content of the sample but does not eliminate pigments or other extracted materials [43]. Despite the risk of losses, the clean-up step has generally been shown to increase the sensitivity and to improve the tailing of the chromatographic peaks by drastically decreasing the occurrence of interfering compounds. Several types of commercial silica gel cartridges have been used which all gave satisfactory results. Best recoveries were obtained with 5% water-deactivated cartridges. Controlled moisture silica gel or alumina eliminates the main part of organic matter

Table 3.1   *Comparison of extraction recoveries obtained in an interlaboratory exercise for butyltin species in coastal sediment [54]. Results correspond to the means of five replicate and standard deviations*

Procedure	MBT	DBT	TBT
Acetic acid HG/GC/AAS	106 ± 13	85 ± 4	118 ± 15
Toluene/tropolone GC/FPD	—	—	98.3 ± 1.3
NaOH/methanol GC/FPD	—	87 ± 3	87 ± 5
SFE/GC/FPD	—	—	82 ± 6
Methanol/tropolone GC/MS	107 ± 5	102 ± 9	105 ± 7
Diethylether/HCl/tropolone	99 ± 5	109 ± 7	101 ± 7

HG/GC/AAS: Hydride generation/gas chromatography/atomic absorption spectrometry.
GC/FPD: Gas chromatography/flame photometric detection.
SFE: Supercritical fluid extraction.
GC/MS: Gas chromatography/mass spectrometry.

extracted from organic-rich sediment but also partly traps the butyltin compounds [47]. An alternative clean-up procedure relies on the saponification of the sample prior to dichloromethane extraction; however, the recovery of TBT in these conditions is low [55].

3.2.4 Derivatization

Derivatization procedures can be employed to separate butyltin species from their matrices and interferences, and to concentrate the species. Some derivatization methods such as hydride generation, for example, are used to generate volatile species which are more easily separated from each other by chromatography. Reactions used currently are mostly centred around the addition of simple groups, involving alkylation with Grignard reagents (e.g. pentylation, ethylation, butylation).

3.2.4.1 Derivatization by Grignard reaction

Many authors use hyphenated techniques which includes a reaction with a Grignard reagent in a solvent medium in order to convert the alkyltin compounds $R_x SnX_{(4-x)}$ into non-polar mixed tetra-alkyltins. The alkylation reaction is then followed by GC separation and Sn detection. This procedure has been applied successfully to various matrices such as water, sediment and biological samples.

Derivatization has to be performed in an aprotic solvent and a drying stage is required prior to the reaction. Peralkylation with a selected R' group Grignard reagent substitutes an alkyl group for the counter-ion to convert the organotin cations into volatile tetra-alkylated derivatives in the solvent medium. The R' groups which have been used include methyl- [24,38], ethyl- [56], butyl- [57], pentyl- [58,59] or hexyl- [60] Grignard reagents. The choice of the size of the alkyl group R' depends both on the volatility of the compounds to be determined and on the separation and detection system. Grignard reaction with hexyl-groups requires a high efficiency capillary GC. Methylation or butylation derivatization have been abandoned by various authors because mixed methyl-butyltin compounds have been claimed to occur in the environment. In addition, the low boiling point of the methyl derivatives may lead to losses during the concentration step. For these reasons, ethylation using tetraethylborate [53,56] and pentylation with e.g. pentyl-magnesium bromide [36,58,61,62] or methyl-magnesium chloride [63] have been successfully applied to the analysis of environmental samples. Hexyltins are also more thermally stable and consequently hexylmagnesium bromide was used to improve the separation between butyltin species [60,64].

Finally, this derivatization technique presents the advantage of limiting losses by evaporation [65]. Water destroys the reagent and the species of interest may be removed from water-based matrices by, for example, extraction of a diethyldithiocarbamate complex into an organic phase prior to derivatization [66].

3.2.4.2 Hydride generation

The reaction of tin and alkyltin compounds with sodium borohydride yielding butyltin hydrides has been known for two decades [67,68]. This derivatization method has been successfully used for the determination of butyltin species [33,69] and it requires a slight acidification of the samples using either acetic acid or nitric acid. Acetic acid ($0.05-0.1 \text{ mol } 1^{-1}$) has been shown to enhance the derivatization yield in comparison to hydrochloric or nitric acid [70]. This reaction also requires that oxygen be purged by a flow of inert gas (e.g. helium). The sodium borohydride may be injected directly in the stirred sample [71] or from a modified hydride system [72].

Hydride generation has been found to be most suitable for the analysis of large volumes of water. However, water samples rich in suspended matter require an extraction step, otherwise very low recoveries (around 15%) are obtained [18]. In addition, inhibition of the reaction due to the presence of hydrocarbons has been demonstrated [42].

With regard to sediment analysis, interferences from metals [33], organics or sulphides have been shown [56,71,73]. Studies of the determination of butyltin species in a sediment heavily polluted in organic matter demonstrated that the inhibition of hydride generation could be lowered by increasing the $NaBH_4$ addition [30,73]. An extensive study has been recently carried out to study the interferences generated by inorganic and organic substances on hydride generation for butyltin determinations in sediment rich in organic matter [74]. Whereas organic compounds (e.g. organic solvent, PCBs, pesticides, humic substances) had only small effects on the signal suppression for DBT and TBT, MBT was severely affected. The most serious interferences encountered for the three butyltin species were associated with the presence of elevated metal content.

Another method involving hydride generation is based on a liquid–liquid extraction of butyltin species followed by derivatization in an aqueous medium or in a solvent [44,75,76]. Butyltin hydrides are gently concentrated by slow evaporation of the solvent and injected into a gas chromatograph. This procedure has been used mainly for DBT and TBT and has been applied to water and sediment samples but most applications were carried out on biological tissues. It requires a clean-up step, generally by passing the extract over a silica microcolumn.

On-column hydride generation has also been used which allowed a direct injection of a solution of butyltin chloride in the gas chromatograph to be performed [56,77]. Derivatization of the extracts is performed directly on top of the chromatographic column [77] or via a packed reactor placed in the injector [44]. Finally, an in-line photolysis coil in a continuous flow system of high-performance liquid chromatography was coupled with hydride generation and flame AAS. Irradiation with ultraviolet light was used to convert TBT into inorganic tin(IV) from which a volatile hydride could be produced after HPLC separation [78].

3.2.5 Separation

The separation of butyltin species can only be performed by techniques which do not destroy the chemical forms. Three basic separation methods exist for speciation analysis: chromatography (e.g. anion exchange, ion pairing reversed phase liquid chromatography (LC) or capillary GC), cold trapping or capillary zone electrophoresis. To our knowledge, the latter technique has not yet been used for the determination of butyltin species.

3.2.5.1 Cold trapping

Cryogenic trapping is often used after hydride generation. The hydride species are carried by an inert gas (helium or nitrogen) to a cryogenic trap (column plunged in liquid nitrogen) and sequentially desorbed according to their respective boiling points [72] after removing the column from liquid nitrogen and progressively heating it (from $c. - 196$ to $200\,°C$). The column is usually a short U-tube (20–40 cm), made of either PTFE, silanized glass or borosilicate glass, and filled with chromatographic material to achieve a better definition of the peaks; the use of a longer column (1 m spiral) allowed a good peak resolution to be obtained [79]. In some cases, only glass wool was used instead of chromatographic material [80]. Improvement in the design of the column and the temperature programme can obviate the use of a water trap [30,72]. A considerable improvement has been made with the possibility of controlling the heating temperature of the column [81].

3.2.5.2 Gas chromatographic separation

Hydrides can be separated by GC [82,83] with a possible pre-concentration step (tenax-GC). Capillary GC has been used after Grignard derivatization either prior to FPD or microwave induced plasma atomic emission spectrometry (MIP-AES) detection [39,66] or MS [63].

3.2.5.3 HPLC separation

Many attempts to set up an HPLC/AAS method for the determination of butyltin species have been made. The HPLC modes were: ion exchange [84], ion pairing [85], micellar and normal. Normal HPLC performed directly on non-polar extracts in the presence of tropolone has been shown to be efficient for the determination of butyltin species in sediment [52]. Gradient elution by a mixture of tropolone in toluene and methanol allowed a complete separation of mono-, di-, tri- and tetrabutyltin to be achieved. An isocratic elution using tropolone was shown to be suitable to resolve the di- and tri-butyltin species; however, a risk of butyltin degradation on the reversed phase material has been observed [88]. HPLC-ETAAS of tropolone complexes of DBT and TBT was successfully applied for water and sediment analysis [49].

HPLC with cyanopropyl-bonded silica column allows a satisfactory separation of butyltin species [86,87]. Ebdon *et al.* [88] have developed a HPLC procedure for the analysis of sea water, comprising a pre-concentration step, HPLC separation and a final step during which the complex tin species morin is formed in an aqueous, micellar medium. DBT and TBT were also separated by a similar procedure, i.e. ion-exchange HPLC and determination by a post-column reaction with morin in a micellar solution [84].

3.2.6 Detection

The detectors used for butyltin determinations are either element specific (e.g. AAS) or non-specific (e.g. FID, FPD, ECD). In general, the compound should arrive alone into the detector to avoid interferences. The choice of the detector depends on the chemical forms to be determined and on the mode of separation used.

3.2.6.1 After cold trapping

A well-known technique involving cold trapping is the detection of butyltin species (after hydride generation) by quartz furnace atomic absorption spectrometry. The hydrides are atomized in the quartz tube heated to 900–1000 °C by an electrical furnace or a classical AAS burner. The atomization rate is enhanced by the addition of hydrogen and oxygen into the quartz furnace. The optimal conditions of flow rate and the geometry of the quartz cell vary widely according to the literature [33,47,89]. In addition, the position of the oxygen and hydrogen inlets seems to be critical as variations of a factor of 20 in sensitivity between different cell designs are not uncommon. Both hollow cathode and electrodeless discharge lamps are available, and several wavelengths can be used for the determination of tin, namely 224.6 nm (the most commonly used), 235.5 nm, 284.0 nm and 286.3 nm. One disadvantage in using a tin-specific detector such as AAS after cold trapping is that species with similar boiling points will arrive at the same time in the detector and will hence not be separated. This problem has been identified for tin species such as dimethylbutyltin and monophenyltin [90]. However, no overlap was noted with respect to butyltin species.

3.2.6.2 After gas chromatography

The electron capture detector is a classical technique in gas chromatography which has been used for the determination of di- and tributyltin chlorides [82,83]. Several developments have been made to obtain quantitative data from biological materials. However, detection limits obtained with the ECD are generally poor and the span of alkyltin determined in one sample is limited. One of the advantages of the detection as chloride is the possible direct concentration of, for example, tributyltin as its chloride salts 'in the field' using a solid phase adsorbent. In estuarine and marine environments, where the toxicity problems of TBT are more acute, the probability of the occurrence of this compound as the chloride salt is high. The direct extraction by adsorption/pre-concentration on C-18 bonded, GC separation and ECD detection could eliminate many analytical steps in the determination protocol for butyltin species [91]. It also combines the sample pre-concentration and may facilitate storage problems mentioned earlier.

Other detectors are often used following gas chromatography, and among them, mass spectrometric detection was developed at an early stage [24]. Although not very sensitive at the outset, this technique has benefited from rapid progress and excellent detection limits were recently obtained in sediment and biological matrices [92,93]. Isotope dilution MS represents a very promising analytical development for the butyltin determinations [94].

Although tin is difficult to excite thermally in flame, several important developments have circumvented this difficulty and flame photometric detection was successfully used after gas chromatography [63,95]. The quantitation is based on the monitoring of the red fluorescence molecular emission of the Sn–H species at 609.5 nm. This type of detector allows an excellent sensitivity to be obtained [63,93]. The best overall sensitivity was obtained with a filterless FPD but this method lacked specificity [96].

To improve the selectivity, an additional scan can be performed in the broad and less sensitive 360–490 nm blue region (SnO band) [97]. Butyltin species were also determined in mussel tissues by GC/FPD using a quartz surface-induced luminescence method [40]. Finally, FPD was recently optimized for SFE [98].

Flame ionization detection was also used, based on the ability of the vapour phase organotins to quench ionization in a hydrogen atmosphere flame [99]. Atomic absorption and atomic emission detection were both successfully used after Grignard reaction and gas chromatography for the determination of butyltin species [66]. Microwave-induced plasma AES was also applied [39].

3.2.6.3   After HPLC separation

Flame AAS is of reduced interest as the detector after HPLC separation owing to its insufficient sensitivity [100]. Extending atom residence times by a slotted-tube atom trap improved the absolute detection limits down to 200 ng which was still too high for many applications [45]. The more sensitive electrothermal AAS has therefore been used instead [79,101]. The design of the interface between the continuous flow of eluent out of the HPLC column and the discontinuous injection of sample in the graphite furnace often hampers a good determination [69]. Difficulties may occur with ion chromatography in which the high salt content of the eluent may lead to serious matrix effects for ETAAS determination and may produce very high background molecular absorption.

Detection after HPLC has also been performed by inductively coupled plasma atomic emission [90,102]. Methods for the nebulization of the solution to be analyzed were developed in order to increase the residence time of the analyte in the plasma and therefore to obtain a better sensitivity [103]. The advantages of this technique in comparison to AAS are the relatively easy coupling between HPLC and ICP and the continuous measurement of tin concentrations.

An improvement has been obtained by coupling HPLC with ICP-MS. In this technique, the detection is obtained after nebulizing the HPLC eluent in an ICP which is itself coupled with a mass spectrometer. On-line monitoring of the HPLC effluent is possible together with a multi-element detection [104,105]. Following separation by HPLC, higher sensitivity was obtained for TBT determination by ICP-MS than by ICP-AES [84]. As indicated above, isotope dilution followed by ICP-MS offers good possibilities for the determination of butyltin species in environmental matrices [94].

High performance thin layer chromatography (HPTLC) has been used to identify tin species in extracts of wood or environmental samples. This technique, although not being very sensitive when a colorimetric detection is carried out, may be useful prior to the quantification of butyltin species by a more sophisticated technique [106]. TBT was determined by coupling HPLC with a sensitive and selective laser-excited atomic fluorescence spectrometry detector [107]; detection limits were comparable to those reported for HPLC-ETAAS [108].

3.2.6.4 Other techniques

Electrochemistry has been applied to the determination of butyltin compounds. Differential pulse anodic stripping voltammetry (DPASV) was used for $DBTCl_2$ in natural water [109,110], while differential pulse detection has been applied to tin species in a liquid chromatographic effluent [111]. Ionspray mass spectrometry/mass spectrometry has also been successfully used after solvent extraction [112], but instrumental requirements will limit the widespread application of such techniques.

3.3 Quality control and evaluation of the method's performance

Basic quality assurance principles should be considered in tin speciation analysis in order to ensure that accurate results with an acceptable uncertainty are obtained. An overview of these principles applied to speciation analysis has been recently published [113]; they involve in particular (i) statistical control principles (e.g. use of control charts), (ii) the evaluation of the method's performance by comparing the procedures with alternative methods or (iii) the use of certified reference materials (CRMs). In the context of this chapter, only points (ii) and (iii) will be discussed.

3.3.1 Calibration

A prerequisite for a good result is a correct calibration. Although calibration is considered as being obvious, experience has shown that this step is often not considered with sufficient care. It is perhaps necessary to stress once again that compounds of well-known stoichiometry should be used and that careful consideration should be given to the choice of the method of calibration: using pure calibrant solutions, matrix-matched solutions or standard additions. Very often yields of reactions (e.g. derivatization) have to be determined which can be considered as being part of the calibration process. In such a case, the calibrant should be taken through the whole determination procedure. However, in order to investigate the possible sources of error, it would be more adequate to calibrate the final determination step with the compound measured in reality (i.e. the derivative) and to run a recovery experiment with the analyte. Although being more time and labour consuming, such verifications are to be performed at regular intervals to detect possible errors and they therefore contribute to the laboratory's quality assurance programme. To date, however, the possibility of obtaining pure calibrants of derivatized tin compounds is rather limited and this need should be considered by chemical companies.

3.3.2   Interlaboratory studies

When all necessary measures have been taken in the laboratory to achieve accurate results, the laboratory should demonstrate its accuracy in intercomparisons, which are also useful for detecting systematic errors. In general, besides the sampling error, three sources of error can be detected in all analyses:

> sample pretreatment (e.g. extraction, separation, clean-up pre-concentration etc.);
> final measurement (e.g. calibration errors, spectral interferences, peak overlap, baseline and background corrections etc.);
> laboratory itself (e.g. training and educational level of workers, care applied to the work, awareness of pitfalls, management, clean bench facilities etc.).

When laboratories participate in an intercomparison, different sample pretreatment methods and techniques of separation and final determination are compared and discussed, as well as the performance of these laboratories. If the results of such an intercomparison are in good and statistical agreement, the collaboratively obtained value is likely to be the best approximation of the truth.

Before conducting an intercomparison the aims should be clearly defined. An intercomparison can be held: (i) to detect the pitfalls of a commonly applied method and to ascertain its performance in practice, or to evaluate the performance of a newly developed method; (ii) to measure the quality of a laboratory or a part of a laboratory (e.g. audits for accredited laboratories); (iii) to improve the quality of a laboratory in collaborative work with mutual learning processes; and (iv) to certify the concentrations of a reference material. In the ideal situation, where the results of all laboratories are under control and accurate, intercomparisons of type (ii) and (iv) will be held only. For the time being, however, types (i) and (iii) play an important role.

Intercomparisons can be organized in a step-by-step approach to test the different parts of analytical procedures, i.e.

> pure calibrant solutions to test the final detection;
> cleaned extracts to test the separation step;
> raw extract to test both the separation and the clean-up steps;
> spiked samples to test the extraction efficiency (as a first approach, knowing that samples with incurred species are better suited);
> natural samples to test the whole procedure.

All methods have their own particular sources of error. For instance, for some procedures, errors may occur due to an incomplete derivatization, a step which is not necessary in other techniques such as high performance liquid chromatography (HPLC). The latter technique, however, may have errors such as incomplete separation, which are not encountered, or to a lesser extent, in the former procedure. If the results of two independent methods are in good agreement, it can be concluded that the risk of systematic error (e.g. insufficient extraction) is limited. This conclusion is stronger when the two methods differ widely, such as derivatization/GC-AAS and HPLC/ICP-MS. If the methods have similarities, such as an extraction step, a comparison of the results would most likely lead to conclusions concerning the accuracy of the method of final determination, and not regarding the analytical result as a whole.

A comparison of methodologies for butyltin determinations was carried out by the National Research Council of Canada [48]. The majority of measurement techniques were based on the selective detection of tin, such as FPD, ICP-AES, ICP-MS, AAS etc. Separation was gas or liquid chromatography or selective extraction involving several organic solvents, sometimes in conjunction with acids and tropolone. For GC, butyltin species were separated and eluted as hydrides, tetraalkylated forms or halides after appropriate derivatization. LC was the second most popular means of separation, in ion exchange or ion pairing chromatography, detection being by ICP-MS or ICP-AES, FAAS or ETAAS. A good agreement was obtained between the different techniques which allowed a CRM to be produced as mentioned below (PACS-1).

Intercomparisons were organized within the BCR-programme (Community Bureau of Reference of the European Commission, Brussels) for improving the quality of tin speciation analyses [17]. This programme involved 15 laboratories from 7 European countries. The first round-robin dealt with the analysis of solutions containing pure analytes (TBT and mixtures of MBT, DBT and triphenyltin). No systematic errors could be detected in the final determination techniques tested at this stage. A second exercise was undertaken in 1989 on the determination of TBT in a spiked sediment. The results of this interlaboratory trial did not reveal any systematic errors in the different analytical methods compared, which encouraged the BCR to organize a certification campaign on TBT in a sediment containing a representative level of TBT (i.e. around 100 ng g^{-1}

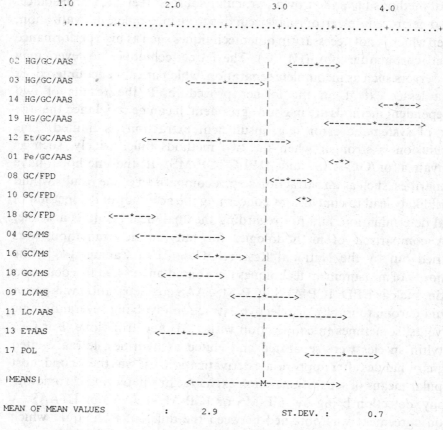

Figure 3.1 Intercomparison of TBT in a spiked sediment (concentrations are given as TBTAc mass fraction in μg g^{-1}). The laboratory codes are indicated along with the methods used (abbreviations defined in Table 3.2). The results plotted correspond to five replicate determinations. MEANS is the average of laboratory means and one standard deviation.

of TBT). Figure 3.1 shows the results obtained in the intercomparison of TBT in a spiked sediment. The coefficient of variation (CV) obtained between laboratories (25%) was considered to reflect the state of the art at that stage and the group of experts recommended to proceed with the organization of a certification campaign. The participants recognized, however, that a better agreement should be achieved for certification. No particular source of error due to a method could be detected and the analytical techniques used in this intercomparison were therefore found suitable for certification.

Interlaboratory studies for the certification of reference materials allowed the investigation of additional techniques used for butyltin determinations under conditions closer to reality, i.e. with samples containing butyltin levels in the range of concentrations usually found in the environment (20–200 ng g^{-1}). The techniques used in the certification involved solvent or acid extraction, derivatization (e.g. Grignard reaction or hydride generation), separation by gas chromatography (GC) or high performance liquid chromatography (HPLC) and various methods of final detection (e.g. quartz furnace atomic absorption spectrometry, flame photometric detection etc.) [54]. Table 3.2 gives an account of the procedures used by each laboratory. The results submitted in the interlaboratory studies were discussed amongst all participants in technical meetings. They were presented in the form of bar-graphs showing the laboratory codes and the methods used, the individual means and standard deviations and the mean of laboratory means with its standard deviation. Figure 3.2a, b shows the sets of results for TBT and DBT in the first interlaboratory study (TBT and DBT in a coastal sediment, CRM 462) after technical scrutiny. Valuable comments were made in the technical discussion which are summarized below.

> For TBT, poorer extraction recoveries were often observed in comparison to recoveries obtained from other sediment materials. Some laboratories found lower recoveries if the spike was allowed to equilibrate longer. The previous interlaboratory exercise on TBT-spiked sediment had identified the need to allow spikes to equilibrate at least overnight to get a realistic assessment of extraction recoveries.
> The evaluation of polarographic methods showed that there were major problems with this sample. The surfactants present made it difficult to detect TBT and polarography was, therefore, a method not recommended for this certification.

Table 3.2 *Summary of the analytical methods used in the interlaboratory study of butyltin species in coastal sediment CRM 462 [54]*

Lab. codes	Sample pre-treatment, derivatization, separation and calibration	Final determination
01	Extraction with acetic acid/hexane/DDTC in pentane. Pentylation with Grignard reagent. Capillary GC separation. TBTOAc and DBTCl$_2$ as calibrants. TPrT as internal standard. Calibration graph	QFAAS Wavelength 286.4 nm addition of air/H$_2$
02	Acetic acid extraction. Hydride generation with NaBH$_4$. Cryogenic trapping (U-tube) and GC separation (packed column). TBTCl and DBTCl$_2$ as calibrants. Standard additions	QFAAS Wavelength 286.4 nm addition of H$_2$/O$_2$
03	Acetic acid extraction. Hydride generation with NaBH$_4$. Cryogenic trapping (U-tube) and GC separation (packed column). TBTCl and DBTC$_2$ as calibrants. Calibration graph. TET as internal standard	QFAAS Wavelength 224.6 nm addition of H$_2$/O$_2$
04	Diethylether/HCl/tropolone extraction. Pentylation with Grignard reagent. Capillary GC separation. TBTO and DBTCl$_2$ as calibrants. DHTC as internal standard. Calibration graph	Mass spectrometry of masses 247, 249, 303 and 305 for TBT, and masses 307 and 319 for DBT
05	HCl/methanol/tropolone extraction. Pentylation with Grignard reagent. Capillary GC separation. TBTCl and DBTCl$_2$ as calibrants. TPrT as internal standard. Calibration graph	Mass spectrometry of masses 301, 303 and 305 for TBT, and masses 305, 317 and 319 for DBT
06	For TBT, supercritical fluid extraction (CO$_2$/HCl/MeOH); for DBT, MeOH/HCl extraction. Ethylation with Grignard reagent. Capillary GC separation. TBTCl and DBTCl$_2$ as calibrants. TPrT as internal standard. Calibration graph	FPD Wavelength 610 nm addition of N$_2$
07	HBr/tropolone extraction. Pentylation with Grignard reagent. Capillary GC separation. TBTCl and DBTCl$_2$ as calibrants. DMTPe$_2$ as internal standard. Calibration graph	FPD Wavelength 610 nm addition of H$_2$/air

Table 3.2 *continued*

Lab. codes	Sample pre-treatment, derivatization, separation and calibration	Final determination
08	NaOH/hexane extraction. Hydride generation with $NaBH_4$. Capillary GC separation. TBTO and $DBTO_2$ as calibrants. TPrT as internal standard. Calibration graph	FPD Wavelength 610 nm addition of N_2
09	Acetic acid extraction. Back-extraction with toluene. HPLC separation. TBTOAc and $DBTAc_2$ as calibrants. Standard additions	ICP-MS of mass 120

ETAAS: electrothermal atomic absorption spectrometry
FPD: flame photometric detection
Et: ethylation
GC: gas chromatography
HG: hydride generation
HPLC: high performance liquid chromatography
ICP-MS: inductively coupled plasma mass spectrometry
MS: mass spectrometry
Pe: pentylation
QFAAS: quartz furnace atomic absorption spectrometry

In some cases (e.g. Labs 04, 07 and 08), the uncertainty of extraction recoveries were not taken into account in the overall uncertainty of the laboratory means which explains the small standard deviations observed.

In the case of DBT, as shown in Figure 3.2b, the overlap obtained between the different sets of results and the range of techniques used was found satisfactory and, as no doubts were thrown on the results presented, the DBT content was agreed to be certified.

A high scatter of results was observed for MBT which prevented certification. Many laboratories reported problems in the extraction step. The addition of complexing agents such as sodium diethyldithiocarbamate may enhance the extraction recovery. However, the addition of complexing agents was found to prevent the hydride generation of a volatile species for MBT. Considering the large variation of results (ranging from 13 to 244 ng g^{-1} as MBT) and the lingering doubts regarding the different techniques used, it was decided not to give any indicative values for MBT which could be misused by laboratories. Further efforts should be made to improve the state of the art of MBT determinations before contemplating certification of this compound at the $ng\,g^{-1}$ level. A similar lack of agreement of MBT results has also been shown in a recent comparison of analytical techniques [53]. Moreover, this study confirmed that the techniques used for DBT and TBT determinations were satisfactory.

Another exercise examining TBT in harbour sediment (CRM 424) was organized. The very low TBT (around $20\,ng\,g^{-1}$) level and the

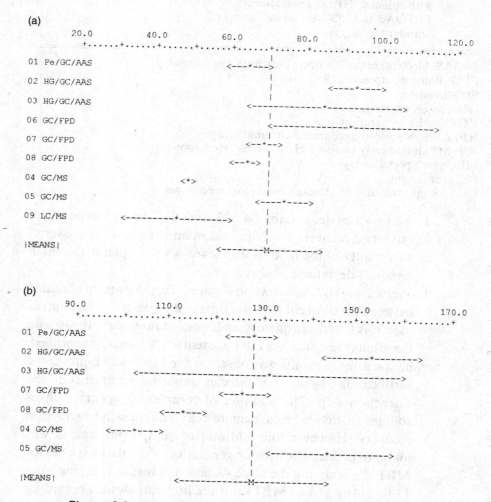

Figure 3.2 Bar graphs of TBT and DBT. The results correspond to five replicate determinations. MEANS is the mean of laboratory means with 95% confidence interval. (a) Tributyltin in $ng\,g^{-1}$ (as TBT); (b) dibutyltin in $ng\,g^{-1}$ (as DBT).

presence of a high content of organic matter created tremendous difficulties for analytical techniques using derivatization [114]. The original data showed a very large scatter, with results ranging from less than $10 \, \text{ng g}^{-1}$ of TBT to more than $150 \, \text{ng g}^{-1}$. Detailed discussions were necessary to explain the sources of discrepancies. Some laboratories reported not detected values: four laboratories using hydride generation/ GC/AAS reported not detected values which corresponded to less than, respectively, $15 \, \text{ng g}^{-1}$, $20 \, \text{ng g}^{-1}$ and $146 \, \text{ng g}^{-1}$. Other laboratories using selective extraction/ETAAS and GC/MS reported values less than, respectively, $49 \, \text{ng g}^{-1}$ and $60 \, \text{ng g}^{-1}$. One laboratory using HG/GC/AAS submitted two results which were below their limit of determination: $15.6 \, \text{ng g}^{-1}$ and $22.4 \, \text{ng g}^{-1}$. Later on, this laboratory submitted five other replicates (means of $(12.0 \pm 3.9) \, \text{ng g}^{-1}$) which confirmed the values found.

The laboratories reported their results of extraction recovery which were generally acceptable. A variety of extraction and clean-up procedures had been used in the exercise which thereby prevented the detection of any substantial systematic errors, for example due to an incomplete extraction of TBT. It was suspected that the main problems were not due to extraction but to possible interferences in the derivatization and/or in the detection step. It was assumed that aromatic compounds could have inhibited the hydride generation but not the ethylation reaction. However, methanolic-HCl used for extraction was not found to extract the oil but still a 35% suppression of hydride generation was observed. The interferences were thought to occur at the atomization stage rather than in the hydride generation step. High concentrations of Fe and Cr could have led to a suppression of the TBT signal, as demonstrated with spiking experiments. It was felt, however, that the methods using derivatization and AAS as final determination were in considerable difficulty with this complicated matrix, due to either unknown interferences or insufficient determination limits. As observed by the participants, the laboratories using gas chromatographic separation and detection either by FPD or MS tended to agree, which would confirm that these methods would be more suited for the determination of TBT in this particular material. This matrix should be reserved for the evaluation of techniques such as GC/FPD and GC/MS. It was also emphasized that such a complicated material should be recommended to very experienced laboratories only. Because of the potential risks of interferences, calibration by standard additions is a prerequisite.

3.3.3   Certified reference materials

The only possibility for any laboratory to verify the analytical performance in a simple manner (i.e. without being obliged to participate systematically in interlaboratory studies) is to use a so-called matrix reference material certified in a reliable manner. The laboratory which measures such an RM by its own procedure and finds a value in disagreement with the certified value is thus warned that its measurement includes an error, for which the source must be identified. Thus, CRMs having well known properties should be used, in particular, to:

, verify the accuracy of results obtained in a laboratory;
monitor the performance of the method (e.g. cusum control charts);
demonstrate equivalence between methods.

The conclusion about the accuracy obtained for the unknown sample must always be a conservative one. If the laboratory finds wrong results for a CRM, it is by no means certain of a good performance for the unknown. However, even if the laboratory finds a value in agreement with the certified value (according to ISO-Guide 33 [115]), it should realize that owing to possible discrepancies in composition (e.g. by way of binding of analytes, fingerprint pattern of interferences, different concentration levels etc.) between the CRM and the unknown, there is a risk that the result of the unknown may be wrong. Therefore, the use of as many relevant CRMs as possible is necessary for good quality assurance.

There are a number of suppliers of RMs and CRMs for the various fields of analysis. Very few CRMs, however, exist at present for the quality control of butyltin determinations in environmental and biological matrices. Table 3.3 lists the CRMs available in 1994 from three certifying bodies: the National Research Council Canada (NRCC), the National Institute for Environmental Science (NIES, Japan) and the BCR (Measurements and Testing Programme, European Commission). A project for the certification of butyltin in mussel tissue is presently running under the auspices of the Measurements and Testing Programme (BCR).

3.4   Conclusions

Increasing efforts are necessary to validate the existing procedures and to develop new techniques for the determination of butyltin species. The two most popular methods presently are Grignard derivatization/FPD and hydride generation/AAS, both of which involve a derivatization step. The comments made in this chapter have stressed, however, that the performance of techniques involving derivatization procedures cannot be fully evaluated in the absence of suitable calibrants to verify their yield. At this stage, the HG/GC/AAS is used frequently for the analysis of water but should be looked at critically for the analysis of solid matrices. Similarly, extraction and separation procedures should be further investigated; for separation, the transfer of technology from the well validated separation procedures used in organic analysis should allow a better understanding of the procedures as applied to speciation and lead to easier and more robust methods. Techniques existing for the determination of butyltin species are certainly very reliable, but further investigations are necessary to validate them fully for use as routine analyses. In this regard, interlaboratory studies using a wide variety of environmental and biological matrices and involving different categories of techniques are necessary. Moreover, an increased production of relevant CRMs is deemed essential. It now appears more important to identify clearly the respective advantages and limitations of a particular technique, rather than the optimum performance, in order to rationalize the determination of butyltin speciation in environmental and biological matrices.

Table 3.3 *Certified reference materials for the quality control of butyltin determinations in environmental and biological matices*

Matrices and RM reference	Butyltin compounds	Certified values	Certifying body
Harbour sediment PACS-1	TBT	$1.27 \pm 0.22\,(\mu g\,g^{-1})$	NRCC [116]
	DBT	$1.16 \pm 0.18\,(\mu g\,g^{-1})$	
	MBT	$0.28 \pm 0.17\,(\mu g\,g^{-1})$	
Coastal sediment CRM 462	TBT	$70 \pm 14\,(ng\,g^{-1})$	BCR [54]
	DBT	$128 \pm 16\,(ng\,g^{-1})$	
Harbour sediment RM 424*	TBT	$020 \pm 5\,(ng\,g^{-1})$	BCR [114]
Fish tissue NIES 11	TBT	$01.3 \pm 0.1\,(\mu g\,g^{-1})$	NIES [117]

*This material was not certified owing to the complexity of the matrix. This RM should be reserved for experienced laboratories and not used as a CRM for routine analysis.
NRCC: National Research Council Canada
BCR: Community Bureau of Reference (now Measurements and Testing Programme, European Commission)
NIES: National Institute for Environmental Studies (Japan)

Once techniques become well understood and controlled, huge efforts will still be necessary to evaluate the risks of errors at the sample collection and storage steps. The comparison of different techniques is again a means to optimize procedures which could possibly be standardized, for example for routine analysis to be performed in monitoring campaigns.

3.5   References

[1] Blunden, S. J. and Chapman, A. (1986) Organotin compounds in the environment, in: *Organometallic Compounds in the Environment*, Craig, P. J. (Ed.), Longman Group Ltd, pp. 111–50.

[2] Laughlin, R. B. and Lindén, O. (1985) *A review of the environmental effects of tributyltin use*, Internal Report, Swedish Environ. Res. Institute, No. B91065, p. 43.

[3] Maguire, R. J. (1987) Environmental aspects of tributyltin, *Applied Organometallic Chemistry*, 1: 475–98.

[4] Waldock, M. J., Thain, J. E. and Waite, M. E. (1987) The distribution and potential effects of TBT in UK estuaries during 1986, *Applied Organometallic Chemistry*, 1: 287–301.

[5] Thompson, J. A. J., Sheffer, M. G., Pierce, R. C., Chau, Y. K., Cooney, J. J., Cullen, W. R. and Maguire, R. J. (1985) *Organotin compounds in the aquatic environment – scientific criteria for assessing their effects on environmental quality*, National Research Council, Ottawa, Canada, ISSN 0316-0114, NRCC No. 22494, p. 284.

[6] Cortez, L., Quevauviller, Ph., Martin, F. and Donard, O. F. X. (1993) A survey of butyltin contamination in Portuguese coastal environment, *Environ. Pollut.*, 82: 57–62.

[7] Leroy, M. J. F., Donard, O. F. X., Astruc, M. and Quevauviller, Ph. (1996) Tin speciation in environmental samples, *Pure and Applied Chemistry*, in press.

[8] Quevauviller, Ph., Bruchet, A. and Donard, O. F. X. (1991) Release of mono- and dibutyltin compounds from PVC materials, *Appl. Organometallic Chemistry*, 5: 125–9.

[9] Donard, O. F. X., Quevauviller, Ph., Ewald, M., Laane, R., Marquenie, J. M. and Ritsema, R. (1989) Organotin compounds in Dutch coastal waters: the Rhine Estuary, *Proc. Heavy Metals in the Environment*, Vernet, J. P. (Ed.), Geneva, Sept. 12–15/89, pp. 526–9.

[10] Quevauviller, Ph., Ewald, M. and Donard, O. F. X. (1989) *Organotin compound distribution in estuarine and coastal waters of the Netherlands*, Rijkswaterstaat Report, DGW 950/Adera 421030.

[11] Quevauviller, Ph. and Donard, O. F. X. (1990) Variability of butyltin determination in water and sediment samples from European coastal environments, *Appl. Organometallic Chemistry*, 4: 353–67.

[12] Quevauviller, Ph., Raoux, C., Etcheber, H. and Donard, O. F. X. (1991) Grain size partitioning of butyltins in estuarine and coastal sediments, *Oceanol. Acta*, SP11: 247–55.

[13] Batley, G. E. and Gardner, D. (1977) Sampling and storage of natural waters for trace metal analysis, *Water Research*, 11: 745–56.

[14] Dooley, C. A. and Homer, V. (1983) *Organotin compounds in the marine environment: uptake and sorption behavior*, Naval Oceans Systems, Technical

Report No. 197, San Diego CA, p. 19.

[15] Blair, W. R., Olson, G. J., Brinckman, F. E., Paule, R. C. and Becker, D. A. (1986) *An international butyltin measurement methods intercomparison: sample preparation and results of analyses*, National Bureau of Standards, Gaithersburg, p. 55.

[16] Young, D. R., Schatzberg, P., Brinckman, F. E., Champ, M. E., Holm, S. E. and Landy, R. B. (1986) Summary report – interagency workshop on aquatic sampling and analysis for organotin compounds, *Proc. Oceans 86 Organotin Symposium*, Washington, D.C., pp. 1135–40.

[17] Quevauviller, Ph., Griepink, B., Maier, E. A., Meinema, H. and Muntau, H. (1990) Analytical quality control of tributyltin determination in the environment, *Proc. Euroanalysis VII*, Vienna, August 24–31/90.

[18] Quevauviller, Ph. and Donard, O. F. X. (1991) Organotin stability during storage of marine waters and sediments, *Fresenius' Zeitung für Analytische Chemie*, **339**: 6–14.

[19] Astruc, M. and Pinel, R. (1982) *L'étain dans l'environnement*, Report CNRS-PIREN, ASP Déchets et Environnement, p. 54.

[20] Harrison, R. M., Hewitt, C. N. and de Mora, S. J. (1985) Environmental analysis using gas chromatography–atomic absorption spectrometry, *Trends in Analytical Chemistry*, **4**: 8–11.

[21] Donard, O. F. X. and Pinel, R. (1989) Tin and germanium, in: *Environmental Analysis Using Chromatography Interfaced with Atomic Spectroscopy*, Harrison, R. M. and Rapsomanikis, S. (Eds), Ellis Horwood Ltd Publishers, London, pp. 189–222.

[22] Donard, O. F. X. and Ritsema, R. (1993) Hyphenated techniques applied to the speciation of organometallic compounds in the environment, in: *Techniques and Instrumentation in Analytical Chemistry – Volume 13, Environmental Analysis*, Barceló (Ed.), Elsevier, Amsterdam, pp. 550–601.

[23] Dirkx, W., Lobinski, R. and Adams, F. C. (1994) Speciation analysis of organotin by GC-AES and GC-AES after extraction and derivatization, in: *Quality Assurance for Environmental Analysis*, Quevauviller, Ph., Maier, E. A. and Griepink, B. (Eds), Elsevier, Amsterdam, pp. 357–409.

[24] Meinema, H. A., Burger-Wiersma, T., Versluis-De Haan, G. and Gevers, E. C. (1978) Determination of trace amounts of butyltin compounds in aqueous systems by gas chromatography/mass spectrometry, *Environmental Science and Technology*, **12**: 288–93.

[25] Schatzberg, P., Adema, C. M., Thomas, W. M. Jr. and Mangum, S. R. (1986) A time-integrating, remotely moored, automated sampling and concentration system for aquatic butyltin monitoring, *Proc. Oceans 86*, Washington, D.C., pp. 1155–9.

[26] Mathias, C. L., Bellama, J. M. and Brinckman, F. E. (1987) Determination of tributyltin in estuarine water using bonded C-18 silica solid phase extraction, hydride derivatisation and GC-FPD, *Proc. Organotin Symposium*, Halifax, pp. 1344–7.

[27] Quevauviller, Ph., Donard, O. F. X., Maier, E. A. and Griepink, B. (1992) Improvement of speciation analysis in environmental matrices, *Mikrochim. Acta*, **109**: 169–90.

[28] Hatfield, M. (1987) Electrically heated quartz atomization cell for hydride generation-atomic absorption spectrometry, *Analytical Chemistry*, **59**: 1887–8.

[29] Randall, L., Donard, O. F. X. and Weber, J. H. (1986) Speciation of n-butyltin compounds by atomic absorption spectrometry with an electrothermal quartz furnace after hydride generation, *Analytica Chimica Acta*, **184**: 197–203.

[30] Desauziers, V., Leguille, F., Lavigne, R., Astruc, M. and Pinel, R. (1989) Butyltin speciation in sediments and biological samples by hydride generation gas chromatography quartz furnace atomic absorption spectrometry: a study of acid leaching procedures, *Applied Organometallic Chemistry*, **3**: 469–74.

[31] Quevauviller, Ph. (1991) *Spéciation de l'étain dans les milieux estuariens et côtiers*, Thesis, University of Bordeaux, No. 607, p. 212.

[32] Tügrul, S., Balkas, T. I. and Goldberg, E. D. (1983) Methyltins in the marine environment, *Marine Pollution Bulletin*, **14**: 297–303.

[33] Donard, O. F. X., Randall, L., Rapsomanikis, S. and Weber, J. H. (1986) Developments in the speciation and determination of alkylmetals (Sn,Pb) using volatilization techniques and chromatography-atomic absorption spectroscopy, *International Journal of Environmental Analytical Chemistry*, **27**: 55–67.

[34] Rapsomanikis, S. and Harrison, R. (1988) Speciation of butyltin compounds in oyster samples, *Applied Organometallic Chemistry*, **2**: 151–62.

[35] Chau, Y. K., Wong, P. T. S. and Bengert, G. A. (1982) Determination of methyltin(IV) and tin(IV) species in water by gas chromatography/atomic absorption spectrometry, *Analytical Chemistry*, **54**: 246–9.

[36] Martin-Landa, I., De Pablod, F. and Marr, I. L. (1989) Determination of organotins in fish and sediments by gas chromatography with flame photometric detection, *Analytical Proceedings*, **26**: 16–18.

[37] Maguire, R. J. and Huneault, H. (1981) Determination of butyltin species in water by gas chromatography with flame photometric detection, *Journal of Chromatography*, **209**: 458–62.

[38] Müller, M. D. (1984) Tributyltin detection at trace levels in water and sediments using GC with flame photometric detection and GC-MS, *Fresenius' Zeitung für Analytische Chemie*, **317**: 32–6.

[39] Lobinski, R., Dirkx, W. M. R., Ceulemans, M. and Adams, F. C. (1992) Optimization of comprehensive speciation of organotin compounds in environmental samples by capillary gas chromatography helium microwave-induced plasma emission spectrometry, *Analytical Chemistry*, **64**: 159–64.

[40] Jiang, G. B., Maxwell, P. S., Siu, K. W. M., Luong, V. T. and Berman, S. S. (1991) Determination of butyltins in mussel by gas chromatography with flame photometric detection using quartz surface-induced luminescence, *Analytical Chemistry*, **63**: 1506–9.

[41] Waldock, M. J., Waite, M. E. and Thain, J. E. (1988) Inputs of TBT to the marine environment from shipping activity in the U.K., *Environ. Technol. Lett.*, **9**: 999–1010.

[42] Mathias, C. L., Bellama, J. M., Olson, G. J. and Brinckman, F. E. (1986) Comprehensive method for determination of aquatic butyltin and butylmethyltin species at ultra trace levels using simultaneous hydridization/extraction with gas chromatography-flame photometric detection, *Environmental Science and Technology*, **20**: 609–15.

[43] Sullivan, J. J., Torkelson, J. D., Welkell, M. M., Hollingworth, T. A., Saxton, W. L., Miller, G. A., Panaro, K. W. and Uhler, A. D. (1988) Determination of tri-n-butyltin and di-n-butyltin in fish as hydride derivatives by reaction gas chromatography, *Analytical Chemistry*, **60**: 626–30.

[44] Ebdon, L., Hill, S. J. and Jones, P. (1985) Speciation of tin in natural waters using coupled high-performance liquid chromatography-flame atomic absorption spectrometry, *Analyst*, **110**: 515–17.

[45] M&T Chemicals (1976) *Standard Test Method AA 33.*

[46] Unger, M. A., McIntyre, W. G., Greaves, J. and Huggett, R. J. (1986) GC determination of butyltins in natural waters by flame photometric detection of hexyl derivatives with mass spectroscopic confirmation, *Chemosphere*, **15**: 461–70.

[47] Lavigne, R. (1989) *Spéciation des ultra-traces d'étain par couplage chromatographie en phase gazeuse-spéctrométrie d'absorption atomique après génération d'hydrure*, Thesis, University of Pau, No. 56, p. 234.

[48] Siu, K. W. and Berman, S. S. (1990) Comparison of methodologies in butyltin species determination, Proc. 3rd Int. Organotin Symposium, Monaco, 17–20 April 1990, pp. 11–16.

[49] Astruc, A., Astruc, M., Pinel, R. and Potin-Gautier, M. (1992) Speciation of butyltin compounds by on-line HPLC-ETAA of tropolone complexes in environmental samples, *Appl. Organometallic Chemistry*, **6**: 39–47.

[50] Lalère, B. (1993) Evaluation of organotin compounds behaviour during microwave digestion procedures, *Symposium on Analytical Sciences*, Deauville, 4–6 May 1993.

[51] Dachs, J., Alzaga, R., Bayona, J. M. and Quevauviller, Ph. (1994) Development of a supercritical fluid extraction procedure for tributyltin determination in sediments, *Anal. Chim. Acta*, **286**: 319–27.

[52] Astruc, M., Lavigne, R., Desauziers, V., Pinel, R. and Astruc, M. (1989) Tributyltin determination in marine sediments: a comparative study of methods, *Applied Organometallic Chemistry*, **3**: 267–71.

[53] Zhang, S., Chau, Y. K., Li, W. C. and Chau, A. S. Y. (1991) Evaluation of extraction techniques for butyltin compounds in sediment, *Applied Organometallic Chemistry*, **5**: 431–4.

[54] Quevauviller, Ph., Astruc, M., Ebdon, L., Desauziers, V., Sarradin, P. M., Astruc, A., Kramer, G. N. and Griepink, B. (1994) Certified reference material (CRM 462) for the quality control of dibutyl- and tributyltin determinations in coastal sediment, *Applied Organometallic Chemistry*, **8**: 629–37.

[55] Luten, J. R., personal communication.

[56] Craig, P. J. and Rapsomanikis, S. (1983) Dismutation of trimethyltin species bound to oxoligand complexes, *Inorganic Chemistry Acta*, **80**: L19–21.

[57] Soderquist, C. J. and Crosby, D. G. (1978) Determination of triphenyltin hydroxide and its degradation products in water, *Analytical Chemistry*, **50**: 1435–9.

[58] Dirkx, W. M. R. and Adams, F. C. (1992) Speciation of organotin compounds in water and sediments by gas chromatography/Atomic absorption spectrometry (GC-AAS), *Mikrochim. Acta*, **109**: 79–81.

[59] Ashby, J. R. and Craig, P. J. (1989) New method for the production of volatile organometallic species for analysis from the environment: some butyltin levels in UK sediments, *The Science of Total Environment*, **78**: 219–32.

[60] Greaves, J. and Unger, M. A. (1988) A selected ion monitoring assay for tributyltin and its degradation products, *Biomedical and Environmental Mass Spectrometry*, **15**: 565–9.

[61] Maguire, R. J. and Tkacz, R. J. (1983) Analysis of butyltin compounds by gas chromatography. Comparison of flame photometric and atomic absorption spectrophotometric detectors, *Journal of Chromatography*, **268**: 99–101.

[62] Short, J. W. (1987) Measuring tri-n-butyltin in salmon by atomic absorption: analysis with and without gas chromatography, *Bulletin of Environmental Contamination and Toxicology*, **39**: 412–16.

[63] Tolosa, I., Bayona, J. M., Albaigés, J., Alencastro, L. F. and Taradellas, J. (1991) Organotin speciation in aquatic matrices by CGC/FPD, ECD and MS and LC/MS, *Fresenius' Journal of Analytical Chemistry*, **339**: 646–53.

[64] Rices, C. D., Espourteille, F. A. and Huggett, R. J. (1987) Analysis of tributyltin in estuarine sediments and oyster tissues, *Crassostrea virginica*, *Applied Organometallic Chemistry*, **1**: 541–4.

[65] Ashby, J. R., Clark, S. and Craig, P. J. (1987) The analysis of organotin compounds from environmental matrices, in: *The Biological Alkylation of Heavy Elements*, Craig, P. J. and Glockling, F. (Eds), pp. 263–90.

[66] Dirkx, W. M. R., Van Mol, W. E., Van Cleuvenbergen, R. J. A. and Adams, F. C. (1989) Speciation of organotin compounds in water by gas chromatography/atomic absorption spectrometry, *Fresenius' Zeitung für Analytische Chemie*, **335**: 769–74.

[67] Cotton, F. A. and Wilkinson, G. (1974) *Advanced Inorganic Chemistry*, Wiley, p. 309.

[68] Hodge, V. F., Seidel, S. L. and Goldberg, E. D. (1979) Determination of tin(IV) and organotin compounds in natural water, coastal sediments and macro algae by atomic absorption spectrometry, *Analytical Chemistry*, **51**: 1256–9.

[69] Astruc, M., Pinel, R. and Astruc, A. (1992) Determination of tributyltin in sediments by hydride generation/GC/QFAAS, *Mikrochim. Acta*, **109**: 73–8.

[70] Pinel, R., Gandjar, I. G., Benabdallah, M. Z., Astruc, A. and Astruc, M. (1984) Dosage de l'étain minéral et organique en traces dans les eaux par spectrophotométrie d'absorption atomique avec génération-décomposition d'hydrures, *Analusis*, **12**: 404–8.

[71] Ulher, A. D., Durell, G. S. and Spellacy, A. M. (1991) Extraction procedure for the measurement of butyltin compounds in biological tissues using toluene, HBr and tropolone, *Bull. Environ. Contam. Toxicol.*, **47**: 217–21.

[72] Donard, O. F. X. (1989) Speciation of trace metals by coupled hydride generation, gas chromatography and atomic absorption detection. A necessary approach? 5. *Colloquium Atomspektrom. Spurenanal.*, Perkin-Elmer, Überlingen, pp. 395–417.

[73] Quevauviller, Ph., Martin, F., Belin, C. and Donard, O. F. X. (1993) Matrix effects in the determination of butyltin compounds in environmental samples by GC/AAS after hydride generation, *Applied Organometallic Chemistry*, **7**: 149–57.

[74] Martin, F. M., Tseng, C. M., Donard, O. F. X., Belin, C. and Quevauviller, Ph. (1994) Interferences generated by organic and inorganic compounds during organotin speciation using hydride generation coupled with cryogenic trapping, gas chromatographic separation and detection by atomic absorption spectrometry, *Anal. Chim. Acta*, **286**: 343–55.

[75] Hattori, Y., Kobayashi, A., Takemoto, S., Takami, K., Kuge, Y., Sugimae, A. and Nakamoto, M. (1984) Determination of trialkyltin, dialkyltin and triphenyltin compounds in environmental water and sediment, *Journal of Chromatography*, **315**: 341–9.

[76] Mathias, C. L., Bellama, J. M., Olson, G. J. and Brinckman, F. E. (1989) Determination of di- and tributyltin in sediment and microbial biofilms using acidified methanol extraction, sodium borohydride derivatization and gas chromatography with flame photometric detection, *International Journal of Environmental Analytical Chemistry*, **35**: 61–8.

[77] Clark, S. and Craig, P. J. (1988) The analysis of inorganic and organometallic antimony, arsenic and tin compounds using an on-column hydride generation method, *Applied Organometallic Chemistry*, **2**: 33–46.

[78] Ebdon, L., Hill, S.J. and Jones, P. (1991) HPLC coupled with in-line photolysis, hydride generation and flame atomic-absorption spectrometry for the speciation of tin in natural waters, *Talanta*, **6**: 607–11.

[79] Weber, J.H., Han, J.S. and François, R. (1988) Speciation of methyl- and butyltin compounds in compartments of the Great Bay Estuary, *Proc. Heavy Metals in the Hydrocycle*, Astruc, M. and Lester, J.N. (Eds), Selper Ltd., London.

[80] Schebeck, L., Andrae, M.O. and Tobschall, H.J. (1991) Methyl- and butyltin compounds in water and sediments of the Rhine river, *Environ. Science and Technology*, **25**: 871–78.

[81] Sarradin, P.M., Leguille, F., Astruc, A., Pinel, R. and Astruc, M. (1994) Optimisation of the atomisation parameters in the speciation of organotin compounds by HG-GC-QFAAS, *J. Anal. At. Spectrosc.*, in press.

[82] Arakawa, Y., Wada, O. and Manabe, M. (1983) Extraction and fluorometric determination of organotin compounds with morin, *Analytical Chemistry*, **55**: 1901–4.

[83] Takahashi, K., Yoshino, T. and Ohyagi, Y. (1987) Photodegradation of tributyltin chloride in sea water by ultra violet light, *J. Chem. Soc. Japan*, **2**: 181–8.

[84] Kleiböhmer, W. and Camman, K. (1989) Separation and determination of organotin compounds by HPLC and fluorimetric detection in micellar systems, *Fresenius, Zeitung für Analytische Chemie*, **335**: 780–4.

[85] Suyani, H., Creed, J., Davidson, T. and Caruso, J. (1989) Inductively coupled plasma mass spectrometry as a detector for micellar liquid chromatography: speciation of alkyltin compounds, *J. of Chromatographic Science*, **27**: 139.

[86] Tsuda, T., Nakanishi, H., Morita, T. and Takebayashi, J. (1986) Simultaneous gas chromatographic determination of dibutyltin and tributyltin compounds in biological and sediment samples, *Journal of the Association of Analytical Chemistry*, **69**: 981–4.

[87] Langseth, W. (1984) Determination of diphenyltin and dialkyltin homologues by HPLC with morin in the eluent, *Talanta*, **31**: 975–8.

[88] Ebdon, L., Hill, S., Ward, R.W. and Robert, W. (1987) Directly coupled chromatography–atomic spectroscopy: Part 2. Directly coupled liquid chromatography–atomic spectroscopy, *Analyst*, **112**: 1–16.

[89] Sarradin, P.M. (1993) *Répartition et évolution du tributylétain dans les sédiuments marins*, Thesis, University of Pau, p. 134.

[90] Quevauviller, Ph., Ritsema, R., Morabito, R., Dirkx, W.M.R., Chiavarini, S., Bayona, J.M. and Donard, O.F.X. (1994) Critical consideration with respect to tin species identification in the environment, *Appl. Organometallic Chemistry*, **8**: 541–9.

[91] Junk, G.A. and Richard, J.J. (1987) Solid phase extraction, GC separation and EC detection of tributyltin chloride, *Chemosphere*, **16**: 61–8.

[92] Morabito, R., Chiavarini, S. and Cremisini, C. (1995) Speciation of organotin compounds in environmental samples by GC-MS, in: *Quality Assurance for Environmental Analysis Within the BCR-programme*, Quevauviller, Ph., Maier, E.A. and Griepink, B. (Eds), Elsevier Science Publ., pp. 411–40.

[93] Müller, M.D. (1987) Comprehensive trace level determination of organotin compounds in environmental samples using high-resolution gas chromatography with flame photometric detection, *Analytical Chemistry*, **59**: 617–23.

[94] Hill, S.J., Brown, A., Rivas, C., Sparkes, S. and Ebdon, L. (1995) High performance liquid chromatography–isotope dilution–inductively coupled

plasma–mass spectrometry for lead and tin speciation in environmental samples, in: *Quality Assurance for Environmental Analysis Within the BCR-programme*, Quevauviller, Ph., Maier, E. A. and Griepink, B. (Eds), Elsevier Science Publ., pp. 389–412.

[95] Aue, W. A. and Flinn, C. G. (1980) Modification of a conventional flame photometric detector for increased tin response, *Analytical Chemistry*, **52**: 1537–8.

[96] Woolins, A. and Cullen, W. R. (1984) Determination of organotin compounds contained in aqueous samples using capillary gas chromatography, *Analyst*, **109**: 1527–9.

[97] Kapila S. and Vogt, C. R. (1980) Some aspects of organic tin analysis by gas chromatography–flame photometry, *Journal of Chromatographic Science*, **18**: 144–7.

[98] Tolosa, I., Dachs, J. and Bayona, J. M. (1992) Tributyltin speciation in aquatic matrices by CGC-FPD and CGC-MS confirmation, *Mikrochim. Acta*, **109**: 87–91.

[99] Hansen, D. R., Gilfoil, T. J. and Hill, H. H. Jr. (1981) Comparison of metal-sensitive flame ionisation and carbon-sensitive flame ionisation detectors for the gas chromatographic determination of organotins, *Analytical Chemistry*, **53**: 857–61.

[100] Kadokami, K., Uehiro, T., Morita, M. and Fuwa, K. (1988) Determination of organotin compounds in water by bonded-phase extraction and high performance liquid chromatography with long-tube atomic absorption spectrometric detection, *Analytical Atomic Spectrometry*, **3**: 187–91.

[101] Jewett, K. L. and Brinckman, F. E. (1981) Speciation of trace di- and triorganotins in water by ion exchange HPLC-GFAA, *Journal of Chromatographic Science*, **19**: 583–93.

[102] Thompson, J. J. and Houk, R. S. (1986) Inductively coupled plasma mass spectrometric detection for multielement flow injection analysis and elemental speciation by reversed-phase liquid chromatography, *Analytical Chemistry*, **58**: 2541–8.

[103] Lafreniere, K. E., Fassel, V. A. and Eckels, D. E. (1987) Elemental speciation via high-performance liquid chromatography combined with inductively coupled plasma atomic emission spectroscopic detection: application of direct injection nebulizer, *Analytical Chemistry*, **59**: 879–87.

[104] Brown, A. A. Ebdon, L. and Hill, S. J. (1994) Development of a coupled liquid chromatography-isotope dilution coupled plasma mass spectrometry method for lead speciation, *Anal. Chim. Acta*, **286**: 391–9.

[105] Suyani, H., Creed, J., Davidson, T. and Caruso, J. (1989) Inductively coupled plasma mass spectrometry and atomic emission spectrometry coupled to high-performance liquid chromatography for speciation and detection of organotin compounds, *Journal of Chromatographic Science*, **27**: 139.

[106] Ohlosson, S. V. and Hintze, W. W. (1983) HPTLC analysis of organotin compounds in preservative solutions and preservative-treated wood, *Journal of High Resolution Chromatography*, **6**(1): 89–94.

[107] Walton, A. P., Wei, G. T., Liang, Z., Michel, R. G. and Morris, M. (1991) Laser excited atomic fluorescence in a flame as a high-sensitivity detector for organomanganese and organotin compounds following separation by high performance liquid chromatography, *Analytical Chemistry*, **63**: 232–40.

[108] Nygren, O., Nillsson, C. and Frech, W. (1988) On-line interfacing of a liquid chromatograph to a continuously heated graphite furnace atomic absorption spectrophotometer for element-specific detection, *Analytical Chemistry*, **60**: 2204–8.

[109] Kitamura, H., Yamada, Y. and Nakamoto, M. (1984) Determination of trace amounts of dibutyltin(IV) dichloride by differential pulse anodic stripping voltammetry, *Chemistry Letters*, pp. 837–40.

[110] MacRehan, W. A. (1981) Differential pulse detection in liquid chromatography and its application to the measurement of organometal cations, *Analytical Chemistry*, **53**: 74–7.

[111] Ochsenkühn-Petropoulou, M., Poulea, G. and Parikassis, G. (1992) Electrochemical speciation of organotin compounds in water and sediments. Application to sea water after ion-exchange separation. *Mikrochimica Acta*, **109**: 93.

[112] Siu, K. W. M., Gardner, G. J. and Berman, S. S. (1989) Ion spray mass spectrometry/mass spectrometry: quantitation of tributyltin in a sediment reference material for trace metals, *Analytical Chemistry*, **61**: 2320–2.

[113] Quevauviller, Ph., Maier, E. A. and Griepink, B. (1996) Quality consideration of results of speciation analysis, in: *Element Speciation in Bioinorganic Chemistry*, Caroli, S. (Ed.), John Wiley & Sons Publ., New York, **8**: 639–44.

[114] Quevauviller, Ph., Astruc, M., Ebdon, L., Kramer, G. N. and Griepink, B. (1994) Interlaboratory study for the improvement of tributyltin determination in harbour sediment (RM 424), *Applied Organometallic Chemistry*, in press.

[115] ISO-Guide 33 (1985) *Certification of reference materials – General and statistical principles.*

[116] National Research Council Canada (1990) *PACS-1. Description sheet*, NRCC, Institute for Environmental Chemistry, Ottawa, K1A 0R6, Canada.

[117] National Institute for Environmental Studies (1990) *NIES certified reference material No 11. Information sheet*, NIES, Yatabe-machi, Tsukuba, Ibaraki, 305, Japan.

4

○ ○ ○ ○ ○ ○ ○ ○ ○ ○ ○ ○ ○ ○ ○ ○ ○ ○ ○ ○

The occurrence, fate and toxicity of tributyltin and its degradation products in fresh water environments

R. James Maguire

4.1 Introduction

The chemistry and toxicity of tin and organotin compounds have been extensively reviewed (e.g., Blunden and Evans, 1990; Maguire, 1987, 1991; World Health Organization, 1990; and references therein). The major uses of organotin compounds are as poly(vinyl chloride) (PVC) stabilizers, industrial catalysts, industrial and agricultural biocides, and wood preserving and antifouling agents. From an environmental point of view, tributyltin compounds used as antifouling agents are of most concern because of their extremely high toxicity to aquatic organisms. This chapter reviews the occurrence, persistence and toxicity of tributyltin, and its degradation products dibutyltin and monobutyltin, in fresh water environments.[1]

4.2 Sources and environmental occurrence

The main use of tributyltin compounds, with direct introduction to water, is as antifouling agents. However, there are other pesticidal uses. In Canada, for example some tributyltin-containing pesticides are also registered for use as material preservatives, joinery wood preservatives, remedial wood preservatives, wood preservative stains, and as slimicides. In some parts of Africa, tributyltin compounds have been applied directly to water in the control of schistosomiasis, and they are used agriculturally against cotton-boll worms (Aboul Dahab, El-Sabrouti and Halim, 1990).

---

[1] Tributyltin has been shown to degrade biologically and abiotically by sequential debutylation, yielding dibutyltin, monobutyltin and inorganic tin (Woggon and Jehle, 1975; Fish, Kimmel, and Casida, 1976; Kimmel, Fish and Casida, 1977; Maguire, Carey and Hale, 1983; François, Short and Weber, 1989). Although the various butyltin species in water are cations or complexes with other solutes, for brevity they are referred to here as tributyltin, dibutyltin and monobutyltin. All alkyl groups are n-alkyl.

Butyltin species are not produced biologically, and therefore their environmental presence is due to anthropogenic input. In general the occurrence of butyltin species in the environment is a result of the use of tributyltin as an antifouling agent. However, the finding of tributyltin in a power plant discharge in Italy (Gabrielides *et al.*, 1990), and in sewage treatment plant influents and effluents in Switzerland (Fent and Muller, 1991) and Canada (Chau, Zhang and Maguire, 1992) probably reflects slimicidal uses of tributyltin. It should be noted that some dibutyltin and monobutyltin compounds are used industrially, primarily as stabilizers for PVC (Maguire, 1991). Consequently, dibutyltin and monobutyltin in industrial plant discharges and in sewage treatment plant influents may arise from non-antifouling uses of tributyltin, or from non-pesticidal uses of dibutyltin and monobutyltin compounds (e.g., through the leaching of organotin-stabilized PVC pipe). It should also be noted that the practices of spreading sewage sludge on land or at sea are other potential routes of entry of butyltin species to the environment.

Methods of analysis for tributyltin, dibutyltin and monobutyltin are reviewed in Chapter 3, and will not be discussed in this Chapter. Tables 4.1, 4.2 and 4.3 summarize the occurrence of tributyltin, dibutyltin and monobutyltin, respectively, in fresh water environments. (There are also some data presented for occurrence in municipal and industrial effluents to fresh water.) Many of these data were collected before antifouling uses of tributyltin were regulated in the various countries. There are far fewer data for fresh water than for seawater. Most of the data are from Canada, Switzerland, England, Germany, the USA, France, Belgium, Portugal, Japan and Italy. These data make a convincing case for the use of tributyltin-containing antifouling paints as the main source of tributyltin in fresh water. Many researchers have noted concentrations of tributyltin, dibutyltin and monobutyltin in the water, sediment and organisms of harbours and marinas that are much higher than those from the more open areas of bays and lakes. In this section, the focus is on tributyltin because it is far more toxic to aquatic organisms than are dibutyltin and monobutyltin (see toxicity section below). The toxicological significance of environmental concentrations of dibutyltin and monobutyltin will be discussed later in the chapter.

The highest concentration of tributyltin found in fresh water (at *c.* 1 m depth) is 7300 ng Sn l^{-1} in a harbour in Antwerp, Belgium (fresh/estuarine water – Dirkx *et al.*, 1989). Tributyltin in fresh waters appears to be present largely in the operationally defined 'dissolved' phase, i.e., that which passes a 0.45 μm filter (Maguire, 1986). The highest concentrations

Table 4.1   *Occurrence of tributyltin in fresh water environments**

Medium	Concentra-tion (ng Sn l^{-1} or ng Sn g^{-1})	Location	Reference
water	n.d.–1190	lakes and rivers in Ontario, Canada	Maguire et al. (1982)
water	0.4–6	rivers and lakes in Switzerland	Muller (1984)
water	n.d.–70	Detriot and St Clair Rivers, Ontario, Canada	Maguire et al. (1985a)
water	n.d.–200	Toronto Harbour, Ontario, Canada	Maguire and Tkacz (1985)
water (overlying sediment)	n.d.–18 100	Toronto Harbour, Ontario, Canada	Maguire and Tkacz (1985)
water	n.d.–2340	Canada	Maguire et al. (1986)
water	2–6	Switzerland	Muller (1987)
water (fresh/estuarine)	2–740	Chesapeake Bay marinas and tributaries, USA	Hall et al. (1988)
water (fresh/estuarine)	36–7300	harbour in Antwerp, Belgium	Dirkx et al. (1989)
water	98–303	marina in Lake Lucerne, Switzerland	Fent (1989a)
water	n.d.–57	marinas in California, USA	Stang and Goldberg (1989)
water	n.d.–1337	Rivers Bure and Yare, England	Waite et al. (1989)
water	1–2	Rhine River, Germany	Schebek, Andreae and Tobschall (1991)
water	n.d.–73	Germany	Schebek et al. (1991)
water	n.d.–308	marinas in Lake Lucerne, Switzerland	Fent and Hunn (1991)
water	n.d.–442	Lake Geneva (Switzerland and France)	Becker et al. (1992)
water	3–671	Switzerland and Germany	Becker and Bringezu (1992)
surface microlayer	n.d.–25 000	lakes and rivers in Ontario, Canada	Maguire et al. (1982)

Table 4.1 *continued*

Medium	Concentration (ng Sn l^{-1} or ng Sn g^{-1})	Location	Reference
surface microlayer	n.d.–33	St Clair River, Ontario, Canada	Maguire *et al.* (1985a)
surface microlayer	n.d.–36 000	lakes and rivers in Ontario, Canada	Maguire and Tkacz (1987)
sediment	n.d.–221	lakes, rivers, harbours and marinas in Ontario, Canada	Maguire (1984)
sediment	1–7	Lakes Constance and Zurich, Switzerland	Muller (1984)
sediment	420	river in Osaka, Japan	Hattori *et al.* (1984)
sediment	n.d.–73	Detroit and St Clair Rivers, Ontario, Canada	Maguire *et al.* (1985a)
sediment (cores)	n.d.–3520	Toronto Harbour, Ontario, Canada	Maguire and Tkacz (1985)
sediment	n.d.–1280	Toronto Harbour, Ontario, Canada	Maguire and Tkacz (1985)
sediment	n.d.–300	Canada	Maguire *et al.* (1986)
sediment	n.d.–0.9	Lake Biwa, Japan	Tsuda *et al.* (1986)
sediment (cores)	n.d.–104	Lake Zurich, Switzerland	Muller (1987)
sediment	n.d.–55	Oshawa and Whitby Harbours, Ontario, Canada	Chau *et al.* (1989)
sediment (fresh water/ estuarine/ marine)	n.d.–213	Portuguese rivers, estuaries	Quevauviller *et al.* (1989)
sediment	9–148	harbours in Lake Ontario, Canada	Scott, Chau and Rais-Firouz (1991)
sediment (cores)	n.d.–838	marinas in Lake Lucerne, Switzerland	Fent and Hunn (1991)
sediment	n.d.–40	Rhine River, Germany	Schebek *et al.* (1991)
sediment	13–182	Germany	Schebek *et al.* (1991)
sediment	n.d.–79	Massachusetts, USA	Wuertz *et al.* (1991)

Table 4.1 *continued*

Medium	Concentration ($ng\,Sn\,l^{-1}$ or $ng\,Sn\,g^{-1}$)	Location	Reference
sediment (spring)	n.d.–182	Severn Sound, Ontario, Canada	Wong and Chau (1992)
sediment (summer)	n.d.–161	Severn Sound, Ontario, Canada	Wong and Chau (1992)
sediment (fall)	n.d.–67	Severn Sound, Ontario, Canada	Wong and Chau (1992)
sediment	14–3452	Switzerland and Germany	Becker and Bringezu (1992)
sediment (fresh water/ estuarine)	n.d.–1291	East Anglia, England	Dowson *et al.* (1992b)
sediment	14–1046	Lake Geneva (Switzerland and France)	Becker *et al.* (1992)
sediment (cores) (fresh water/ estuarine)	n.d.–3500	England	Dowson *et al.* (1993)
clam (*Anodonta cygnaea*)	100–688	Lake Geneva (Switzerland and France)	Becker *et al.* (1992)
zebra mussel (*Dreissena polymorpha*)	1148–3834	marinas in Lake Lucerne, Switzerland	Fent and Hunn (1991)
zebra mussel (*Dreissena polymorpha*)	601–3828	Lake Geneva (Switzerland and France)	Becker *et al.* (1992)
Tateboshi gai (shellfish)	11	Lake Biwa, Japan	Tsuda *et al.* (1986)
yellow perch (*Perca flavescens*) (whole)	10	Jordan Harbour, Lake Ontario, Canada	Maguire *et al.* (1986)
white sucker (*Catostomus commersoni*) (whole)	n.d.–20	harbours in Lake Ontario, Canada	Maguire *et al.* (1986)
round crucian carp and largemouth black bass (muscle)	1–9	Lake Biwa, Japan	Tsuda *et al.* (1986)

Table 4.1 *continued*

Medium	Concentration (ng Sn l^{-1} or ng Sn g^{-1})	Location	Reference
lake chub (*Leuciscus cephalus*) (muscle)	70	marina in Lake Murten, Switzerland	Fent and Hunn (1991)
fish (unidentified) (whole)	16–179	harbours in Lake Ontario, Canada	Scott *et al.* (1991)
young-of-the-year fish (unidentified) (whole)	4–16	Midland Bay, Severn Sound, Ontario, Canada	Wong and Chau (1992)
pike (*Esox lucius*) (whole)	n.d.–98	Severn Sound, Ontario	Wong and Chau (1992)
power plant discharge	4982	Leghorn, Italy	Gabrielides *et al.* (1990)
STP effluent	6	Switzerland	Muller (1987)
sewage sludge	195–2730	Switzerland	Muller (1987)
untreated wastewater	26–89	Zurich, Switzerland	Fent and Muller (1991)
tertiary STP effluent	n.d.–2	Zurich, Switzerland	Fent and Muller (1991)
sewage sludge	115–361	Zurich, Switzerland	Fent and Muller (1991)
STP influent	n.d.–55 200	five Canadian cities	Chau *et al.* (1992)
STP effluent	n.d.	five Canadian cities	Chau *et al.* (1992)
sewage sludge	n.d.–278	five Canadian cities	Chau *et al.* (1992)

*Concentrations in sediment and sludge are by dry weight, and in biota are by wet weight. Water samples are whole water. Abbreviations used: n.d., not determined; STP, sewage treatment plant.

Table 4.2   *Occurrence of dibutyltin in fresh water environments**

Medium	Concentration ($ng\,Sn\,l^{-1}$ or $ng\,Sn\,g^{-1}$)	Location	Reference
water	4–630	Lake Michigan, USA	Hodge, Seidel and Goldberg (1979)
water	n.d.–3700	lakes and rivers in Ontario, Canada	Maguire *et al.* (1982)
water	n.d.–100	Toronto Harbour, Ontario, Canada	Maguire and Tkacz (1985)
water (overlying sediment)	n.d.–720	Toronto Harbour, Ontario, Canada	Maguire and Tkacz (1985)
water	n.d.–100	Detroit and St Clair Rivers, Ontario, Canada	Maguire *et al.* (1985a)
water	n.d.–1360	Canada	Maguire *et al.* (1986)
water (fresh/ estuarine)	n.d.–10	Canada	Maguire *et al.* (1986)
water	2–16	Switzerland	Muller (1987)
water (fresh/ estuarine)	3–340	Chesapeake Bay marinas and tributaries, USA	Hall *et al.* (1988)
water (fresh/ estuarine)	48–15 700	harbour in Antwerp, Belgium	Dirkx *et al.* (1989)
water	n.d.–113	Rivers Bure and Yare, England	Waite *et al.* (1989)
water	n.d.–31	marina, Lake Lucerne, Switzerland	Fent (1989a)
water	n.d.–15	marinas in California, USA	Stang and Goldberg (1989)
water	n.d.–52	marinas in Lake Lucerne, Switzerland	Fent and Hunn (1991)
water	0.5–2	Rhine River, Germany	Schebek *et al.* (1991)
water	1–16	Germany	Schebek *et al.* (1991)
water	n.d.–124	Lake Geneva (Switzerland and France)	Becker *et al.* (1992)
surface microlayer	n.d.– 1 330 000	lakes and rivers in Ontario, Canada	Maguire *et al.* (1982)

Table 4.2 *continued*

Medium	Concentration (ng Sn l^{-1} or ng Sn g^{-1})	Location	Reference
surface microlayer	n.d.–5	St Clair River, Ontario, Canada	Maguire *et al.* (1985a)
surface microlayer	n.d.–365 000	Ontario lakes and rivers	Maguire and Tkacz (1987)
sediment	n.d.–180	lakes, rivers, harbours and marinas in Ontario, Canada	Maguire (1984)
sediment	180	river in Osaka, Japan	Hattori *et al.* (1984)
sediment	n.d.–260	Toronto Harbour, Ontario, Canada	Maguire and Tkacz (1985)
sediment (cores)	n.d.–530	Toronto Harbour, Ontario, Canada	Maguire and Tkacz (1985)
sediment	n.d.–36	Detroit and St Clair Rivers, Ontario	Maguire *et al.* (1985a)
sediment	n.d.–580	Canada	Maguire *et al.* (1986)
sediment (fresh water/ estuarine)	n.d.–60	Canada	Maguire *et al.* (1986)
sediment (cores)	n.d.–55	Lake Zurich, Switzerland	Muller (1987)
sediment	n.d.–21	Oshawa and Whitby Harbours, Ontario, Canada	Chau *et al.* (1989)
sediment	n.d.–97	harbours in Lake Ontario, Canada	Scott *et al.* (1991)
sediment (cores)	n.d.–73	marinas in Lake Lucerne, Switzerland	Fent and Hunn (1991)
sediment (cores)	n.d.–26	marina in Lake Lucerne, Switzerland	Fent, Hunn and Sturm (1991)
sediment	10–50	Rhine River, Germany	Schebek *et al.* (1991)
sediment	15–44	Germany	Schebek *et al.* (1991)
sediment (spring)	n.d.–135	Severn Sound, Ontario, Canada	Wong and Chau (1992)
sediment (summer)	n.d.–175	Severn Sound, Ontario, Canada	Wong and Chau (1992)

Table 4.2 *continued*

Medium	Concentration (ng Sn l^{-1} or ng Sn g^{-1})	Location	Reference
sediment (fall)	n.d.–36	Severn Sound, Ontario, Canada	Wong and Chau (1992)
sediment	4–441	Lake Geneva (Switzerland and France)	Becker *et al.* (1992)
sediment (fresh water/ estuarine)	0.8–220	East Anglia, England	Dowson *et al.* (1992b)
sediment	13	Main River, Germany	Cai, Rapsonanikis and Andreae (1993)
sediment (cores) (fresh water/ estuarine)	n.d.–2000	England	Dowson *et al.* (1993)
macrophyte (*Cladophora* sp.)	91	Wye River mouth, Ontario, Canada	Wong and Chau (1992)
clam (*Anodonta cygnaea*)	22–107	Lake Geneva (Switzerland and France)	Becker *et al.* (1992)
zebra mussel (*Dreissena polymorpha*)	240–1239	marinas in Lake Lucerne, Switzerland	Fent and Hunn (1991)
zebra mussel (*Dreissena polymorpha*)	61–2122	Lake Geneva (Switzerland and France)	Becket *et al.* (1992)
round crucian carp and largemouth black bass (muscle)	n.d.–2.1	Lake Biwa, Japan	Tsuda *et al.* (1986)
lake chub (*Leuciscus cephalus*) (muscle)	10	marina in Lake Murten, Switzerland	Fent and Hunn (1991)
fish (unidentified) (whole)	4–109	harbours in Lake Ontario, Canada	Scott *et al.* (1991)
young-of-the- year fish (unidentified) (whole)	3	Midland Bay, Severn Sound, Ontario, Canada	Wong and Chau (1992)

Table 4.2 *continued*

Medium	Concentration (ng Sn l^{-1} or ng Sn g^{-1})	Location	Reference
pike (*Esox lucius*) (whole)	n.d.–33	Severn Sound, Ontario	Wong and Chau (1992)
power plant discharge	48	Leghorn, Italy	Gabrielides *et al.* (1990)
landfill leachate	243	Switzerland	Fent (1991)
STP effluent	10	Switzerland	Muller (1987)
sewage sludge	195–2730	Switzerland	Muller (1987)
STP influent	127–191	Zurich, Switzerland	Fent (1989b, 1994)
sewage sludge	600	Zurich, Switzerland	Fent (1989b, 1994)
organotin production plant effluent	8570	Germany	Schebek *et al.* (1991)
STP effluent (domestic effluent only)	5–8	Germany	Schebek *et al.* (1991)
STP effluent (40% domestic–60% industrial effluent)	3–48	Germany	Schebek *et al.* (1991)
untreated wastewater	65–523	Zurich, Switzerland	Fent and Muller (1991)
tertiary STP effluent	2–7	Zurich, Switzerland	Fent and Muller (1991)
sewage sludge	209–454	Zurich, Switzerland	Fent and Muller (1991)
STP influent	n.d.–2400	five Canadian cities	Chau *et al.* (1992)
STP effluent	n.d.–2000	five Canadian cities	Chau *et al.* (1992)
sewage sludge	n.d.–305	five Canadian cities	Chau *et al.* (1992)
STP influent	6	Bordeaux, France	Donard, Quevauviller and Bruchet (1993)
STP primary effluent	5	Bordeaux, France	Donard *et al.* (1993)
STP primary clarifier	9	Bordeaux, France	Donard *et al.* (1993)
STP activated sludge	n.d.	Bordeaux, France	Donard *et al.* (1993)
STP effluent	4	Bordeaux, France	Donard *et al.* (1993)

*Concentrations in sediment and sludge are by dry weight, and in biota are by wet weight. Water samples are whole water. Abbreviations used: n.d., not determined; STP, sewage treatment plant.

Table 4.3 *Occurrence of monobutyltin in fresh water environments**

Medium	Concentration ($ng\,Sn\,l^{-1}$ or $ng\,Sn\,g^{-1}$)	Location	Reference
water	9–510	Lake Michigan, USA	Hodge et al. (1979)
water	n.d.–5800	harbours and marinas in Ontario, Canada	Maguire et al. (1982)
water	30–200	tributaries of Baltimore Harbor, MD, USA	Jackson et al. (1982)
water	n.d.–90	Toronto Harbour, Ontario, Canada	Maguire and Tkacz (1985)
water (overlying sediment)	n.d.–290	Toronto Harbour, Ontario, Canada	Maguire and Tkacz (1985)
water	n.d.–115	Detroit and St Clair Rivers, Ontario, Canada	Maguire et al. (1985a)
water	n.d.–1890	Canada	Maguire et al. (1986)
water (fresh/ estuarine)	n.d.–10	Canada	Maguire et al. (1986)
water	5–21	Switzerland	Muller (1987)
water	n.d.–77	Rivers Bure and Yare, England	Waite et al. (1989)
water	n.d.–9	marinas in California, USA	Stang and Goldberg (1989)
water	n.d.–41	marina, Lake Lucerne, Switzerland	Fent (1989a)
water (fresh/ estuarine)	14–1200	harbour in Antwerp, Belgium	Dirkx et al. (1989)
water	n.d.–51	marinas in Lake Lucerne, Switzerland	Fent and Hunn (1991)
water	0.7–2.3	Rhine River, Germany	Schebek et al. (1991)
water	0.9–6	Germany	Schebek et al. (1991)
surface microlayer	n.d.–100	St Clair River, Canada	Maguire et al. (1985a)
surface microlayer	n.d.–66 800	lakes and rivers in Ontario, Canada	Maguire and Tkacz (1987)
sediment (core)	n.d.–15	Palace Moat, Tokyo, Japan	Tugrul, Balkas and Goldberg (1983)

Table 4.3 *continued*

Medium	Concentration ($ng\,Sn\,l^{-1}$ or $ng\,Sn\,g^{-1}$)	Location	Reference
sediment	n.d.–390	lakes, rivers, harbours and marinas in Ontario, Canada	Maguire (1984)
sediment	n.d.–80	Toronto Harbour, Ontario, Canada	Maguire and Tkacz (1985)
sediment (cores)	n.d.–60	Toronto Harbour, Ontario, Canada	Maguire and Tkacz (1985)
sediment	n.d.–35	Detroit and St Clair Rivers, Canada	Maguire et al. (1985a)
sediment	n.d.–4730	Canada	Maguire et al. (1986)
sediment (fresh water/ estuarine)	n.d.–10	Canada	Maguire et al. (1986)
sediment (cores)	n.d.–23	Lake Zurich, Switzerland	Muller (1987)
sediment	n.d.–45	Oshawa and Whitby Harbours, Ontario, Canada	Chau et al. (1989)
sediment (fresh water/ estuaries/ marine)	n.d.–1428	Portuguese rivers, estuaries	Quevauviller et al. (1989)
sediment	n.d.–59	harbours in Lake Ontario, Canada	Scott et al. (1991)
sediment (cores)	n.d.–105	marinas in Lake Lucerne, Switzerland	Fent and Hunn (1991)
sediment (cores)	n.d.–25	marina in Lake Lucerne, Switzerland	Fent et al. (1991)
sediment	n.d.–60	Rhine River, Germany	Schebek et al. (1991)
sediment	27–34	Germany	Schebek et al. (1991)
sediment (spring)	n.d.–63	Severn Sound, Ontario, Canada	Wong and Chau (1992)
sediment (summer)	n.d.–96	Severn Sound, Ontario, Canada	Wong and Chau (1992)
sediment (fall)	n.d.–32	Severn Sound, Ontario, Canada	Wong nd Chau (1992)

Table 4.3 *continued*

Medium	Concentration (ng Sn l^{-1} or ng Sn g^{-1})	Location	Reference
sediment (fresh water/ estuarine)	n.d.–50	East Anglia, England	Dowson *et al.* (1992b)
sediment	19	Main River, Germany	Cai *et al.* (1993)
sediment (cores) (fresh water/ estuarine)	n.d.–300	England	Dowson *et al.* (1993)
lake chub (*Leuciscus cephalus*) (muscle)	2	marina in Lake Murten, Switzerland	Fent and Hunn (1991)
fish (unidentified) (whole)	n.d.–47	harbours in Lake Ontario, Canada	Scott *et al.* (1991)
power plant discharge	<68	Leghorn, Italy	Gabrielides *et al.* (1990)
landfill leachate	112	Switzerland	Fent (1991)
STP effluent	8	Switzerland	Muller (1987)
sewage sludge	84–2520	Switzerland	Muller (1987)
organotin production plant effluent	2130	Germany	Schebek *et al.* (1991)
STP effluent (domestic effluent only)	9	Germany	Schebek *et al.* (1991)
STP effluent (40% domestic–60% industrial effluent)	9–19	Germany	Schebek *et al.* (1991)
untreated wastewater	92–288	Zurich, Switzerland	Fent and Muller (1991)
tertiary STP effluent	3–12	Zurich, Switzerland	Fent and Muller (1991)
sewage sludge	68–530	Zurich, Switzerland	Fent and Muller (1991)
STP influent	142–250	Zurich, Switzerland	Fent (1989b, 1994)
sewage sludge	1200	Zurich, Switzerland	Fent (1989b, 1994)
STP influent	1900–20 600	five Canadian cities	Chau *et al.* (1992)

Table 4.3 *continued*

Medium	Concentration (ng Sn l^{-1} or ng Sn g^{-1})	Location	Reference
STP effluent	700–14 500	five Canadian cities	Chau *et al.* (1992)
sewage sludge	n.d.–440	five Canadian cities	Chau *et al.* (1992)
STP influent	13	Bordeaux, France	Donard *et al.* (1993)
STP primary effluent	7	Bordeaux, France	Donard *et al.* (1993)
STP primary clarifier	10	Bordeaux, France	Donard *et al.* (1993)
STP activated sludge	6	Bordeaux, France	Donard *et al.* (1993)
STP effluent	10	Bordeaux, France	Donard *et al.* (1993)

*Concentrations in sediment and sludge are by dry weight, and in biota are by wet weight. Water samples are whole water. Abbreviations used: n.d., not determined; STP, sewage treatment plant.

of tributyltin found in sediment were about 3500 ng Sn g^{-1} dry weight in Toronto Harbour in Canada (Maguire and Tkacz, 1985), the Elbe River in Germany (Becker and Brigezu, 1992) and southeastern England (fresh/estuarine water – Dowson, Bubb and Lester, 1993). There are few data on tributyltin concentrations in fresh water organisms. Tributyltin has been found in zebra mussels (*Dreissena polymorpha*) in Lake Lucerne and Lake Geneva in Switzerland at concentrations up to 3800 ng Sn g^{-1} wet weight (Fent and Hunn, 1991; Becker *et al.*, 1992). To date, surveys for tributyltin in fresh water organisms have not revealed effects such as those observed in marine organisms (e.g., shell deformities in oysters and 'imposex' in dogwhelks).

High concentrations of tributyltin have also been observed in the surface microlayer of fresh water (Maguire *et al.*, 1982, 1985a; Maguire and Tkacz, 1987). Surface microlayers of natural waters (thickness < 300 µm) have long been of interest in environmental chemistry because they are often enriched in metals, lipophilic contaminants, nutrients, dissolved and particulate organic matter, and microorganisms (MacIntyre, 1974; Liss, 1975; Armstrong and Elzerman, 1982; Eisenreich, 1982; Meyers and Kawka, 1982). In particular, contamination of the surface microlayer by high concentrations of toxic substances relative to subsurface water may pose hazards to organisms which spend part or all of their lives at the air–water interface (e.g., Von Westernhagen *et al.*, 1987). Surface microlayer enrichment is also an important phenomenon in the cycling of contaminants from water to the atmosphere as jet and film drops ejected from air bubbles bursting at the surface of water (MacIntyre, 1974; Liss, 1975; Blanchard and Syzdek, 1975; Piotrowicz *et al.*, 1979). In addition, the possibility exists for enhanced sunlight photodegradation of chemicals in surface microlayers since there is little attenuation of sunlight compared with the attenuation experienced in penetration to greater depths in the water column. The highest concentration of tributyltin reported for a fresh water surface microlayer was $36\,000\,\text{ng}\,\text{Sn}\,\text{l}^{-1}$ in a lake in Ontario, Canada (Maguire and Tkacz, 1987), a concentration that is higher than acute toxicity thresholds for many fresh water organisms. It has occasionally been observed that the concentration of tributyltin in the surface microlayer was so much greater than that in subsurface water that the microlayer contained a significant amount of tributyltin relative to that in the whole depth of subsurface water (Maguire *et al.*, 1982; Maguire and Tkacz, 1987). However, these large enrichments were rare, and probably transient because turbulent conditions would mix the surface microlayer with subsurface water. There are few studies of the temporal variability of concentrations of toxic substances in surface microlayers. Pellenbarg and Church (1979) have shown that microlayer enrichments can be significant over long periods. They observed that the microlayer in a Delaware salt marsh carried an average of 10% of the copper, 19% of the zinc and 23% of the iron in terms of the total metal flux, including the dissolved and particulate compartments. In contrast, Maguire and Tkacz (1988) have shown that over a one-year period the surface microlayer did not contribute significantly to the loading of chlorinated hydrocarbons from the Niagara River to Lake Ontario. It is concluded that there is not enough information on the temporal stability of butyltin concentrations in surface microlayers to assess their hazard to aquatic biota.

In addition to tributyltin, dibutyltin and monobutyltin species, other related compounds such as tetrabutyltin, tributylmethyltin, dibutyl-dimethyltin and butyltrimethyltin have been found in fresh water environments (Maguire, 1984; Maguire *et al.*, 1986). These compounds have only been found rarely, and at low concentrations. Tetrabutyltin was likely a contaminant of tributyltin-containing antifouling paint formulations, and the methylated butyltin compounds likely resulted from the natural methylation of the butyltin species.

4.3   Persistence and fate

There are many factors that influence the distribution and persistence of a chemical in aquatic environments, such as its physical and chemical properties, and ecosystem-specific properties like the nature and concentration of dissolved and suspended material, nature and concentration of microbial populations, temperature, degree of insolation, etc. Important physical, chemical and biological removal mechanisms for aquatic ecosystems are (i) volatilization and adsorption to suspended solids and sediment, (ii) chemical and photochemical degradation or transformation, and (iii) uptake and transformation by microorganisms, respectively. A more detailed description of the way in which physical–chemical properties and ecosystem-specific properties determine the fate of chemicals has been given by Howard (1989). Most of the data in the literature on the persistence of organotin compounds refer to the aquatic environment, and degradation pathways and kinetics for tributyltin are in general the same for dibutyltin and monobutyltin.

Volatilization of tributyltin from fresh water is not likely to be a significant pathway. There was negligible volatilization of tributyltin from distilled water over a period of at least two months (Maguire et al., 1983), and from fresh water–sediment mixtures over a period of eleven months (Maguire and Tkacz, 1985). Adsorption to sediment is fairly strong, and desorption appears to be slow (Maguire and Tkacz, 1985). Chemical degradation of tributyltin in fresh water is also a slow process, with a half-life greater than eleven months in a sterilized fresh water–sediment mixture. The fast degradation of tributyltin observed in certain fine-grained marine sediments (Stang, Lee and Seligman, 1992) (half-lives <7 days) has not been observed in fresh water sediments. Sunlight photodegradation of tributyltin appears to be the fastest of the purely physical or chemical routes of degradation in, or dissipation from, fresh water, but even then the half-life is greater than three months (Maguire et al., 1983). Sunlight photodegradation of tributyltin is likely to be less important with increasing depth in fresh water compared with seawater because of the greater transparency of seawater. However, in fresh water the rate of photodegradation can be accelerated by dissolved organic material such as fulvic acid (Maguire et al., 1983), which indicates the potential influence that an ecosystem-specific property can have on persistence. There is some evidence that the photolysis of tributyltin in water proceeds through a sequential debutylation pathway (Woggon and Jehle, 1975; Maguire et al., 1983).

Tributyltin in water appears to be much more susceptible to biological degradation than purely chemical or photochemical degradation. There are reports of degradation in fish (Ward *et al.*, 1981) and by microorganisms such as bacteria, algae and fungi (e.g., Slesinger and Dressler, 1978; Barug, 1981; Orsler and Holland, 1982; Maguire *et al.*, 1984, 1985b; Maguire and Tkacz, 1985; Olson and Brinckman, 1986; Stang and Seligman, 1986; Seligman, Valkirs and Lee, 1986) in natural waters and sediments, as well as in more concentrated cultures of microorganisms. Although there are some exceptions, in general it appears that aquatic organisms degrade tributyltin through a sequential debutylation pathway in a manner analogous to mammalian metabolism (Fish *et al.*, 1986; Kimmel *et al.*, 1977). This mechanism probably involves hydroxylated butyltin intermediates which are unstable and which lose butene, yielding dibutyltin, monobutyltin or tin species. Such hydroxylated intermediates have been demonstrated in mammalian microsomal preparations (Fish *et al.*, 1986; Kimmel *et al.*, 1977; Ishizaka, Suzuki and Saito, 1989), from microsomal preparations from crab stomach and fish liver (Lee, 1986), and estuarine algae (Lee, Valkirs and Seligman 1989). These hydroxylated intermediates are likely also produced in the metabolism of tributyltin by fresh water organisms, although that has not yet been demonstrated. As mentioned above, biological methylation of tributyltin and its degradation products is possible, but it appears to be a minor transformation pathway.

Because biological degradation in water and sediment appears to be the most important factor limiting the persistence of tributyltin in aquatic ecosystems, it is expected that ecosystem-specific characteristics such as temperature and the kinds and concentrations of tributyltin-degrading microorganisms will be important determinants of the persistence of tributyltin in any particular location. Table 4.4 summarizes the persistence of tributyltin, dibutyltin and monobutyltin in fresh water environments, and compares these data with data from estuarine and marine environments. There is general agreement between results obtained for fresh water and those obtained for estuarine water and seawater. It appears that tributyltin is only of low-to-moderate persistence in fresh water (i.e., half-lives of several days to a few months), depending upon the presence or absence of tributyltin-degrading microorganisms, temperature, the degree of insolation (enhancing algal degradation), and the season of the year. In sediment, however, tributyltin appears to be at least moderately persistent, with measured half-lives on the order of months and estimated half-lives from sediment core data on the order of years. There are fewer data on the persistence of dibutyltin and monobutyltin in fresh water (or other) environments. Those data indicate that dibutyltin and monobutyltin have roughly the same persistence as tributyltin in water and sediment.

4.4   General toxicity summary

The toxicity of tin compounds has been studied extensively (e.g., Luijten, 1972; Piver, 1973; Hall and Pinkney, 1985; Snoeij, Penninks and Seinen, 1987; Cooney and Wuertz, 1989; Boyer, 1989; and references therein). Organotin compounds are more toxic than inorganic tin compounds. Progressive introduction of organic groups to the tin atom in any $R_nSn^{(4-n)+}$ series produces maximal biological activity against all species when $n = 3$, i.e., for the triorganotin compounds (Wong et al., 1982; Walsh et al., 1985; Vighi and Calamari, 1985; Laughlin et al., 1985; Dooley and Kenis, 1987; Salazar and Salazar, 1989; Josephson, Lindsay and Stuiber, 1989; Nagase et al., 1991). However, within the class of triorganotin compounds there are considerable variations in toxicity with the nature of the organic substituents (Davies and Smith, 1980). For insects, trimethyltin compounds are the most toxic; for mammals, the triethyltin compounds; for Gram-negative bacteria, the tri-n-propyltin compounds; for Gram-positive bacteria, yeasts, fungi and fish, the tri-n-butyltin compounds. Further increase in the n-alkyl chain length produces a sharp drop in toxicity. Triphenyltin compounds are particularly toxic to phytoplankton (Wong et al., 1982), while tricyclohexyltin compounds show high acaricidal activity (Davies and Smith, 1980).

The variation of X within any particular series of R_3SnX compounds usually has little effect on biological activity (Polster and Halacka, 1971; Davies and Smith, 1980; Nagase et al., 1991). In addition, some mammalian toxicity data for dialkyltin and monoalkyltin compounds have indicated only a modest variation in toxicity with the nature of the X substituent (World Health Organization, 1980). However, Nagase et al. (1991) have demonstrated a ten-fold variation in acute toxicity to the red killifish (*Oryzias latipes*) with the nature of X in the series Bu_2SnX_2.

The mechanism of toxic action of triorganotin compounds appears to be through disruption of oxidative phosphorylation, by (a) secondary responses caused by discharge of a hydroxyl–chloride ion gradient across mitochondrial membranes, (b) interaction with the basic energy conservation system involved in the synthesis of ATP, and (c) an interaction with mitochondrial membranes to cause swelling and disruption (Selwyn, 1976; Aldridge, 1976). (The biological effects of tetraorganotin compounds in mammals appear to be caused principally by their biologically and

Table 4.4 *Persistence of tributyltin, dibutyltin and monobutyltin in aquatic environments*

Species	Medium	Half-life	Location/source of medium	Reference
Tributyltin	fresh water	4 months (in dark)	Toronto Harbour, Ontario, Canada	Maguire and Tkacz (1985)
		est. few weeks to few months in sunlight, depending upon season	Toronto Harbour, Ontario, Canada	Maguire (1986)
		6 days at 20 °C, (subdued light)	southern England	Thain, Waldock and Waite (1987)
		11 days (in dark)	Daini Neya River, Osaka Prefecture, Japan	Hattori et al. (1988)
		26 days in winter (in lab window)	Osaka, Japan	Watanabe, Sakai and Takatsuki (1992)
	fresh water/sediment mixture	5 months (in dark)	Toronto Harbour, Ontario, Canada	Maguire and Tkacz (1985)
	fresh water/estuarine sediment	0.9–5.2 years (estimated from sediment core data)	southeast England	Dowson et al. (1993)
	estuarine water	1–2 weeks at 22–28 °C (in dark) (faster under incandescent light)	Annapolis and Baltimore Harbors, MD, USA	Brinckman et al. (1988)
		'quite stable' at 4 °C (in dark)	Annapolis and Baltimore Harbors, MD, USA	Brinckman et al. (1988)

estuarine sediment	3.8 years (estimated from sediment core data)	Georges River, NSW, Australia	Kilby and Batley (1993)
seawater	6–13 days (in sunlight) 7–19 days (in dark) est. 60 days at 5 °C, (subdued light)	San Diego Harbor, CA, USA southern England	Seligman et al. (1986) Thain et al. (1987)
	6 days (in sunlight)	San Diego Bay, CA, USA	Seligman et al. (1988)
	13 days in one location no degradation over 15 days in two other locations (in dark)	Osaka Bay, Japan	Hattori et al. (1988)
	4–19 days (in sunlight) 7–26 days in autumn or winter, depending on source (in lab window)	various USA locations Osaka, Japan	Seligman et al. (1989) Watanabe et al. (1992)
seawater/sediment microcosm	9 days (in sunlight)	Narragansett Bay, RI, USA	Adelman, Hinga and Pilsor (1990)
marine sediment	1.85 years (estimated from sediment core data)	Tamaki Estuary, Auckland, New Zealand	de Mora, King and Miller (1989)
	>8–15 years (estimated from sediment core data)	Arcachon Bay, France	Astruc et al. (1990)

Table 4.4 *continued*

Species	Medium	Half-life	Location/source of medium	Reference
Dibutyltin	fresh water	5 days (in dark)	Daini Neya River, Japan	Hattori *et al.* (1988)
	fresh water/estuarine sediment	10 days (in lab window)	Osaka, Japan	Watanabe *et al.* (1992)
		1.5–3.0 years (estimated from sediment core data)	southeast England	Dowson *et al.* (1993)
	estuarine sediment	6 years (estimated from sediment core data)	Georges River, NSW, Australia	Kilby and Batley (1993)
	seawater	est. 90 days at 5 °C (subdued light)	southern England	Thain *et al.* (1987)
		8 days in one location, no degradation over 15 days in water from two other locations	Osaka, Japan	Hattori *et al.* (1988)
		4–10 days in filtered solutions (5–10 μm)	Osaka Bay, Japan	Watanabe *et al.* (1992)
		1–7 days (in lab window)	New Hampshire, USA	François *et al.* (1989)
	seawater/sediment microcosm	12–18 days at 20 °C (in sunlight)	Narragansett Bay, RI, USA	Adelman *et al.* (1990)

Monobutyltin	seawater	2–14 days (in lab window)	New Hampshire, USA	François et al. (1989)
	fresh water/estuarine sediment	1.8–3.7 years (estimated from sediment core data)	southeast England	Dowson et al. (1993)
	seawater/sediment microcosm	no degradation over 50 days at 20 °C (in sunlight)	Narragansett Bay, RI, USA	Adelman et al. (1990)

abiotically produced triorganotin degradation products (Davies and Smith, 1980).) Dialkyltin compounds also show the same trend of decreasing toxicity with increasing alkyl chain length observed with trialkyltin compounds, and they are considerably less toxic than trialkyltin species. The mechanism of toxic action may be through interfering with α-keto acid oxidation, possibly by binding to dithiol groups (Cain, Hyams and Griffiths, 1977; Penninks and Seinen, 1980; Davies and Smith, 1980; Penninks, Verschuren and Seinen, 1983). Dimethyltin dichloride inhibits RNA polymerase from *Escherichia coli* (Yamada, 1977). (Dialkyltin compounds more strongly inhibit the polymerase reaction than trialkyltin compounds.) Dibutyltin dichloride and dioctyltin dichloride are immunosuppressive compounds that impair thymus-dependent immunity in animals (Vos, 1977; Li, Dahl and Hill, 1982; Smialowicz *et al.*, 1988). Little work has been identified on the mechanism of toxic action of monoorganotin compounds. These compounds appear to be far less toxic than the triorganotin or even the diorganotin compounds.

4.5   Toxicity of butyltin species to fresh water organisms

The toxicity of organotin compounds to aquatic organisms has received a good deal of attention in the past 10 years (see reviews by Hall and Pinkney, 1985, and Cooney and Wuertz, 1989). Wong *et al.* (1982) showed a direct relationship between the toxicity of organotin compounds to algae and their octanol–water partition coefficients. Cooney *et al.* (1989) tested the effects of various organotin compounds on 29 yeasts representing 10 genera. Yeasts varied in their sensitivity from strain to strain, but tributyltin was the most toxic compound tested. Mono- and dimethyltin were the least toxic. Triphenyltin, dibutyltin, monobutyltin, trimethyltin, triethyltin and diethyltin showed intermediate toxicity. Eng *et al.* (1988) showed that the total surface area parameter is an excellent predictor of toxicity provided that the toxicity is primarily related to hydrophobicity, as it is for diorganotin compounds, rather than to electronic or steric effects. Organisms included in this study were neuroblastoma cells, fibroblasts, algae (*Ankistrodesmus falcatus*), water fleas (*Daphnia magna*), and mud crabs (*Rhithropanopeus harrisii*).

Table 4.5 summarizes the results of acute toxicity tests of tributyltin, dibutyltin and monobutyltin on fresh water organisms, and their significance is discussed below. In contrast to the situation for tributyltin, only limited information was identified on the toxicity of dibutyltin and monobutyltin to fresh water organisms. Some of the studies summarized in Table 4.5 also contain information on accumulation, tissue distribution and metabolism of tributyltin, as do other articles (e.g., Fent, 1991; Schwaiger et al., 1992; Fent and Hunn, 1993), but these aspects will not be discussed in this chapter.

A complication in studies of the toxicity of monobutyltin and dibutyltin compounds may be the presence of tributyltin compounds that can be orders of magnitude more toxic. Only one article was found on this point in the fresh water toxicity literature. Wester et al. (1990) used dibutyltin dichloride of 98% purity in a study of the effects of dibutyltin and tributyltin in the Japanese medaka (Oryzias latipes).[2] The no-observed-effects-concentrations (NOEC) using histopathology were $0.06 \mu g \, Sn \, l^{-1}$ for tributyltin chloride and $< 125 \mu g \, Sn \, l^{-1}$ for dibutyltin dichloride, a difference of approximately 2000. However, considerable amounts of tributyltin were found in dibutyltin-exposed fish, and subsequent analysis indicated that the dibutyltin dichloride used contained tributyltin chloride at a concentration (0.33%) that could have accounted for the observed toxicity of dibutyltin. This point should be kept in mind in considering the toxicity of monobutyltin and dibutyltin compounds in the studies in Table 4.5 and those summarized below. It may be that toxicity thresholds reported for monobutyltin and dibutyltin compounds reflect the toxicity of a tributyltin contaminant and should be considered as upper limits.

In this section, the most sensitive fresh water organism to each butyltin species is identified, with emphasis being placed on ecologically significant acute toxicity tests, where possible (e.g., LC_{50} values chosen in preference to enzyme activity tests). Information on chronic toxicity is also provided where available, and the most sensitive organism in chronic toxicity tests is also identified.

4.5.1   Tributyltin
4.5.1.1   Acute toxicity
For tributyltin, a coelenterate (Hydra sp.) was the most sensitive fresh water organism in acute toxicity tests. The 48 h EC_{50} (response to prodding) was $0.4 \mu g \, Sn \, l^{-1}$ (Brooke et al., 1986). No data were identified on the acute toxicity of tributyltin in sediment to benthic fresh water organisms.

---

[2] Oryzias latipes has been described as either Japanese medaka or red killifish. The designations by the original authors are retained here.

Table 4.5 *Acute toxicity of tributyltin, dibutyltin and monobutyltin compounds to fresh water organisms**

Compound	Species	Size/age	Static/flow-through	Temp. (°C)	pH	Hardness (mg l^{-1} CaCO$_3$)	Response	Conc. (µg Sn l^{-1})	Nominal/measured	Reference
Tributyltin										
chloride	bacteria (mixed culture)		stat	24			30 min IC$_{50}$ (dehydrogenase activity)	1400	n	Liu and Thomson (1986)
chloride	bacteria (various sp.)		stat	20	7.2		1 week EC$_{50}$ (growth)	<10–>10000	n	Wuertz et al. (1991)
chloride	cyanobacterium (*Plectonema boryanum*)		stat	22			20 min IC$_{50}$ (photosynthesis)	145	n	Avery et al. (1991)
chloride	cyanobacterium (*Anabaena cylindrica*)		stat	22			20 min IC$_{50}$ (photosynthesis)	120	n	Avery et al. (1991)
chloride	cyanobacterium (*Anabaena cylindrica*)		stat	22			3 h IC$_{50}$ (nitrogenase activity)	120	n	Avery et al. (1991)
chloride	alga (*Ankistrodesmus falcatus*)		stat	20	8		4 h IC$_{50}$ (primary productivity)	7.4	n	Wong et al. (1982)
chloride	alga (*Scenedesmus quadricauda*)		stat	20	8		4 h IC$_{50}$ (primary productivity)	6	n	Wong et al. (1982)
chloride	alga (*Anabaena flos-aquae*)		stat	20	8		4 h IC$_{50}$ (primary productivity)	5	n	Wong et al. (1982)
chloride	alga (*Scenedesmus obliquus*)		stat	25			96 h EC$_{50}$ (growth)	1.3	n	Huang et al. (1993)
chloride	algae (Lake Ontario, Canada)		stat	20	8		4 h IC$_{50}$ (primary productivity)	1	n	Wong et al. (1982)
chloride	algae (13 sp.)		stat	20	6.5–7.5		14 d EC$_{100}$ (growth)	23–740	n	Blanck, Wallin and Wangberg (1984)
bis(tri-n-butyltin) oxide	water flea (*Daphnia magna*)						48 h EC$_{50}$ (immobilization)	7	m	Steinhauser et al. (1985)
bis(tri-n-butyltin) oxide	water flea (*Daphnia magna*)						48 h LC$_{50}$	14	n	Foster (1981)
chloride	water flea (*Daphnia magna*)		stat	20	7.5	200	24 h LC$_{50}$	5	n	Vighi and Calamari (1985)

Compound	Species	Size/weight	Test	Temp (°C)	Hardness	pH	Endpoint	Value		Reference
bis(tri-n-butyltin) oxide	water flea (*Daphnia magna*)	<24hr	flow	24	52	7.5	48 h EC_{50} (immobilization)	1.8	n	Brooke *et al.* (1986)
bis(tri-n-butyltin) oxide	mosquito larvae (*Culex* sp.)		stat	17	52	7.6	96 h EC_{50} (prodding response)	4.2	n	Brooke *et al.* (1986)
acetate	mosquito (*Culex pipiens pipiens*)	larva	stat	25		6.5	24 h LC_{90}	99	n	Gras and Rioux (1965)
acetate	mosquito (*Culex pipiens pipiens*)	larva	stat	25		6.5	24 h LC_{50}	235	n	Gras and Rioux (1965)
bis(tri-n-butyltin) oxide	mosquito (*Culex pipiens pipiens*)	larva	stat	25		6.5	24 h LC_{50}	76	n	Gras and Rioux (1965)
bis(tri-n-butyltin) oxide	mosquito (*Culex pipiens pipiens*)	larva	stat	25		6.5	24 h LC_{90}	138	n	Gras and Rioux (1965)
bis(tri-n-butyltin) oxide	annelid (*Lumbriculus variegatus*)	9 mg	flow	18	52	7.1	48 h EC_{50} (prodding response)	4.5	n	Brooke *et al.* (1986)
bis(tri-n-butyltin) oxide	coelenterate (*Hydra* sp.)	<1 mg	stat	26	51	7.5	48 h EC_{50} (body/tentacle length)	0.4	n	Brooke *et al.* (1986)
bis(tri-n-butyltin) oxide	amphipod (*Gammarus pseudolimnaeus*)	30 mg	flow	18	52	7.1	96 h LC_{50}	1.5	n	Brooke *et al.* (1986)
bis(tri-n-butyltin) oxide	snail (*Biomphalaria glabrata*)	10 mm	stat			5.5	24 h LC_{50}	10–20	n	Hopf *et al.* (1967)
acetate	snail (*Biomphalaria glabrata*)	10 mm	stat			5.5	24 h LC_{50}	20–60	n	Hopf *et al.* (1967)
pentachlorophenate	snail (*Biomphalaria glabrata*)	10 mm	stat			5.5	24 h LC_{50}	20–80	n	Hopf *et al.* (1967)
bis(tri-n-butyltin) oxide	snail (*Biomphalaria glabrata*)	10–26 mm	stat				6 h LC_{50}	82	n	Seiffer and Schoof (1967)
acetate	snail (*Biomphalaria glabrata*)	10–26 mm	stat				6 hr LC_{50}	58	n	Seiffer and Schoof (1967)
bis(tri-n-butyltin) oxide	Asiatic clam larvae (*Corbicula fluminea*)	3 weeks	stat				24 h LC_{50}	419	n	Foster (1981)
bis(tri-n-butyltin) oxide	European frog (*Rana temporaria*)	larva	ren (1 d)				5 d LC_{40}	6	n	Laughlin and Linden (1982)
fluoride	European frog (*Rana temporaria*)	larva	ren (1 d)				5 d LC_{50}	12	n	Laughlin and Linden (1982)

Table 4.5 *(continued)*

Compound	Species	Size/age	Static/flow-through	Temp. (°C)	Hardness (mg l⁻¹ CaCO₃)	pH	Response	Conc. (μg Sn l^{-1})	Nominal/measured	Reference
chloride	minnow (*Phoxinus phoxinus*)	embryo	ren (1 d)	21	120	8	96 h LC$_{100}$	2.7–3.8	m	Fent (1992)
bis(tri-*n*-butyltin) oxide	*Tilapia rendalli*	6–20 g	flow	25	350	7.6	24 h EC$_{50}$ (loss of positive rheotaxis)	11	n	Chliamovitch and Kuhn (1977)
bis(tri-*n*-butyltin) oxide	guppy (*Lebistes reticulatus*)						7 d LC$_{50}$	16	n	Polster and Halacka (1971)
acetate	guppy (*Lebistes reticulatus*)						7 d LC$_{50}$	12	n	Polster and Halacka (1971)
oleate	guppy (*Lebistes reticulatus*)						7 d LC$_{50}$	14	n	Polster and Halacka (1971)
benzoate	guppy (*Lebistes reticulatus*)						7 d LC$_{50}$	10	n	Polster and Halacka (1971)
chloride	guppy (*Lebistes reticulatus*)						7 d LC$_{50}$	9	n	Polster and Halacka (1971)
laurate	guppy (*Lebistes reticulatus*)						7 d LC$_{50}$	12	n	Polster and Halacka (1971)
bis(tri-*n*-butyltin) oxide	red killifish (*Oryzias latipes*)	2 cm	ren (1 d)	20			48 h LC$_{50}$	14	n	Nagase *et al.* (1991)
acetate	red killifish (*Oryzias latipes*)	2 cm	ren (1 d)	20			48 h LC$_{50}$	27	n	Nagase *et al.* (1991)
chloride	red killifish (*Oryzias latipes*)	2 cm	ren (1 d)	20			48 h LC$_{50}$	13	n	Nagase *et al.* (1991)
fluoride	red killifish (*Oryzias latipes*)	2 cm	ren (1 d)	20			48 h LC$_{50}$	24	n	Nagase *et al.* (1991)
methoxide	red killifish (*Oryzias latipes*)	2 cm	ren (1 d)	20			48 h LC$_{50}$	8	n	Nagase *et al.* (1991)
ethoxide	red killifish (*Oryzias latipes*)	2 cm	ren (1 d)	20			48 h LC$_{50}$	9	n	Nagase *et al.* (1991)
chloride	bluegill sunfish (*Lepomis macrochirus*)	cells	stat	34			24 h Neutral Red$_{50}$ assay (cytotoxicity)	66	n	Babich and Borenfreund (1988)
bis(tri-*n*-butyltin) oxide	bluegill sunfish (*Lepomis macrochirus*)						96 h LC$_{50}$	48	n	Foster (1981)
bis(tri-*n*-butyltin) oxide	channel catfish (*Ictalurus punctatus*)	1.4 g	flow	18	52	7.1	96 h LC$_{50}$	2.3	n	Brooke *et al.* (1986)

Compound	Species	Size/stage	Test	Temp	Hardness	pH	Endpoint	Value	n/m	Reference
bis(tri-*n*-butyltin) oxide	fathead minnow (*Pimephales promelas*)	0.07 g	flow	24	52	7.5	96 h LC$_{50}$	1.1	n	Brooke *et al.* (1986)
bis(tri-*n*-butyltin) oxide	rainbow trout (*Oncorhynchus mykiss*)	yearling	ren (1 d)	18	250		48 h LC$_{50}$	4	n	Alabaster (1969)
bis(tri-*n*-butyltin) oxide	rainbow trout (*Oncorhynchus mykiss*)	10–20 g	flow	16	350	7.6	24 h EC$_{50}$ (loss of positive rheotaxis)	6	n	Chiamovitch and Kuhn (1977)
chloride	rainbow trout (*Oncorhynchus mykiss*)	yolk sac fry	flow	14	94–102	6.8	12 d LC$_{100}$	1.8	n	Seinen *et al.* (1981)
bis(tri-*n*-butyltin) oxide	rainbow trout (*Oncorhynchus mykiss*)	0.2 g	flow	16	51	7.1	96 h LC$_{50}$	1.6	n	Brooke *et al.* (1996)
bis(tri-*n*-butyltin) oxide	rainbow trout (*Oncorhynchus mykiss*)	1.5 g	flow	16	135	7.8	96 h LC$_{50}$	1.4	m	Martin *et al.* (1989)
bis(tri-*n*-butyltin) oxide	lake trout (*Salvelinus namaycush*)	5.9 g	flow	16	135	7.8	96 h LC$_{50}$	5.2	m	Martin *et al.* (1989)
Dibutyltin dichloride	bacteria (mixed culture)		stat	24			30 min IC$_{50}$ (dehydrogenase activity)	19 000	n	Liu and Thomson (1986)
dichloride	bacterium (*Pseudomonas putida*)						4 h EC$_{10}$ (growth)	975	m	Steinhauser *et al.* (1985)
dichloride	cyanobacterium (*Plectonema boryanum*)		stat	22			20 min IC$_{50}$ (photosynthesis)	1300	n	Avery *et al.* (1991)
dichloride	cyanobacterium (*Anabaena cylindrica*)		stat	22			20 min IC$_{50}$ (photosynthesis)	1067	n	Avery *et al.* (1991)
dichloride	cyanobacterium (*Anabaena cylindrica*)		stat	22			3 h IC$_{50}$ (nitrogenase activity)	356	n	Avery *et al.* (1991)

Table 4.5 (continued)

Compound	Species	Size/age	Static/flow-through	Temp. (°C)	Hardness (mg l^{-1} CaCO$_3$)	pH	Response	Conc. (µg Sn l^{-1})	Nominal/measured	Reference
dichloride	alga (*Ankistrodesmus falcatus*)		stat	20		8	4 h IC$_{50}$ (primary productivity)	2652	n	Wong et al. (1982)
dichloride	alga (*Scenedesmus obliquus*)		stat	25			96 h EC$_{50}$ (growth)	6.5	n	Huang et al. (1993)
dilaurate	water flea (*Daphnia magna*)		stat	20	200	7.5	24 h LC$_{50}$	350	n	Vighi and Calamari (1985)
dilaurate	water flea (*Daphnia magna*)			25			48 h EC$_{50}$ (immobilization)	124	m	Steinhauser et al. (1985)
dichloride	mosquito (*Culex pipiens pipiens*)	larva	stat	25		6.5	24 h LC$_{50}$	238	n	Gras and Rioux (1965)
dichloride	mosquito (*Culex pipiens pipiens*)	larva	stat	25		6.5	24 h LC$_{90}$	2067	n	Gras and Rioux (1954)
dilaurate	mosquito (*Culex pipiens pipiens*)	larva	stat	25		6.5	24 h LC$_{50}$	134	n	Gras and Rioux (1965)
dilaurate	mosquito (*Culex pipiens pipiens*)	larva	stat	25		6.5	24 h LC$_{90}$	1647	n	Gras and Rioux (1965)
dichloride	guppy (*Poecilia reticulata*)	3–4 weeks	ren (2 d)	23			3 month EC$_{50}$ (histopathological effects)	<125	n	Wester and Canton (1987)
diacetate	red killifish (*Oryzias latipes*)	2 cm	ren (1 d)	20			48 h LC$_{50}$	1270	n	Nagase et al. (1991)
dichloride	red killifish (*Oryzias latipes*)	2 cm	ren (1 d)	20			48 h LC$_{50}$	2260	n	Nagase et al. (1991)
oxide	red killifish (*Oryzias latipes*)	2 cm	ren (1 d)	20			48 h LC$_{50}$	401	n	Nagase et al. (1991)
maleate	red killifish (*Oryzias latipes*)	2 cm	ren (1 d)	20			48 h LC$_{50}$	4500	n	Nagase et al. (1991)
dilaurate	red killifish (*Oryzias latipes*)	2 cm	ren (1 d)	20			48 h LC$_{50}$	220	n	Nagase et al. (1991)
dichloride	golden orfe (*Leuciscus idus melanotus*)						24 h LC$_{50}$	234	m	Steinhauser et al. (1985)
dilaurate	golden orfe (*Leuciscus idus melanotus*)		stat	34			24 h LC$_{50}$	377	m	Steinhauser et al. (1985)
dichloride	bluegill sunfish (*Lepomis macrochirus*)	cells					24 h Neutral Red$_{50}$ assay (cytotoxicity)	100	n	Babich and Borenfreund (1988)

Compound	Organism		Regime	Temp (°C)	Test/endpoint	Value		Reference
Monobutyltin trichloride	bacteria (mixed culture)		stat	24	30 min IC_{50} (dehydrogenase activity)	>126000	n	Liu and Thomson (1986)
thiostannonic acid	bacterium (Pseudomonas putida)				4 h EC_{10} (growth)	53	m	Steinhauser et al. (1985)
trichloride	cyanobacterium (Plectonema boryanum)		stat	22	20 min IC_{50} (photosynthesis)	>11 865	n	Avery et al. (1991)
trichloride	cyanobacterium (Anabaena cylindrica)		stat	22	20 min IC_{50} (photosynthesis)	>11 865	n	Avery et al. (1991)
trichloride	alga (Ankistrodesmus falcatus)		stat	20	4 h IC_{50} (primary productivity)	10 500	n	Wong et al. (1982)
trichloride	water flea (Daphnia magna)		stat	20	24 h LC_{50}	20 580	n	Vighi and Calamari (1985)
trichloride	bluegill sunfish (Lepomis macrochirus)	cells	stat	34	24 h Neutral Red_{50} assay (cytotoxicity)	9255	n	Babich and Borenfreund (1988)
trichloride	red killifish (Oryzias latipes)	2 cm	ren (1 d)	20	48 h LC_{50}	16000	n	Nagase et al. (1991)

*Abbreviations used: EC_{50}, IC_{50}, concentration which effectively inhibits 50% of function/growth; LC_{xx}, concentration that kills xx% of test organisms; ren, renewal (tests solutions renewed at specified intervals); rt, room temperature.

4.5.1.2 Chronic toxicity

Brooke *et al.* (1986) determined the toxicity of tributyltin to the fathead minnow (*Pimephales promelas*) in an early life-stage test (33 days) with flow-through conditions and measured concentrations. Mean fish weights at the end of the test were significantly ($p \leq 0.01$) less than the control fish at exposures of tributyltin at $0.18\,\mu g\,Sn\,l^{-1}$ and above. Mean standard length at the end of the test was significantly ($p \leq 0.05$) reduced at the lowest exposure concentration ($0.03\,\mu g\,Sn\,l^{-1}$). Mean exposures $\geq 0.06\,\mu g\,Sn\,l^{-1}$ resulted in length reductions at a higher level of significance ($p \leq 0.01$). The no-effect concentration determined was $\leq 0.03\,\mu g\,Sn\,l^{-1}$, and the acute/chronic toxicity ratio was > 32.5.

Brooke *et al.* (1986) also determined the chronic toxicity of tributyltin to the water flea (*Daphnia magna*) in a 21-day static, toxicant renewal test with measured concentrations. Survival was not affected at exposure concentrations $< 0.21\,\mu g\,Sn\,l^{-1}$. The number of young produced per adult surviving the 21-day exposure and the number of young produced per adult per reproductive day were significantly ($p < 0.01$) reduced at exposure concentrations $\geq 0.08\,\mu g\,Sn\,l^{-1}$. The no-effect concentration range was 0.04–$0.08\,\mu g\,Sn\,l^{-1}$, with a calculated minimum acceptable toxicant concentration (MATC) of $0.06\,\mu g\,Sn\,l^{-1}$. The acute/chronic toxicity ratio was 30.7.

Bodar, von Donselaar and Herwig (1990) studied histopathological effects on the ultrastructural organization of digestive tract and storage cells after a 7-day exposure to tributyltin. The ultrastructural integrity of storage cells appeared to be seriously affected at a tributyltin concentration of $0.41\,\mu g\,Sn\,l^{-1}$, and there was a reduction in food consumption rate caused primarily by a significant ($p < 0.001$) reduction of the frequency of filtering limb movements. In addition, conspicuous lipid accumulations were observed in posterior gut epithelium and in a specific hemolymphatic cell type.

The NOEC (thymus atrophy, liver vacuolation or hyperplasia of the hemopoietic tissue) in guppies (*Poecilia reticulata*) was $0.002\,\mu g\,Sn\,l^{-1}$ for an exposure period of 3 months (Wester and Canton, 1987). It was concluded that long-term low-dose exposure of guppies to tributyltin caused thymus atrophy, increase in granulocytes, accumulation of glycogen and fat in the liver, and changes in cornea, retina and skin.

The NOEC for tributyltin in the Japanese medaka (*Oryzias latipes*) (histopathological effects) was $0.06\,\mu g\,Sn\,l^{-1}$ (Wester *et al.*, 1990). The histopathological effects included vacuolation of hepatocytes, tubulonephrosis and glomerulopathy, vacuolation of the retinal pigment epithelium, keratitis and inflammation of oral and skin epithelium, hyperplasia and swelling of the gas gland epithelium and thyroid activation.

The comparative toxicity of various organotin compounds in early life stages of rainbow trout (*Oncorhynchus mykiss*) was investigated by de Vries *et al.* (1991). Beginning with yolk sac fry, trout were continuously exposed for 110 days. Tributyltin chloride caused acute mortality at a concentration of $1.9 \mu g \, Sn \, l^{-1}$, and the NOEC (mortality) was $0.02 \mu g \, Sn \, l^{-1}$. The lowest-observed-effect-concentration (LOEC) for mortality was $0.07 \mu g \, Sn \, l^{-1}$. At the end of the exposure period, resistance to infection was examined by an intraperitoneal challenge with *Aeromonas hydrophila*, a bacterium pathogenic to fish. Resistance to the bacterial challenge was decreased with tributyltin chloride at $0.07 \mu g \, Sn \, l^{-1}$. This might be indicative of a suppressed immune function or generally diminished fish health. These authors and Schwaiger *et al.* (1992) have reported a number of histopathological effects of low concentrations of tributyltin on rainbow trout.

Walker *et al.* (1989) found that length, fecundity and hatching success of Japanese medaka (*Oryzias latipes*) were significantly reduced during a 90-day exposure to tributyltin at $0.1 \mu g \, Sn \, l^{-1}$. Embryo viability was significantly reduced at $0.9 \mu g \, Sn \, l^{-1}$. Fry survival was not affected at $0.1–0.9 \mu g \, Sn \, l^{-1}$, and no abnormal embryos or fry were detected.

Embryonic exposure of minnows (*Phoxinus phoxinus*) to tributyltin at $2.7–4.3 \mu g \, Sn \, l^{-1}$ over 9 days resulted in severe body axis malformations, inhibition of coordinated movements due to disruption of neuromuscular functioning, necrosis of muscle fibres, changes in kidneys and effects on the eyes of embryos and hatched larvae. Up to 50% of larvae exposed at $21 \, °C$ to an initial concentration of tributyltin of $0.34 \mu g \, Sn \, l^{-1}$ for 6 days showed abnormalities (deformations, erratic swimming, paralysis) but no mortality. These abnormalities were not observed at the same concentration, but at $16 \, °C$ (Fent, 1992; Fent and Meier, 1992).

With the assumption that it is preferable to consider LOEC values for mortality over NOEC values, the most sensitive fresh water organism identified in chronic toxicity tests with tributyltin was rainbow trout (*Oncorhynchus mykiss*), with a 110-day LOEC (mortality) of $0.07 \mu g \, Sn \, l^{-1}$ (de Vries *et al.*, 1991).

No data were identified on the chronic toxicity of tributyltin in sediment to benthic fresh water organisms.

4.5.2   Dibutyltin

4.5.2.1   Acute toxicity

For dibutyltin, the alga *Scenedesmus obliquus* was the most sensitive fresh water organism in acute toxicity tests. The 96-hour EC_{50} value (growth) was $6.5 \, \mu g \, Sn \, l^{-1}$ (Huang *et al.*, 1993). No data were identified on the acute toxicity of dibutyltin in sediment to benthic fresh water organisms.

4.5.2.2   Chronic toxicity

Exposure of clams (*Anodonta anatina*) to $15 \, \mu g \, Sn \, l^{-1}$ dibutyltin dichloride for 7 months (weekly static renewal) caused decreases in weight and carbohydrate stores, but no mortality (Holwerda and Herwig, 1986). The NOEC for histopathological effects in guppies (*Poecilia reticulata*) was $< 125 \, \mu g \, Sn \, l^{-1}$ for an exposure period of one month (Wester and Canton, 1987). Wester *et al.* (1990) did not observe thymic atrophy in the Japanese medaka after exposure to butyltin compounds, nor did Seinen *et al.* (1981) observe it in rainbow trout, although it has been observed in the rat and the guppy. Thus the thymotoxicity of butyltin compounds appears to be a species-specific phenomenon. In the Japanese medaka, dibutyltin dichloride affected the liver and gas gland, probably as a result of impaired glycogen breakdown. In addition, it was toxic to kidney and retina. The NOEC for dibutyltin in the Japanese medaka (histopathological effects) was $< 125 \, \mu g \, Sn \, l^{-1}$ (Wester *et al.*, 1990). De Vries *et al.* (1991) investigated the comparative toxicity of various organotin compounds in early life stages of rainbow trout (*Oncorhynchus mykiss*). Beginning with yolk sac fry, trout were continuously exposed for 110 days. Dibutyltin dichloride, with a no-lethal-effect-level of $19 \, \mu g \, Sn \, l^{-1}$ for trout yolk sac fry, was about 1000 times less toxic than tributyltin chloride. The LOEC for mortality was $95 \, \mu g \, Sn \, l^{-1}$. At the end of the exposure period, resistance to infection was examined by an intraperitoneal challenge with *Aeromonas hydrophila*, a bacterium pathogenic to fish. Resistance to the bacterial challenge was decreased with dibutyltin dichloride at about $100 \, \mu g \, Sn \, l^{-1}$. This might be indicative of a suppressed immune function or generally diminished fish health.

The most sensitive fresh water organism identified in chronic toxicity tests with dibutyltin was rainbow trout (*Oncorhynchus mykiss*), with a LOEC (mortality) of $95 \, \mu g \, Sn \, l^{-1}$ (de Vries *et al.*, 1991).

No data were identified on the chronic toxicity of dibutyltin in sediment to fresh water benthic organisms.

4.5.3   Monobutyltin

4.5.3.1   Acute toxicity

For monobutyltin, the bacterium *Pseudomonas putida* was the most sensitive fresh water organism identified in acute toxicity tests. Its 4-hour EC_{10} (growth) was $53 \mu g \, Sn \, l^{-1}$ (Steinhauser *et al.*, 1985), and was 200–2000 times more sensitive than acute toxicities to other bacteria tested. This relatively high toxicity should be verified. No data were identified on the acute toxicity of monobutyltin in sediment to benthic fresh water organisms.

4.5.3.2   Chronic toxicity

No data were identified on the chronic toxicity of monobutyltin to fresh water organisms in water or sediment.

4.6   Assessment of the toxicity of concentrations of butyltin species in fresh water environments

It is evident from the data in Table 4.1 that maximal concentrations of tributyltin in many fresh water locations (subsurface) exceeded acute or chronic toxicity thresholds for the most sensitive organisms, and exceeded, for example, the United Kingdom's environmental quality standard (EQS) for tributyltin in fresh water of $8 \, ng \, Sn \, l^{-1}$ (Dowson, Bubb and Lester, 1992a). The EQS is defined as the maximum concentration of a pollutant allowed over a particular period or geographical area (Cleary, 1991). The fresh water acute toxicity threshold for dibutyltin in the most sensitive organism was also exceeded in Antwerp Harbour (Dirkx *et al.*, 1989). It should be noted, however, that many of the data summarized in Table 4.1 were collected before the regulation of tributyltin-containing antifouling paints in many countries.

It is of interest to compare mean values for butyltin concentrations with toxicity thresholds. Canadian data were used in the following example. In this assessment, effects thresholds for the most sensitive fresh water biota were estimated by dividing the LOEC in toxicity tests by various factors that account for the limited data available (Table 4.6) (such factors were used recently in an assessment of the toxicity of non-pesticidal organotin compounds in the Canadian environment – Maguire *et al.*, 1993). Emphasis was placed on ecologically relevant test results (e.g., mortality rather than enzyme assay endpoints). The estimated effects thresholds (EET) were then compared with the mean environmental concentrations observed in fresh water in Canada. If the EET/EC ratio was ≤1 for a

Table 4.6 *Comparisons between estimated effects thresholds for fresh water biota and concentrations of butyltin species in fresh water in Canada*

Species	Endpoint for most sensitive fresh water organism	Factors applied	Estimated effects threshold (EET)	Environmental concentrations (EC)*	EET/EC
tributyltin	110d LOEC for mortality in rainbow trout, *Oncorhynchus mykiss* 0.07 µg Sn l^{-1} (de Vries et al. (1991)	10[a]	0.007 µg Sn l^{-1}	0.22 µg Sn l^{-1} (mean across Canada where detected; detection frequency = 23%)	0.03
dibutyltin	96 h EC$_{50}$ (growth) for *Scenedesmus obliquus* 6.5 µg Sn l^{-1} (Huang et al., 1993)	20[b] × 4[c]	0.08 µg Sn l^{-1}	0.15 µg Sn l^{-1} (mean across Canada where detected; detection frequency = 24%)	0.5
monobutyltin	4h EC$_{10}$ (growth) for *Pseudomonas putida* 53 µg Sn l^{-1} (Steinhauser et al., 1985)	20[b] × 4[c]	0.66 µg Sn l^{-1}	0.22 µg Sn l^{-1} (mean across Canada where detected; detection frequency = 24%)	3

*For tributyltin, the highest concentration observed in fresh water (subsurface) in Canada was 2340 ng Sn l^{-1} (excluding a high concentration found just above the sediment in a core from Toronto Harbour). The mean and median concentrations derived from original data in all studies (n = 64, n.d. values for an additional 210 samples not included in calculations, detection limit 0.1–10 ng Sn l^{-1}) were 220 ng Sn l^{-1} and 50 ng Sn l^{-1}, respectively. For dibutyltin, the highest concentration observed in fresh water (subsurface) in Canada was 3700 ng Sn l^{-1}. The mean and median concentrations derived from original data in all studies (n = 65, n.d. values for an additional 209 samples not included in calculations, detection limit 0.1–10 ng Sn l^{-1}) were 148 ng Sn l^{-1} and 25 ng Sn l^{-1}, respectively. For monobutyltin, the highest concentration observed in fresh water (subsurface) in Canada was 5800 ng Sn l^{-1}. The mean and median concentrations derived from original data in all studies (n = 65, n.d. values for an additional 209 samples not included in calculations, detection limit 0.1–10 ng Sn l^{-1}) were 216 ng Sn l^{-1} and 27 ng Sn l^{-1}, respectively.
[a] To convert from chronic LOEC to a NOEC and to account for differences between laboratory and field conditions and species sensitivity.
[b] To convert from acute LC$_{50}$, IC$_{50}$, or EC$_{50}$ to chronic NOEC and to account for the extrapolation of laboratory results to the field and differences in species sensitivity.
[c] To compensate for limited toxicity data or use of non-standard toxicity tests.

given species, then the potential exists (existed) for that compound to cause harmful effects to aquatic biota.

The environmental concentrations listed in Table 4.6 are assumed to overestimate the true mean concentrations in Canada because most sampling was conducted at sites where maximum concentrations would be expected (e.g., harbours and marinas), and the detection frequencies at these sites were low, about 24%. Non-detections were not included in the calculations of the mean environmental concentrations.

Comparison of the estimated effects thresholds to mean concentrations of the butyltin species in fresh water in Canada indicates that tributyltin and dibutyltin have (or had) the potential to cause harmful effects to fresh water biota, but not monobutyltin. As discussed above, no assessment was possible for the butyltin species in surface microlayers of fresh water.

No data were identified on the toxicity of sediment-associated butyltin species to fresh water benthic organisms. Consequently, an assessment can not be made of the toxicity of residues of butyltin species in sediment. It should be noted that bioturbation and resuspension of butyltin-containing sediments by storms or dredging may reintroduce such species to the water compartment.

4.7   Research recommendations

Several recommendations follow from the data gaps identified above.

(1) Continued monitoring of butyltin residues in fresh water, sediment and biota to determine the effects of the regulation of tributyltin-containing paints in various countries. It is desirable to augment such monitoring with determinations of triphenyltin and copper, which are other antifoulants and potential replacements for tributyltin.

(2) Determination of the toxicity of butyltin species in sediment to benthic fresh water organisms.

(3) More information is required on the acute and chronic fresh water toxicity of dibutyltin and monobutyltin, in particular verification of the reported high toxicity of these species to some algae and bacteria, respectively.

(4) Development of suitable fresh water biological indicators of butyltin contamination.

(5) More data on the persistence of butyltin species in fresh water sediments, and verification of common assumptions made in persistence determinations by sediment core research (e.g.,

knowledge of sedimentation rates, accounting for mixing effects such as bioturbation/storm mixing/current effects, determination of historical fluxes of tributyltin to fresh water).

(6) Determination of the temporal stability of high concentrations of butyltin species in surface microlayers.

4.8 References

Aboul Dahab, O., El-Sabrouti, M. A. and Halim, Y. (1990). Tin compounds in sediments of Lake Maryut, Egypt. *Environ. Pollut.* **63**, 329–44.

Adelman, D., Hinga, K. R. and Pilson, M. E. Q. (1990). Biogeochemistry of butyltins in an enclosed marine ecosystem. *Environ. Sci. Technol.* **24**, 1027–32.

Alabaster, J. S. (1969). Survival of fish in 164 herbicides, insecticides, fungicides, wetting agents and miscellaneous substances. *Inter. Pest Contr.* **11**, 29–35.

Aldridge, W. N. (1976). The influence of organotin compounds on mitochondrial functions. In *Organotin Compounds: New Chemistry and Applications*, ed. J. J. Zuckerman, Advances in Chemistry Series 157, American Chemical Society, Washington, DC, USA, pp. 186–96.

Armstrong, D. E. and Elzerman, A. W. (1982). Trace metal accumulation in surface microlayers. *J. Great Lakes Res.* **8**, 282–7.

Astruc, M., Lavigne, R., Pinel, R., Leguille, F., Desauziers, V., Quevauviller, P. and Donard, O. F. X. (1990). Speciation of tin in sediments of Arcachon Bay (France). In *Metals Speciation, Separation, and Recovery*, eds J. W. Patterson and R. Passino, Lewis Publ., Chelsea, MI 48118, USA, ISBN 0-87371-268-4, pp. 263–74.

Avery, S. V., Miller, M. E., Gadd, G. M., Codd, G. A. and Cooney, J. J. (1991). Toxicity of organotins towards cyanobacterial photosynthesis and nitrogen fixation. *FEMS Microbiol. Lett.* **84**, 205–10.

Babich, H. and Borenfreund, E. (1988). Structure-activity relationships for diorganotins, chlorinated benzenes and chlorinated anilines established with bluegill sunfish BF-2 cells. *Fundament. Appl. Toxicol.* **10**, 295–301.

Barug, D. (1981). Microbial degradation of bis(tributyltin) oxide. *Chemosphere* **10**, 1145–54.

Becker, E. C. and Bringezu, S. (1992). Contamination of surface waters by organotin compounds – concentrations effects, quality objectives, use limitations. *Z. Wasser-Abwasser-Forsch.* **25**, 40–6.

Becker, K., Merlini, L., de Bertrand, N., de Alencastro, L. F. and Tarradellas, J. (1992). Elevated levels of organotins in Lake Geneva: bivalves as sentinel organism. *Bull. Environ. Contam. Toxicol.* **48**, 37–44.

Blanchard, D. C. and Syzdek, L. D. (1975). Electrostatic collection of jet and film drops. *Limnol. Oceanogr.* **20**, 762–73.

Blanck, H., Wallin, G. and Wangberg, S.-A. (1984). Species-dependent variation in algal sensitivity to chemical compounds. *Ecotoxicol. Environ. Safety* **8**, 339–51.

Blunden, S. J. and Evans, C. J. (1990). Organotin compounds. In *The Handbook of Environmental Chemistry, Vol. 3, Part E, Anthropogenic Compounds*, ed. O. Hutzinger, Springer-Verlag, New York, NY, USA, ISBN 0-387-51423-6, pp. 1–44.

Bodar, C. W. M., van Donselaar, E. G. and Herwig, H. J. (1990). Cytopathological investigations of digestive tract and storage cells in *Daphnia magna* exposed to cadmium and tributyltin. *Aquat. Toxicol.* **17**, 325–38.

Boyer, I. J. (1989). Toxicity of dibutyltin, tributyltin and other organotin compounds to humans and to experimental animals. *Toxicol.* **55**, 253–98.

Brinckman, F. E., Olson, G. J., Blair, W. R. and Parks, E. J. (1988). Implications of molecular speciation and topology of environmental metals: uptake mechanisms and toxicity of organotins. In *Aquatic Toxicology and Hazard Assessment: 10th Vol.*, eds W. J. Adams, G. A. Chapman and W. G. Landis, American Society for Testing and Materials, Philadelphia, PA, USA, ASTM STP 971, pp. 219–32.

Brooke, L. T., Call, D. J., Poirier, S. H., Markee, T. P., Lindberg, C. A., McCauley, D. J. and Simonson, P. G. (1986). *Acute toxicity and chronic effects of bis(tri-n-butyltin) oxide to several species of freshwater organisms.* Center for Lake Superior Environmental Studies report, University of Wisconsin-Superior, Superior, WI 54880, USA, 20 pp.

Cai, Y., Rapsomanikis, S. and Andreae, M. O. (1993). Determination of butyltin compounds in sediment using gas chromatography–atomic absorption spectrometry: comparison of sodium tetrahydroborate and sodium tetraethylborate derivatization methods. *Anal. Chim. Acta* **274**, 243–51.

Cain, K., Hyams, R. L. and Griffiths, D. E. (1977). Studies on energy-linked reactions: inhibition of oxidative phosphorylation and energy-linked reactions by dibutyltin dichloride. *Fed. Eur. Biochem. Soc. Lett.* **82**, 23–8.

Chau, Y. K., Wong, P. T. S., Bengert, G. A. and Yaromich, J. (1989). Bioaccumulation of butyltin compounds by mussels in harbours. *Chem. Speciation. Biol. Avail.* **1**, 151–6.

Chau, Y. K., Zhang, S. and Maguire, R. J. (1992). Occurrence of butyltin species in sewage and sludge in Canada. *Sci. Total. Environ.* **121**, 271–91.

Chliamovitch, Y.-P. and Kuhn, C. (1977). Behavioural, hematological and histological studies on acute toxicity of bis(tri-n-butyltin) oxide on *Salmo gairdneri* Richardson and *Tilapia rendalli* Boulenger. *J. Fish. Biol.* **10**, 575–85.

Cleary, J. J. (1991). Organotin in the marine surface microlayer and subsurface waters of south-west England: relation to toxicity thresholds and the UK environmental quality standard. *Mar. Environ. Res.* **32**, 213–22.

Cooney, J. J., de Rome, L., Laurence, O. and Gadd, G. M. (1989). Effects of organotin and organolead compounds on yeasts. *J. Industr. Microbiol.* **4**, 279–88.

Cooney, J. J. and Wuertz, S. (1989). Toxic effects of tin compounds on microorganisms. *J. Industr. Microbiol.* **4**, 375–402.

Davies, A. G. and Smith, P. J. (1980). Recent advances in organotin chemistry. *Adv. Inorg. Chem. Radiochem.* **23**, 1–77.

de Mora, S. J., King, N. G. and Miller, M. C. (1989). Tributyltin and total tin in marine sediments: profiles and the apparent rate of TBT degradation. *Environ. Technol. Lett.* **10**, 901–8.

de Vries, H., Penninks, A. H., Snoeij, N. J. and Seinen, W. (1991). Comparative toxicity of organotin compounds to rainbow trout (*Oncorhynchus mykiss*) yolk sac fry. *Sci. Total Environ.* **103**, 229–43.

Dirkx, W. M. R., Van Mol, W. E., Van Cleuvenbergen, R. J. A. and Adams, F. C. (1989). Speciation of organotin compounds in water by gas chromatography/atomic absorption spectrometry. *Fresenius Z. Anal. Chem.* **335**, 769–74.

Donard, O. F. X., Quevauviller, P. and Bruchet, A. (1993). Tin and organotin speciation during wastewater and sludge treatment processes. *Water Res.* **27**, 1085–9.

Dooley, C. A. and Kenis, P. (1987). Response of bioluminescent bacteria to alkyltin compounds. *Proc. Inter. Organotin Symp. Oceans '87 Conf.*, Halifax, NS, Canada, Sept. 28–Oct. 1, Vol. 4, pp. 1517–24, IEEE Service Center, 445, Hoes Lane, Piscataway, NJ, USA 08854.

Dowson, P. H., Bubb, J. M. and Lester, J. N. (1992a). Organotin distribution in sediments and waters of selected east coast estuaries in the UK. *Mar. Pollut. Bull.* **24**, 492–8.

Dowson, P. H., Pershke, D., Bubb, J. M. and Lester, J. N. (1992b). Spatial distribution of organotins in sediments of lowland river catchments. *Environ. Pollut.* **76**, 259–66.

Dowson, P. H., Bubb, J. M. and Lester, J. N. (1993). Depositional profiles and relationships between organotin compounds in freshwater and estuarine sediment cores. *Environ. Monitor. Assess.* **28**, 145–60.

Eisenreich, S. J. (1982). Atmospheric role in trace metal exchange at the air-water surface. *J. Great Lakes Res.* **8**, 243–56.

Eng, G., Tierney, E. J., Bellama, J. M. and Brinckman, F. E. (1988). Correlation of molecular total surface area with organotin toxicity for biological and physicochemical applications. *Appl. Organometal. Chem.* **2**, 171–5.

Fent, K. (1989a). Teratogenic effects of tributyltin on embryos of the freshwater fish *Phoxinus phoxinus. Neue Zurcher Zeit.* **106**, 95–6.

Fent, K. (1989b). Organotin speciation in municipal wastewater and sewage sludge: ecotoxicological consequences. *Mar. Environ. Res.* **28**, 477–83.

Fent, K. (1991). Bioconcentration and elimination of tributyltin chloride by embryos and larvae of minnows, *Phoxinus phoxinus. Aquat. Toxicol.* **20**, 147–58.

Fent, K. (1994). Organotins in municipal wastewater and sewage sludge. In *Organotins: Fate and Effects in the Environment*, eds M. A. Champ and P. F. Seligman, Elsevier Publ., Amsterdam, in press.

Fent, K. (1992). Embryotoxic effects of tributyltin on the minnow *Phoxinus phoxinus. Environ. Pollut.* **76**, 187–94.

Fent, K. and Hunn, J. (1991). Phenyltins in water, sediment and biota of freshwater marinas. *Environ. Sci. Technol.* **25**, 956–63.

Fent, K. and Hunn, J. (1993). Uptake and elimination of tributyltin in fish-yolk-sac larvae. *Mar. Environ. Res.* **35**, 65–71.

Fent, K., Hunn, J. and Sturm, M. (1991). Organotins in Lake Sediment. *Naturwissenschaften* **78**, 219–21.

Fent, K. and Meier, W. (1992). Tributyltin-induced effects on early life stages of minnows, *Phoxinus phoxinus. Arch. Environ. Contam. Toxicol.* **22**, 428–38.

Fent, K. and Muller, M. D. (1991). Occurrence of organotins in municipal wastewater and sewage sludge and behavior in a treatment plant. *Environ. Sci. Technol.* **25**, 489–93.

Fish, R. H., Kimmel, E. C. and Casida, J. E. (1976). Bioorganotin chemistry: reactions of tributyltin derivatives with a cytochrome P-450 dependent monooxygenase enzyme system. *J. Organometal. Chem.* **118**, 41–54.

Foster, R. B. (1981). Use of Asiatic clam larvae in aquatic hazard evaluations. In *Ecological Assessment of Communities of Indigenous Aquatic Organisms*, eds J. M. Bates and C. I. Weber, Amer. Soc. for Test. Materials, Philadelphia, PA, USA, ASTM STP 730, pp. 280–8.

François, R., Short, F. T. and Weber, J. H. (1989). Accumulation and persistence of tributyltin in eelgrass (*Zostera marina* L.) tissue. *Environ. Sci. Technol.* **23**, 191–6.

Gabrielides, G. P., Alzieu, C., Readman, J. W., Bacci, E., Aboul Dahab, O. and

Salihoglu, I. (1990). MED POL survey of organotins in the Mediterranean. *Mar. Pollut. Bull.* **21**, 233–7.

Gras, G. and Rioux, J.-A. (1965). Relation entre la structure chimique et l'activité insecticide des composés organiques de l'étain (essai sur les larves de *Culex pipiens pipiens* L.). *Arch. Inst. Pasteur Tunis* **42**, 9–22.

Hall, L.W., Jr., Bushong, S.J., Johnson, W.E. and Hall, W.S. (1988). Spatial and temporal distribution of butyltin compounds in a northern Chesapeake Bay marina and river system. *Environ. Monitor. Assess.* **10**, 229–44.

Hall, L.W., Jr. and Pinkney, A.E. (1985). Acute and sublethal effects of organotin compounds on aquatic biota: an interpretive literature evaluation. *CRC Crit. Rev. Toxicol.* **14**, 159–209.

Hattori, Y., Kobayashi, A., Nonaka, K., Sugimae, A. and Nakamoto, M. (1988). Degradation of tributyltin and dibutyltin compounds in environmental water. *Water Sci. Technol.* **20**, 71–6.

Hattori, Y., Kobayashi, A., Takemoto, S., Takami, K., Kuge, Y., Sugimae, A. and Nakamoto, M. (1984). Determination of trialkyltin, dialkyltin and triphenyltin compounds in environmental water and sediments. *J. Chromatogr.* **315**, 341–9.

Hodge, V.F., Seidel, S.L. and Goldberg, E.D. (1979). Determination of tin (IV) and organotin compounds in natural waters, coastal sediments and macro algae by atomic absorption spectrometry. *Anal. Chem.* **51**, 1256–9.

Holwerda, D.A. and Herwig, H.J. (1986). Accumulation and metabolic effects of di-n-butyltin dichloride in the freshwater clam, *Anodonta anatina*. *Bull. Environ. Contam. Toxicol.* **36**, 756–62.

Hopf, H.S., Duncan, J., Beesley, J.S.S., Webley, D.J. and Sturrock, R.F. (1967). Molluscicidal properties of organotin and organolead compounds with particular reference to triphenyllead acetate. *Bull. World Health Org.* **36**, 955–61.

Howard, P.H. (1989). *Handbook of Environmental Fate and Exposure Data for Organic Chemicals, Vol. 1. Large Production and Priority Pollutants.* Lewis Publ., Chelsea, MI 48118, USA.

Huang, G., Bai, Z., Dai, S. and Xie, Q. (1993). Accumulation and toxic effects of organometallic compounds on algae. *Appl. Organometal. Chem.* **7**, 373–80.

Ishizaka, T., Suzuki, T. and Saito, Y. (1989). Metabolism of dibutyltin dichloride in male rats. *J. Agric. Food Chem.* **37**, 1096–101.

Jackson, J.A., Blair, W.R., Brinckman, F.E. and Iverson, W.P. (1982). Gas chromatographic speciation of methylstannanes in the Chesapeake Bay using purge and trap sampling with a tin-selective detector. *Environ. Sci. Technol.* **16**, 110–19.

Josephson, D.B., Lindsay, R.C. and Stuiber, D.A. (1989). Inhibition of trout gill and soybean lipoxygenases by organotin compounds. *J. Environ. Sci. Health* **B24**, 539–58.

Kilby, G.W. and Batley, G.E. (1993). Chemical indicators of sediment chronology. *Austr. J. Mar. Freshwater Res.* **44**, 635–47.

Kimmel, E.C., Fish, R.H. and Casida, J.E. (1977). Bioorganotin chemistry. Metabolism of organotin compounds in microsomal monooxygenase systems and in mammals. *J. Agric. Food Chem.* **25**, 1–9.

Laughlin, R.B., Jr., Johannesen, R.B., French, W., Guard, H. and Brinckman, F.E. (1985). Structure-activity relationships for organotin compounds. *Environ. Toxicol. Chem.* **4**, 343–51.

Laughlin, R. and Linden, O. (1982). Sublethal responses of the tadpoles of the European frog *Rana temporaria* to two tributyltin compounds. *Bull. Environ. Contam. Toxicol.* **28**, 494–9.

Lee, R. F. (1986). Metabolism of bis(tributyltin) oxide by estuarine animals. *Proc. Organotin Symp. Oceans '86 Conference*, Washington, DC, USA, Sept. 23–25, Vol. 4, pp. 1182–8. IEEE Service Center, 445 Hoes Lane, Piscataway, NJ 08854, USA.

Lee, R. F., Valkirs, A. O. and Seligman, P. F. (1989). Importance of microalgae in the biodegradation of tributyltin in estuarine waters. *Environ. Sci. Technol.* **23**, 1515–18.

Li, A. P., Dahl, A. R. and Hill, J. O. (1982). *In vitro* cytotoxicity and genotoxicity of dibutyltin dichloride and dibutylgermanium dichloride. *Toxicol. Appl. Pharmacol.* **64**, 482–5.

Liss, P. S. (1975). Chemistry of the sea surface microlayer. In *Chemical Oceanography*, eds J. P. Riley and G. Skirrow, Academic Press, New York, NY, USA, pp. 193–243.

Liu, D. and Thomson, K. (1986). Biochemical responses of bacteria after short exposure to alkyltins. *Bull. Environ. Contam. Toxicol.* **36**, 60–6.

Luijten, J. G. A. (1972). Applications and biological effects of organotin compounds. In *Organotin Compounds*, ed. A. K. Sawyer, Marcel Dekker, New York, NY, USA, pp. 93:–76.

MacIntyre, F. (1974). The top millimetre of the ocean. *Scient. Amer.* **230**, 62–77.

Maguire, R. J. (1984). Butyltin compounds and inorganic tin in sediments in Ontario. *Environ. Sci. Technol.* **18**, 291–4.

Maguire, R. J. (1986). Review of the occurrence, persistence and degradation of tributyltin in fresh water ecosystems in Canada. *Proc. Organotin Symp. Oceans '86 Conference*. Washington, DC, USA, Sept. 23–25, Vol. 4, pp. 1252–5. IEEE Service Center, 445 Hoes Lane, Piscataway, NJ 08854, USA.

Maguire, R. J. (1987). Review of environmental aspects of tributyltin. *Appl. Organometal. Chem.* **1**, 475–98.

Maguire, R. J. (1991). Aquatic environmental aspects of non-pesticidal organotin compounds. *Water Pollut. Res. J. Can.* **26**, 243–360.

Maguire, R. J., Carey, J. H. and Hale, E. J. (1983). Degradation of the tri-n-butyltin species in water. *J. Agric. Food Chem.* **31**, 1060–5.

Maguire, R. J., Chau, Y. K., Bengert, G. A., Hale, E. J., Wong, P. T. S. and Kramar, O. (1982). Occurrence of organotin compounds in Ontario lakes and rivers. *Environ. Sci. Technol.* **16**, 698–702.

Maguire, R. J., Liu, D. L. S., Thomson, K. and Tkacz, R. J. (1985b). *Bacterial degradation of tributyltin*. National Water Research Institute Report 85-82, Department of the Environment, Burlington, Ont. L7R 4A6, Canada, 20 pp.

Maguire, R. J., Long, G., Meek, M. E. and Savard, S. (1993). *Canadian Environmental Protection Act priority substances list assessment report on non-pesticidal organotin compounds*. Department of the Environment, Commercial Chemicals Evaluation Branch, Ottawa, Ont. K1A 0H3, Canada, 32 + pp. (ISBN 0-662-20719-X).

Maguire, R. J. and Tkacz, R. J. (1985). Degradation of the tri-n-butyltin species in water and sediment from Toronto Harbour. *J. Agric. Food Chem.* **33**, 947–53.

Maguire, R. J. and Tkacz, R. J. (1987). Concentration of tributyltin in the surface microlayer of natural waters. *Water Pollut. Res. J. Can.* **22**, 227–33.

Maguire, R. J. and Tkacz, R. J. (1988). Chlorinated hydrocarbons in the surface microlayer and subsurface water of the Niagara River, 1985–86. *Water Pollut.*

Res. J. Can. **23**, 292–300.

Maguire, R. J., Tkacz, R. J. and Sartor, D. L. (1985a). Butyltin species and inorganic tin in water and sediment of the Detroit and St. Clair Rivers. *J. Great Lakes Res.* **11**, 320–7.

Maguire, R. J., Tkacz, R. J., Chau, Y. K., Bengert, G. A. and Wong, P. T. S. (1986). Occurrence of organotin compounds in water and sediment in Canada. *Chemosphere* **15**, 253–74.

Maguire, R. J., Wong, P. T. S. and Rhamey, J. S. (1984). Accumulation and metabolism of tri-n-butyltin cation by a green alga, *Ankistrodesmus falcatus. Can. J. Fish. Aquat. Sci.* **41**, 537–40.

Martin, R. C., Dixon, D. G., Maguire, R. J., Hodson, P. V. and Tkacz, R. J. (1989). Acute toxicity, uptake, depuration and tissue distribution of tri-n-butyltin in rainbow trout, *Salmo gairdneri. Aquat. Toxicol.* **15**, 37–52.

Meyers, P. A. and Kawka, O. E. (1982). Fractionation of hydrophobic organic materials in surface microlayers. *J. Great Lakes Res.* **8**, 288–98.

Muller, M. D. (1984). Tributyltin detection at trace levels in water and sediments using GC with flame photometric detection and GC-MS. *Fresenius Z. Anal. Chem.* **317**, 32–6.

Muller, M. D. (1987). Comprehensive trace level determination of organotin compounds in environmental samples using high resolution gas chromatography with flame photometric detection. *Anal. Chem.* **59**, 617–23.

Nagase, H., Hamasaki, T., Sato, T., Kito, H., Yoshioka, Y. and Ose, Y. (1991). Structure-activity relationships for organotin compounds on the red killifish *Oryzias latipes. Appl. Organometal. Chem.* **5**, 91–7.

Olson, G. J. and Brinckman, F. E. (1986). Biodegradation of tributyltin by Chesapeake bay microorganisms. *Proc. Organotin Symp. Oceans '86 Conference,* Washington, DC, USA, Sept. 23–25, Vol. 4, pp. 1196–201. IEEE Service Center, 445 Hoes Lane, Piscataway, NJ 08854, USA.

Orsler, R. J. and Holland, G. E. (1982). Degradation of tributyltin oxide by fungal culture filtrates. *Inter. Biodeterior. Bull.* **18**, 95–8.

Pellenbarg, R. E. and Church, T. M. (1979). The estuarine surface microlayer and trace metal cycling in a salt marsh. *Science* **203**, 1010–12.

Penninks, A. H. and Seinen, W. (1980). Toxicity of organotin compounds. IV. Impairment of energy metabolism of rat thymocytes by various dialkyltin compounds. *Toxicol. Appl. Pharmacol.* **56**, 221–31.

Penninks, A. H., Verschuren, P. M. and Seinen, W. (1983). Di-n-butyltin dichloride uncouples oxidative phosphorylation in rat liver mitochondria. *Toxicol. Appl. Pharmacol.* **70**, 115–20.

Piotrowicz, S. R., Duce, R. A., Fasching, J. L. and Weisel, C. P. (1979). Bursting bubbles and their effect on the sea-to-air transport of Fe, Cu and Zn. *Mar. Chem.* **7**, 307–24.

Piver, W. T. (1973). Organotin compounds: industrial applications and biological investigation. *Environ. Health Perspect.* **4**, 61–79.

Polster, M. and Halacka, K. (1971). Contribution in the hygienic and toxicological problems associated with some organotin compounds when used as antimicrobial agents. *Ernahrungsforschung* **16**, 527–35.

Quevauviller, P., Lavigne, R., Pinel, R. and Astruc, M. (1989). Organotins in sediments and mussels from the Sado estuarine system (Portugal). *Environ. Pollut.* **57**, 149–66.

Salazar, M. H. and Salazar, S. M. (1989). *Acute effects of (bis)tributyltin oxide on marine organisms. Summary of work performed 1981 to 1983.* Naval Ocean Systems Center Tech. Rep. 1299, 60+ pp., San Diego, CA 92152-5000, USA.

Schebek, L., Andreae, M. O. and Tobschall, H. J. (1991). Methyl- and butyltin compounds in water and sediments of the Rhine River. *Environ. Sci. Technol.* **25**, 871–8.

Schwaiger, J., Bucher, F., Ferling, H., Kalbus, W. and Negele, R.-D. (1992). A prolonged toxicity study on the effects of sublethal concentrations of bis(tri-n-butyltin) oxide (TBTO): histopathological and histochemical findings in rainbow trout (*Oncorhynchus mykiss*). *Aquat. Toxicol.* **23**, 31–48.

Scott, B. F., Chau, Y. K. and Rais-Firouz, A. (1991). Determination of butyltin species by GC/atomic emission spectroscopy. *Appl. Organometal. Chem.* **5**, 151–7.

Seiffer, E. A. and Schoof, H. F. (1967). Tests of 15 experimental molluscicides against *Australorbis glabratus. Publ. Health Rep.* **82**, 833–9.

Seinen, W., Helder, T., Vernij, H., Penninks, A. and Leeuwangh, P. (1981). Short term toxicity of tri-n-butyltin chloride in rainbow trout (*Salmo gairdneri* Richardson) yolk sac fry. *Sci. Total Environ.* **19**, 155–66.

Seligman, P. F., Grovhoug, J. G., Valkirs, A. O., Stang, P. M., Fransham, R., Stallard, M. O., Davidson, B. and Lee, R. F. (1989). Distribution and fate of tributyltin in the United States marine environment. *Appl. Organometal. Chem.* **3**, 31–47.

Seligman, P. F., Valkirs, A. O. and Lee, R. F. (1986). Degradation of tributyltin in San Diego Bay, California, waters. *Environ. Sci. Technol.* **20**, 1229–35.

Seligman, P. F., Valkirs, A. O., Stang, P. M. and Lee, R. F. (1988). Evidence for rapid degradation of tributyltin in a marina. *Mar. Pollut. Bull.* **19**, 531–4.

Selwyn, M. J. (1976). Triorganotin compounds as ionophores and inhibitors of ion translocating ATPases. In *Organotin Compounds: New Chemistry and Applications*, ed. J. J. Zuckerman, Advances in Chemistry Series 157, American Chemical Society, Washington, DC, USA, ISBN 0-8412-0343-1, pp. 205–26.

Slesinger, A. E. and Dressler, I. (1978). The environmental chemistry of three organotin chemicals. *Proc. Organotin Workshop*, ed. M. L. Good, New Orleans, LA, USA, Feb. 17–19, pp. 115–62.

Smialowicz, R. J., Riddle, M. M., Rogers, R. R., Rowe, D. G., Luebke, R. W., Fogelson, L. D. and Copeland, C. B. (1988). Immunologic effects of perinatal exposure of rats to dioctyltin dichloride. *J. Toxicol. Environ. Health* **25**, 403–22.

Snoeij, N. J., Penninks, A. H. and Seinen, W. (1987). Biological activity of organotin compounds – an overview. *Environ. Res.* **44**, 335–53.

Stang, P. M. and Goldberg, E. D. (1989). Butyltins in California river and lake marina waters. *Appl. Organometal. Chem.* **3**, 183–7.

Stang, P. M., Lee, R. F. and Seligman, P. F. (1992). Evidence for rapid, non-biological degradation of tributyltin compounds in autoclaved and heat-treated fine-grained sediments. *Environ. Sci. Technol.* **26**, 1382–7.

Stang, P. M. and Seligman, P. F. (1986). Distribution and fate of butyltin compounds in the sediment of San Diego Bay. *Proc. Organotin Symp. Oceans '86 Conference*, Washington, DC, USA, Sept. 23–25, Vol. 4, pp. 1256–61. IEEE Service Center, 445 Hoes Lane, Piscataway, NJ 08854, USA.

Steinhauser, K. G., Amann, W., Spath, A. and Polenz, A. (1985). Investigations on the aquatic toxicity of organotin compounds. *Vom Wasser* **65**, 203–14.

Thain, J. E., Waldock, M. J. and Waite, M. E. (1987). Toxicity and degradation studies of tributyltin (TBT) and dibutyltin (DBT) in the aquatic environment. In *Proc. Inter. Organotin Symp. Oceans '87 Conf.*, Halifax, NS, Sept. 29–Oct. 1, Vol. 4, pp. 1398–404. IEEE Service Center, 445 Hoes Lane, Piscataway, NJ 08854,

USA.

Tsuda, T., Nakanishi, H., Morita, T. and Takebayashi, J. (1986). Simultaneous gas chromatographic determination of dibutyltin and tributyltin compounds in biological and sediment samples. *J. Assoc. Off. Anal. Chem.* **69**, 981–4.

Tugrul, S., Balkas, T. I. and Goldberg, E. D. (1983). Methyltins in the marine environment. *Mar. Pollut. Bull.* **14**, 297–303.

Vighi, M. and Calamari, D. (1985). QSARs for organotin compounds on *Daphnia magna. Chemosphere* **14**, 1925–32.

Von Westernhagen, H., Landolt, M., Kocan, R., Furstenberg, G., Janssen, D. and Kremling, K. (1987). Toxicity of sea-surface microlayer: effects on herring and turbot embryos. *Mar. Environ. Res.* **23**, 273–90.

Vos, J. G. (1977). Immune suppression as related to toxicology. *CRC Crit. Rev. Toxicol.* **5**, 67–101.

Waite, M. E., Evans, K. E., Thain, J. E. and Waldock, M. J. (1989). Organotin concentrations in the Rivers Bure and Yare, Norfolk Broads, England. *Appl. Organometal. Chem.* **3**, 383–91.

Walker, W. W., Heard, C. S., Lotz, K., Lytle, T. F., Hawkins, W. E., Barnes, C. S., Barnes, D. H. and Overstreet, R. M. (1989). Tumorigenic, growth, reproductive and developmental effects in medaka exposed to bis(tri-n-butyltin) oxide. *Proc. Oceans '89 Conference*, Seattle, WA, USA, Sept. 18–21, Vol. 2, pp. 516–24. IEEE Service Center, 445 Hoes Lane, NJ 08854, USA.

Walsh, G. E., McLaughlan, L. L., Lores, E. M., Louie, M. K. and Deans, C. H. (1985). Effects of organotins on growth and survival of two marine diatoms, *Skeletonema costatum* and *Thalassiosira pseudonana. Chemosphere* **14**, 383–92.

Ward, G. S., Cramm, G. C., Parrish, P. R., Trachman, H. and Slesinger, A. (1981). Bioaccumulation and chronic toxicity of bis(tributyltin) oxide (TBTO): tests with a saltwater fish. In *Aquatic Toxicology and Hazard Assessment: Fourth Conf.*, eds D. R. Branson and K. L. Dickson, Amer. Soc. for Test. and Mater., PA, USA, ASTM STP 737, pp. 183–200.

Watanabe, N., Sakai, S. and Takatsuki, H. (1992). Examination for degradation paths of butyltin compounds in natural waters. *Water Sci. Technol.* **25**, 117–24.

Wester, P. W. and Canton, J. H. (1987). Histopathological study of *Poecilia reticulata* (guppy) after long-term exposure to bis(tri-n-butyltin) oxide (TBTO) and di-n-butyltin dichloride (DBTC). *Aquat. Toxicol.* **10**, 143–65.

Wester, P. W., Canton, J. H., Van Iexsel, A. A. J., Kranjc, E. I. and Vaessen, H. A. M. G. (1990). The toxicity of bis(tri-n-butyltin) oxide (TBTO) and di-n-butyltin dichloride (DBTC) in the small fish species *Oryzias latipes* (medaka) and *Poecilia reticulata* (guppy). *Aquat. Toxicol.* **16**, 53–72.

Woggon, H. and Jehle, D. (1975). The residue analysis of biocidal organotin compounds using anodic stripping voltammetry. *Die Nahrung* **19**, 271–5.

Wong, P. T. S. and Chau, Y. K. (1992). *Occurrence of butyltin compounds in Severn Sound, Ontario.* National Water Research Institute Report 92-119, Department of the Environment, Burlington, Ont. L7R 4A6, Canada, 26 pp.

Wong, P. T. S., Chau, Y. K., Kramar, O. and Bengert, G. A. (1982). Structure-toxicity relationship of tin compounds on algae. *Can. J. Fish Aquat. Sci.* **39**, 483–8.

World Health Organization (1980). *Tin and organotin compounds: a preliminary review.* United Nations Environment Programme – World Health Organization, Environmental Health Criteria **15**, Geneva, Switzerland, 109 pp., ISBN 92-4-154075-3.

World Health Organization (1990). *Tributyltin Compounds.* United Nations Environment

Programme – World Health Organization, Environmental Health Criteria 116, Geneva, Switzerland, 273 + pp., ISBN 92-4-157116-0.

Wuertz, S., Miller, C. E., Pfister, R. M. and Cooney, J. J. (1991). Tributyltin-resistant bacteria from estuarine and freshwater sediments. *Appl. Environ. Microbiol.* 57, 2783–9.

Yamada, J. (1977). The mode of RNA polymerase inhibition by alkyltin compounds. *Bull. Fac. Educ. Yamaguchi Univ.* 27, 73–8.

5

○ ○

The distribution and fate of tributyltin in the marine environment

Graeme Batley

5.1 Introduction

Earlier chapters have outlined the extensive usage of tributyltin (TBT) in marine antifouling paints which commenced in Europe around 1960, almost a decade after its biocidal properties were first reported (Bennett, 1983). Almost 30 years later it has been banned from use on small craft, to protect shellfish, such as oysters, mussels and scallops, and gastropods, from its impacts in confined waterways. The legacy of the unknown quantities of TBT that have leached from many square kilometres of painted surfaces in waters worldwide, has been that sediments in the vicinity of marinas, dockyards, wharves and other areas of high boating activity now contain measurable concentrations of TBT. This derives not only from paint scrapings, but from the preferential partitioning of dissolved TBT onto colloidal and particulate surfaces, which ultimately accumulate in bottom sediments. Once resident here, the concern is for its availability to sediment-dwelling biota, and its potential remobilisation to overlying waters.

There has been extensive research into the fate of TBT in the marine environment, and this has been reviewed by a number of authors (e.g. Laughlin & Linden, 1985; Clark, Sterritt & Lester, 1988; Muller, Renberg & Rippen, 1989; Stewart & de Mora, 1990). This includes degradation and removal processes in the water column and in sediments, as well as biotic transformations. Reactions will involve adsorption, photolysis, and both microbial and abiotic decomposition processes. The prognosis is that these processes will eventually result in the complete removal of TBT from the aquatic environment, and there are now sufficient data to permit an estimation of the time that this might take. A concern is that the products of these processes may also have undesirable impacts on aquatic biota. This chapter examines current knowledge on the distribution and behaviour of TBT once it enters the estuarine and marine environment.

5.2   Sources of tributyltin in estuarine and marine waters

Marine antifouling paint preparations represent by far the greatest source of TBT to the marine environment. As the active biocidal ingredient, TBT is present either alone or in combination with copper, at concentrations as high as 3% Sn dry weight in the paint formulations. Modern TBT-based paints are copolymer formulations based on either tributyltin oxide (TBTO) or tributyltin fluoride. Freshly painted surfaces are designed to reach a constant TBT leach rate of $1.6 \mu g$ Sn cm^{-2} per day, although the leach rate of freshly painted surfaces will be as high as $6 \mu g$ Sn cm^{-2} per day, reducing in several weeks to the desired constant rate.

The stripping of old paint from boat hulls, carried out on slipways or drydocks by surface blasting with water or abrasive slag fines, can be a significant source of TBT to marine waters. It is only recently that efforts have been made to contain hydroblasting wastes and prevent their uncontrolled entry into natural waterways.

As discussed in Chapter 2, TBT is also used in cooling waters to prevent slime formation, as fungicides, bacteriocides and insecticides, and as a preservative for wood and other products. These sources are unlikely to be significant contributors of TBT to marine water systems except as a component of leachates, or industrial or sewage effluents.

5.3   Tributyltin concentrations in marine waters

5.3.1   Pre-banning

Since its first use in marine antifouling paints in the 1970s, TBT and its degradation products have been extensively monitored in marine waterways worldwide, and the data are well-documented in the scientific literature (e.g. Champ & Pugh, 1987; Maguire, 1987; Seligman *et al.*, 1987, 1989). There have been isolated examples where concentrations have exceeded 300 ng Sn l^{-1} (Cleary & Stebbing, 1985), most likely extremely close to point sources, but in general concentrations below 100 ng Sn l^{-1} are a more common occurrence.

Table 5.1 lists some typical data for dissolved TBT recorded prior to the imposition of bans on the use of TBT-based paints on boats under 25 m in length, in the late 1980s. Not surprisingly, the highest concentrations have been found near marinas or dockyard areas.

A large commercial ship leaching TBT at the constant leach rate will contribute more than 200 g of TBT to the waters in its immediate vicinity during a three-day stay in port. If freshly painted, this amount could be as high as 600 g. In a large marina or dockyard or complex, this can result in a significant dissolved TBT concentration, and typical measurements in such environments have shown concentrations in the range 100–200 ng $Sn l^{-1}$ and occasionally as high as 600 ng $Sn l^{-1}$, as shown by the data in Table 5.1.

In small estuaries, marinas and moorings for smaller pleasure craft can also contribute significantly to a dissolved TBT load. Typically these are in the range 10–70 ng $Sn l^{-1}$ (Batley et al., 1989).

The application of classical steady-state tidal prism models has enabled prediction of dissolved TBT concentrations similar to those measured (Batley et al., 1989).

The steady-state concentration C_{ss}, can be expressed as:

$$C_{ss} = gy/v(1 - y)$$

where g is the load in nanograms introduced per tidal cycle, y is the

Table 5.1 *Tributyltin concentrations in some marine and estuarine waters*

Site	Concentration $(ng\,Sn\,l^{-1})$	Reference
San Diego Bay, USA	20–380	Valkirs et al., 1986a,b
Chesapeake Bay, USA	20–170	Hall et al., 1987
Chesapeake Bay, USA	30	Matthias et al., 1986
Chesapeake Bay, USA	1–27	Olson & Brinckman, 1986
Torquay Harbour, Devon, UK	<4–210	Cleary & Stebbing, 1985
Torquay Marina, Devon, UK	120–243	Cleary & Stebbing, 1987a
River Tamar, Plymouth, UK	10–558	Cleary & Stebbing, 1987a
Poole Harbour, Dorset, UK	2–139	Langston et al., 1987
Marinas in Poole Harbour, UK	234–646	Langston et al., 1987
River Avon, UK	4–110	Ebdon et al., 1988
Southampton, UK	<1–275	Waldock et al., 1988
Oleron Island, France	16–615	Waldock et al., 1988
Toulon Harbour, France	2–342	Alzieu et al., 1989
Georges River, NSW, Australia	8–40	Batley et al., 1989
Lakes Entrance, Victoria, Australia	249	Batley & Scammell, 1991
Tamaki Estuary, New Zealand	n.d.–321	King et al., 1989
Leghorn, Northern Tyrrhenian Coast, Italy	164–332	Bacci & Gaggi, 1989
Antwerp Harbour, Belgium	<1–182	Dirkx et al., 1993
Marinas, Hong Kong	90–1000	Lau, 1991

fraction remaining after one tidal cycle, and v is the low-tide volume.

$$y = [1 - p/(p + v)](1 - d)$$

where p is the tidal prism or exchange volume and d is the decay rate of TBT per tidal cycle (%/100).

Both seasonal and short-term variability from tidal influences have been reported. Clavell, Seligman & Stang (1986) showed that tidal exchange resulted in variations from as low as 7 to values such as 136 ng $Sn\,l^{-1}$, differences of up to 20-fold, near the mouth of a yacht basin in San Diego Bay (USA) (Figure 5.1). Seligman *et al.* (1986) demonstrated temporal variations in the Elizabeth River, Virginia (USA), with peaks attributed to December painting of cruise ships.

A study by Ebdon *et al.* (1988) of the temporal variability of TBT concentrations in six estuaries in south and southwest England, also reflected the changes in boating activity. Dissolved TBT concentrations were at their maximum in May dropping to below the detection limits until the autumn. In some instances, peaks in concentration occurred in August or September (Figure 5.2).

The soluble product leached from painted surfaces will be either TBTO or the TBT cation. Early studies suggested that the most logical species of TBT in sea water was TBTOH, but subsequently both this species and TBTCl were confirmed (Laughlin, Guard & Coleman, 1986). Measurements using nuclear magnetic resonance spectroscopy (NMR) showed that in chloroform extracts of sea water spiked with TBTO, TBTO, TBTCl and

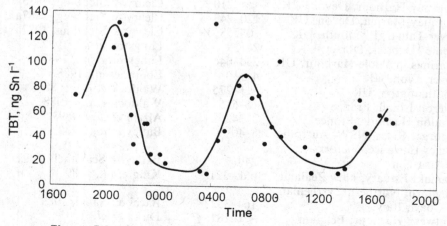

Figure 5.1 Variation in dissolved TBT in San Diego Bay (USA) over a tidal cycle. (From Clavel *et al.*, 1986.)

$TBTCO_3$ were present. The speciation of TBTO at pH 8 in sea water was suggested to involve the following equilibria:

$$(Bu_3Sn)_2O + H_2O \rightarrow 2Bu_3SnOH$$
$$Bu_3SnOH + H_3O^+ \rightarrow Bu_3SnOH_2^+$$
$$Bu_3SnOH_2^+ + HOCO_2^- \rightarrow Bu_3SnOCO_2^- + H_3O^+$$
$$Bu_3SnOH_2^+ + Cl^- \rightarrow Bu_3SnCl + H_2O$$

The solubility of TBTO in sea water has been measured at $8\,mg\,Sn\,l^{-1}$ (Schatzberg, 1987), but this concentration will not be reached in natural waters.

5.3.2 Post-banning

The imposition of bans on the use of TBT has resulted in significant reductions in dissolved TBT concentrations, often to below the analytical detection limits. This is particularly so in enclosed waterways and estuaries dominated by the small craft affected by the legislation. Typical of the results obtained elsewhere in the world, are the data reported by Huggett et al. (1992) for waters in San Diego Bay, USA, and supported by reductions in impacts on biota, particularly oysters (Alzieu et al., 1990; Waldock et al., 1988; Batley, Brockbank & Scammell, 1992).

In a San Diego Bay yacht basin, TBT concentrations of around $82\,ng$ $Sn\,l^{-1}$ prior to a 1987 ban, decreased to around $10\,ng\,Sn\,l^{-1}$ in 1990, and $5\,ng\,Sn\,l^{-1}$ in 1991 (Huggett et al., 1992) (Figure 5.3). Mean dissolved TBT concentrations at a marina site, Hampton Road, Virginia decreased over the same period from $102\,ng\,Sn\,l^{-1}$ to $20\,ng\,Sn\,l^{-1}$. Similar decreases have been observed in samples from both Europe and Japan, and have been reported in a summary of monitoring studies produced by CEFIC (1994).

In estuaries and harbours where large vessels are the major source of TBT, concentrations in waters, sediments and biota have not shown the same reductions. The major continuing source appears to be drydocks and other facilities where ship-building, repair and repainting occurs, as the areas affected by traffic of larger vessels show relatively low TBT concentrations (CEFIC, 1994).

5.4 Tributyltin in the surface microlayer

Substantial evidence has been accumulated to demonstrate the enrichment of TBT in the surface microlayer. This is not surprising given that the unburnt gasoline and other organic compounds that accumulate as

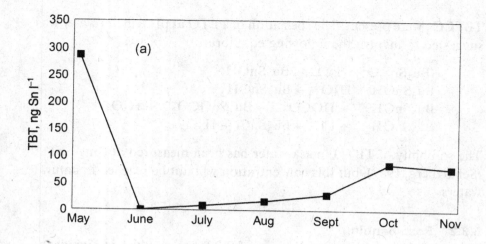

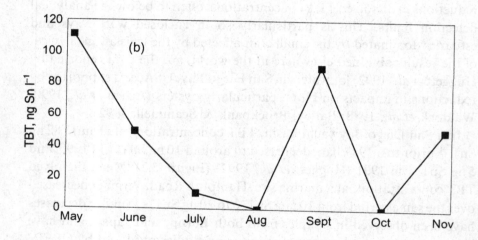

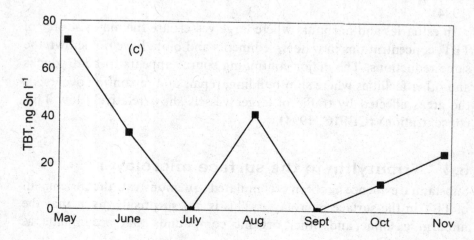

surface films are not dissimilar from the solvents used in analytical procedures to extract TBT from waters. The hydrophobic tendencies of TBT species can be implied from their octanol–water partition coefficients (K_{ow}). Laughlin et al. (1986) reported measurements for log K_{ow} for TBTO in sea water showing a dependency on salinity. Log values decreased from 3.8 in deionised water to 3.7 at 25 psu, increasing again with increasing salinity as the equilibrium shifted to TBTCl. Typically log values higher than 3 classify a compound as lipophilic, i.e. likely to accumulate in the fatty tissue of aquatic organisms (Connell, 1988).

Reported microlayer enrichment factors for TBT are quite variable, but this could be a combination of the variability in the concentration of hydrophobic organic compounds in the surface layer and the inability of sampling techniques to sample reproducibly to the same depth. Sampling techniques have included the use of Teflon sheets, glass plates and stainless steel mesh frames. The potential for adsorptive losses on these surfaces is unknown. More recently, rotating drum samplers have been used and, in our laboratory, a sampler having a polycarbonate drum was used to minimise losses.

Typical data, shown in Table 5.2, exhibit a range of enrichment factors, the highest near marinas, presumably where the concentrations of other surface contaminants are high. These values are nowhere near the magnitude of those measured for freshwater systems (Maguire & Tkacz, 1987) which ranged from 70 to 47 300. This may reflect both a different

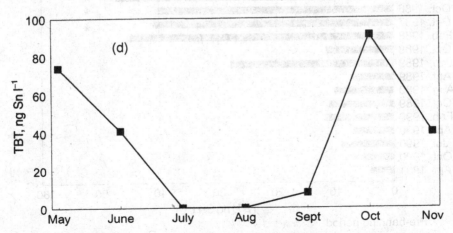

Figure 5.2 Seasonal fluctuations in TBT concentrations at four sites (designated (a)–(d)) in the south and southwest of England. (From Ebdon et al., 1988.)

solubility of TBT in freshwaters and a difference in the nature of the surface film organics. It is important to note that often results for both subsurface and surface waters are based on unfiltered samples, and this may make if difficult to compare enrichment factors.

Enrichment of TBT in the surface microlayer has been suggested as the reason that intertidal oysters are more susceptible to TBT accumulation than subtidal oysters, mussels and scallops (Batley & Scammell, 1990).

Table 5.2 *TBT in the surface microlayer of estuarine and marine waters*

Site	Surface microlayer concentration ($ng\,Sn\,l^{-1}$)	Enrichment factor	Reference
Chesapeake Bay, USA	2450[a]	11	Matthias *et al.*, 1988
Chesapeake Bay, USA	16–480[a]	2.2–25	Hall *et al.*, 1987
Southwest England	170–835[a]	1.9–27	Cleary & Stebbing, 1987a,b
Green Bay, NH, USA	n.d.–125	>40	Donard *et al.*, 1986

[a]Unfiltered samples.

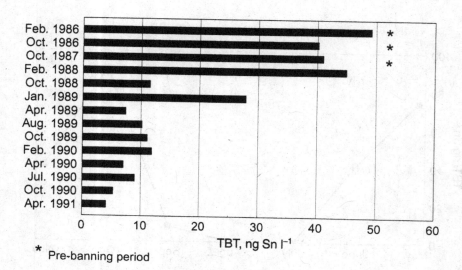

* Pre-banning period

Figure 5.3 Changes in TBT concentrations in San Diego Bay (USA) after legislation restricting its use. (From Huggett *et al.*, 1992.)

5.5 Degradation processes of dissolved TBT

5.5.1 Degradation routes and kinetics

The degradation of tributyltin in marine waters can proceed by both biotic and abiotic routes. In either case, the transformation pathway is the same, involving sequential debutylation via dibutyltin (DBT) and monobutyltin (MBT) to inorganic tin as tin(IV). Numerous studies of degradation kinetics have been reported. By far the most comprehensive are those of Seligman and coworkers (Seligman *et al.*, 1986a, b; Lee, Valkirs & Seligman, 1987). A number of laboratory microcosm experiments have been conducted with ambient, unfiltered and filtered sea water from quite diverse harbours and estuaries, under both light and dark conditions, using samples spiked with TBTCl and radiolabelled TBTO (butyl-1-^{14}C). Results are summarised in Table 5.3.

These results showed that the kinetics of degradation were first order, with a dependence of half-life on both temperature and concentration, but in the concentration ranges typical of contaminated waters, the half-life lay between 6 and 9 days. Degradation in the dark, simulating near-bottom conditions, yielded half-lives that were 40% longer, suggesting the role that photosynthetic organisms such as diatoms may play. Filtration, treatment with formalin, and autoclaving significantly reduced the rate of TBT degradation, while addition of nutrients to increase phytoplankton density caused a rapid increase in decay rate, thus implicating microbial debutylation as the major degradation pathway (Seligman *et al.*, 1986c; Watanabe *et al.*, 1992). Filtration through 0.45 µm filters had a greater effect than coarse filtration using 5–10 µm filters. The temperature effect was also important, with no TBT degradation observed at 5 °C compared to about 60% at 28 °C after 6 days in the light (Olson & Brinckman, 1986).

The reduced degradation rate observed at very high TBT concentrations is believed to result from possible toxic effects of TBT to some microorganisms. Variations in degradation rates will reflect differences in microbial populations (Lee *et al.*, 1987), as shown in Figure 5.4. More recent studies by the same authors have reinforced these findings (Lee *et al.*, 1989).

A range of microbial species including algae, fungi and bacteria have been shown to biodegrade TBT (Barug, 1981; Olson & Brinckman, 1986); however, there is clear evidence for the involvement of photosynthetic organisms. Lee *et al.* (1989) showed that diatoms were able to degrade TBT rapidly whereas dinoflagellates, chrysophytes and chlorophytes

Table 5.3 *Degradation half-lives for TBT in unfiltered estuarine waters*

Site	TBT (ng Sn l^{-1})	Treatment	Half-life (days)	Reference
San Diego Bay, South Bay, CA, USA	820	Light, 17°C	12	Seligman, Valkirs & Lee, 1986c
San Diego Bay, South Bay, CA, USA	820	Dark, 17°C	15	Seligman, Valkirs & Lee, 1986c
San Diego Bay, Shelter Island, USA	820	Light 17°C	7	Seligman, Valkirs & Lee, 1986c
San Diego Bay, Shelter Island, USA	820	Dark, 17°C	9	Seligman, Valkirs & Lee, 1986c
San Diego Bay, Shelter Island, USA	440	Light, 17°C	7	Seligman, Valkirs & Lee, 1986c
Skidaway Island, GA, USA	205	Light, 29°C	5	Seligman, Valkirs & Lee, 1986c
Skidaway Island, GA, USA	370	Dark, 29°C	9	Seligman, Valkirs & Lee, 1986c
Skidaway Island, GA, USA	533	Light, 12°C	7	Seligman, Valkirs & Lee, 1986c
Skidaway Island, GA, USA	656	Dark, 12°C	11	Seligman, Valkirs & Lee, 1986c
Skidaway River, GA, USA	205	Light, 29°C	6	Lee, Valkirs & Seligman, 1987
Skidaway River, GA, USA	370	Dark, 29°C	9	Lee, Valkirs & Seligman, 1987
Skidaway River, GA, USA	615	Light, 29°C	6	Lee, Valkirs & Seligman, 1987
Skidaway River, GA, USA	615	Dark, 29°C	10	Lee, Valkirs & Seligman, 1987
Skidaway River, GA, USA	615	Light, 19°C	9	Lee, Valkirs & Seligman, 1987
Skidaway River, GA, USA	615	Light, 11°C	8	Lee, Valkirs & Seligman, 1987
Skidaway River, GA, USA	615	Dark, 11°C	13	Lee, Valkirs & Seligman, 1987
Skidaway River, GA, USA	164	Light, 28°C	3	Lee, Valkirs & Seligman, 1989
Skidaway River, GA, USA	164	Dark, 28°C	7	Lee, Valkirs & Seligman, 1989
Toba Bay marina, Japan	2900	Light, 15°C	26	Watanabe, Sakai & Takatsuki, 1992
Toba Bay, Japan	1300	Light, 20°C	7	Watanabe, Sakai & Takatsuki, 1992
Toba Bay, Japan	1600	5 µm filter, 25°C	5	Watanabe, Sakai & Takatsuki, 1992
Toba Bay, Japan	1100	5 µm filter, 25°C	38	Watanabe, Sakai & Takatsuki, 1992
Toba Bay, Japan	1100	autoclaved	66	Watanabe, Sakai & Takatsuki, 1992

were markedly slower in TBT degradation. They presented evidence for the formation of (hydroxybutyl)dibutyltins in sunlit natural waters. More stable hydroxybutyltins were detectable, but (β-hydroxybutyl)dibutyltin was unstable and rapidly degraded to DBT. No hydroxybutyltin species were found in samples kept in the dark. They observed that estuarine waters containing algae and [^{14}C]TBT (400 ng Sn l^{-1}), after 4 days in the light, showed a reduction in TBT content to 62%, accompanied by a conversion of 25% to DBT, 12% to (hydroxybutyl)dibutyltins, and <1% to MBT.

The half-life for the degradation of DBT to MBT is appreciably longer than the debutylation of TBT; Adelman, Hinga & Pilson (1990) measured values of 12–18 days at 20 °C in an enclosed mesocosm. The half-life for complete mineralisation of butyl groups, as measured by the formation of $^{14}CO_2$, has been measured at between 50 and 75 days (Seligman *et al.*, 1986b).

Olson & Brinckman (1986) found occasional evidence for tetrabutyltin formation during the incubation of TBT in Chesapeake Bay waters at 28 °C, and they postulated a redistribution mechanism with TBT. The reaction was not readily duplicated and its origin is unclear.

Methyltin species have also been reported in coastal waters (Jackson *et al.*, 1982; Weber, Han & François, 1988; Dirkx *et al.*, 1993), typically at concentrations of 2–9 ng Sn l^{-1} for monomethyltin, and 2–4 ng Sn l^{-1} for dimethyltin (Dirkx *et al.*, 1993). This could represent an anthropogenic source, but there is evidence for the biologically-mediated methylation of

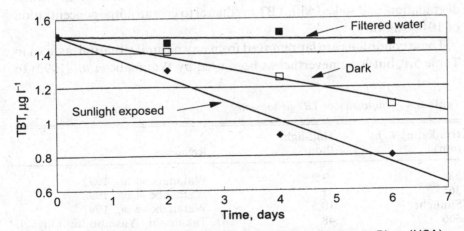

Figure 5.4 Degradation of ^{14}C-TBT added to Skidaway River (USA) water under light and dark conditions. (From Lee *et al.*, 1987).

tin(IV) (and tin(II)) in sediments forming a series of methyltin products $(Me_nSn^{(4-n)+})$ (Guard, Cobet & Coileman, 1981; Hallas, Means & Cooney, 1982; Donard, Short & Weber, 1987).

5.5.2 Evaporative losses

Evaporation has been considered as a source of TBT losses in degradation and mesocosm experiments in several studies (Adelman *et al.*, 1990; Watanabe *et al.*, 1992). Calculations predicted a half-life between 1.2 and 3585 days, on the basis of measured vapour pressures and water solubilities of TBTO and TBT acetate, and assuming the mass transfer coefficients at the sea water surface. This wide range results from the range of vapour pressures reported for TBTO; however, TBTCl is likely to be more important. Watanabe *et al.* (1992) suggested that if, like TBT acetate, TBTCl has a high vapour pressure, then the half-life for evaporative loss could be quite low, although in their experiments they assumed a half-life of 70 days.

5.5.3 Photolytic degradation

Experiments by Maguire, Carey & Hale (1983) on the photolysis of TBT in fresh water showed a half-life exceeding 89 days. These results have been challenged by Watanabe *et al.* (1992), because they were conducted in an air-tight Pyrex tube, whereas direct surface irradiation with adequate air contact was required.

Results for the photodegradation of TBT in seawater, presented in Table 5.4, show surprisingly rapid degradation of TBT. These data were obtained using seawater filtered to eliminate the effect of microbial degradation, and spiked with TBT in ethanol to give an initial concentration of 10 000 ng Sn l^{-1}.

These conditions are far removed from any natural situation, as seen in Table 5.1, but have nevertheless been used by Watanabe *et al.* (1992) to

Table 5.4 *Photolysis of TBT in seawater*

Irradiating light (nm)	Half-light (hours)	Reference
254	3.8	Watanabe *et al.*, 1992
366	31	Watanabe *et al.*, 1992
Sunlight	10.5	Watanabe *et al.*, 1992
366	48	Takahashi, Yashino & Ohyagi, 1987

estimate the relative importance of photolytic and microbial pathways (Table 5.5).

They considered the cases of sea waters that were 'rich' in suspended matter, and biologically active, as well as 'clean' open ocean water. Calculations were made for both winter and summer conditions, assuming 5 h of sunlight per day in summer and 1/10 of this in winter. They concluded that biodegradation was the major control of TBT decomposition in estuarine waters with a half-life of 6 days (in summer), in agreement with the findings of other studies described earlier. Photolysis was dominant in open ocean waters, but the half-life was almost three times longer (Watanabe *et al.*, 1992).

5.5.4 TBT degradation products in marine waters
Where measurements of DBT and MBT have been reported, their concentrations have been quite variable, although generally lower than the measured TBT concentrations (e.g. Hall *et al.*, 1987; Valkirs, Stallard & Seligman, 1987; Batley *et al.*, 1989). This is to be expected in sites receiving a continuous fresh input of TBT, but where this is not the case, values exceeding the TBT concentration could be expected. The distribution of these degradation products will be complicated by differences in their partitioning behaviour on suspended particulates, which will be discussed later.

5.6 Sediment/water partitioning
The equilibrium partitioning of contaminants between sediment and water phases is traditionally described by a partition coefficient, K_D, whose value is generally determined from laboratory studies with natural waters and sediments. Measurements on selected field samples of suspended sediment and water concentrations of a contaminant will yield an apparent partitioning or sorption coefficient, which will approach the equilibrium sediment–water partition coefficient, K_D, only in cases where the sediments are in equilibrium with the overlying waters.

On the basis of its octanol–water partition coefficient, it would be expected that TBT would favourably partition to particulate organic matter in suspended sediments in marine waters. Measurements of apparent K_D values for TBT partitioning in field samples have confirmed this, but, not surprisingly, have shown considerable variation because of the variable organic content of the sediment (Table 5.6).

For the partitioning of hydrophobic contaminants, it is more appropriate

Table 5.5 Degradation rates for TBT removal processes in seawater[a]

| Process | Degradation rate (d^{-1}) | | | |
| | Summer | | Winter | |
	SS-rich[b]	Open Ocean	SS-rich[b]	Open Ocean
Biodegradation	0.103 (86.1%)	0.0082 (19.8%)	0.0165 (90.4%)	0.0002 (36.8%)
Photolysis	0.0165 (13.8%)	0.0330 (79.9%)	0.00165 (8.8%)	0.0033 (60.6%)
Evaporation	0.0014 (0.1%)	0.00014 (0.3%)	0.00014 (0.8%)	0.00014 (2.6%)
Total	0.112	0.0413	0.0186	0.0054
Half-life, days	5.8	16.8	37	127

[a]From Watanabe et al. (1992).
[b]Suspended sediment-rich.

to report K_{OW} values, where the sediment concentration is expressed on the basis of organic carbon (Di Toro *et al.*, 1991).

The binding of a polar molecule such as TBT may be a function both of its hydrophobicity, associated with the bulky butyl groups, and its polarity, and on this basis organic carbon content of sediments may not be the only control on adsorption. Sediment particle size, or more correctly surface area of the sediments, is likely to be important. A low surface area sandy substrate is likely to have a low binding capacity, compared to clay–silt particles. The wide spread of apparent K_D values in Table 5.6 reflects a combination of such factors, but there is insufficient information to enable interpretation of the data. Differences in TBT degradation rates between the dissolved and particulate phases, can also be important. Since solution-phase degradation is fastest, samples where degradation has proceeded further are likely to show apparently higher values for K_D.

An equilibrium partitioning coefficient in such a dynamic environment as an estuary will only be applicable if the physical and chemical partitioning is not kinetically limited. More detailed studies have been undertaken to characterise fully the sorptive behaviour. A universal finding is that equilibration is rapidly achieved, in times less than several hours after addition of TBT to a mixture of water and sediment (Harris & Cleary, 1987). Adsorption kinetics appear to be first order.

Laboratory studies have been carried out on a range of natural sediments and waters. For sediments from Chesapeake Bay, Unger, MacIntyre & Huggett (1987, 1988) measured partition coefficients from 110 to 8200 (Figure 5.5) for both adsorption and desorption, but using sediment concentrations of up to $13\,500\,\mu g\ Sn\,g^{-1}$. The adsorption process appeared to be reversible. Normalisation of these values to organic carbon content gave values for K_{OC} with an equally wide range, between 12 000 and 200 000, indicating that carbon content is not the only factor controlling TBT adsorption.

For enclosure experiments in Pearl Harbour, Hawaii, K_D values between 6250 and 55 400 were measured (Stang & Seligman, 1987). Experiments were carried out using an *in situ* dome positioned over TBT-containing sediments. No TBT was observed to desorb from the bottom sediments, but is adsorbed at the daily rate of $0.23\,ng\ Sn\,cm^{-2}$. The system appeared to be at or near equilibrium. This is consistent with the data of Unger *et al.* (1988), where adsorption isotherms were determined by vigorous mixing of sediments with water. The implication is that during dredging of TBT-contaminated sediments, it is likely that

Table 5.6 *Sediment–water partitioning of TBT*

Site	Dissolved TBT (ng Sn l^{-1})	Apparent K_D (l kg^{-1})	TBT (%)	Reference
San Diego Bay, USA	57	227	46	Valkirs et al., 1986a,b
San Diego Bay, USA	44	1 590	62	Valkirs et al., 1986a,b
San Diego Bay, USA	213	38 900	—	Valkirs et al., 1987
San Diego Bay, USA	193	4 610	—	Valkir et al., 1987
Sarah Creek, Chesapeake Bay, USA	7	2 500	—	Unger et al., 1988
Sarah Creek, Chesapeake Bay, USA	9	5 500	—	Unger et al., 1988
Sarah Creek, Chesapeake Bay, USA	4	2 600	—	Unger et al., 1988

some TBT will be released either by dissolution or resuspension, but the extent of this release and the potential for biological impact, as shown by measurements on sediment interstitial waters, is likely to be small, as will be shown later. In many highly contaminated sites, the sediment contains paint flakes which will be a source of more TBT than desorption from sediments.

Randall & Weber (1986) carried out a comprehensive series of equilibrium sorption experiments using an artificial sorbent, hydrous iron oxide, in the presence of dissolved fulvic acid. Three iron oxide concentrations 10, 100 and 1000 mg l^{-1} were used at three salinities 5, 20 and 35 psu, and three pH values 6.2, 7.2 and 8.2, each in the presence of 10 mg l^{-1} of fulvic acid.

Adsorption of the fulvic acid imparted a net negative charge to the otherwise positively charged iron oxide, and although there was precipitation of all fulvic acid at the two higher iron oxide concentrations, it was in excess in the case of the 10 mg l^{-1} iron oxide. Particulate iron/fulvic acid affected the measured K_D values. Increasing the iron oxide from 10 to 1000 mg l^{-1}, and effectively increasing the fulvic content of the particulates, increased K_D from 21 000 to 35 000 at 35 psu salinity and from 1500 to 13 000 at 5 psu.

Adsorption was reduced with increasing pH, presumably because of competition for TBT between the carboxylate groups on the fulvic acid, and hydroxide and carbonate anions.

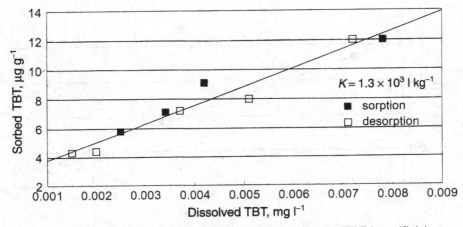

Figure 5.5 Adsorption and desorption isotherms for TBT in artificial seawater on sediment from a tidal creek in Chesapeake Bay, USA. (From Unger *et al.*, 1988).

Increases in salinity have been shown in several studies to result in a decrease in TBT adsorption (Harris & Cleary, 1987; Unger *et al.*, 1987, 1988; Dirkx *et al.*, 1993). For sediments from Carter Creek in Chesapeake Bay (USA), equilibrated with artificial sea water with added TBT, Unger *et al.* (1987) reported a linear decrease in K_D ranging from $1400 \, l \, kg^{-1}$ at zero salinity to 700 at 35 psu (Figure 5.6). These results are in apparent disagreement with the results of Randall & Weber (1986) and Harris & Cleary (1987), although the former study used a very high particulate concentration. The trend is the same as that observed for cadmium and zinc on Rhine River sediments (Salomons & Förstner, 1984).

In the study of Randall & Weber (1986), using artificial sediments at only two salinities, 5 and 35 psu, greater adsorption occurred at the higher salinity, supposedly through competition of Na^+ with TBT for the carboxylate ions on fulvic acid, i.e. a salting out effect. Harris & Cleary (1987) found that the percentage of TBT associated with a fixed concentration ($60 \, mg \, l^{-1}$) of sediments from the Crouch and Tamar Estuaries (UK) increased with salinity, but in a sigmoidal relationship reaching a plateau at around 16 psu. The K_D values calculated from these were extremely low ($2.2 \, l \, kg^{-1}$) and inconsistent with those reported previously, and would imply negligible to very low sediment binding. The physical characteristics of the sediments were not specified.

The decrease in sorption with salinity found by Unger *et al.* (1987, 1988), may result from ion exchange competition of sea water cations with

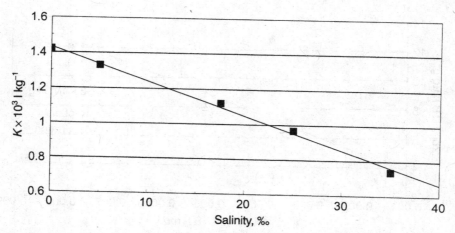

Figure 5.6 The effect of salinity on the sorption coefficient for TBT on a sediment from a tidal creek in Chesapeake Bay, USA. (From Unger *et al.*, 1987.)

sorbed TBT, or from changes in the TBT species in solution.

The adsorptive behaviour of TBT will be dependent as much on solution speciation as it is on the composition of the sediment phase. Differences in behaviour may ultimately be the result of using artificial water samples free of natural complexing agents, unrepresentative sediment concentrations and sediments of differing chemical compositions. It would appear that equilibrium log K_D values are typically in the range 2–5, but that for application to a field model for TBT partitioning, site specific adsorption experiments are required.

The role of colloids in binding dissolved TBT has not yet been considered, but on the basis of observed sediment–water partitioning, they may play an important role as a solution species. The biological availability of TBT in this form may well differ from truly soluble species.

The partitioning of DBT and MBT have received only minor attention. In their study with synthetic sediments, Randall & Weber (1986) reported that the adsorption of butyltin cations followed the order MBT > TBT > DBT, which they suggested was indicative of the differing importance of charge and hydrophobicity as controls. The *in situ* adsorption/desorption studies of Stang & Seligman (1987) showed that apparent K_D values for both DBT and MBT were substantially lower than those for TBT. While TBT values ranged from 6250 to 55 440, those for DBT lay between 2070 and 26 080, and for MBT between 1760 and 28 750. The reasons for the differing behaviour of MBT are not clear, and additional studies on natural sediments are required.

5.7 Tributyltin concentrations in estuarine and marine sediments

As for waters, there have been extensive studies of TBT distribution and behaviour in estuarine sediments. Some typical concentrations are listed in Table 5.7. As expected a wide range of concentrations has been reported, with high concentrations specifically related to TBT usage in the immediate vicinity. In the instances of highest concentrations, it is likely that the samples contain TBT as paint flakes rather than adsorbed to particles sedimenting from the water column. No data were available for marine sediments, and unless these were specifically receiving estuarine inputs, it is unlikely that their TBT content would be significant.

Both DBT and MBT are also found in estuarine sediments, and these dominate where TBT has entered the sediment by adsorption, because they are more persistent.

A study of sediment pore water concentrations (Astruc *et al.*, 1990) showed that although total dissolved tin was often significant at 20–200 ng Sn l^{-1}, TBT was rarely detectable. The degradation products were present, principally MBT and to a lesser extent DBT, but only in the surface layers, and at concentrations below 7 ng Sn l^{-1}.

5.8   Degradation processes of TBT in estuarine sediments

The degradation of TBT in estuarine sediments leads to the same products, DBT and MBT as are formed in solution; however, the rate is considerably slower. There have been few studies on the degradation processes.

Stang *et al.* (1992) examined the degradation of TBT (120 ng Sn l^{-1}) incubated for up to 7 days with sterile sea water and a range of sediment types. The percentages of degradation products measured on the sediments, with and without sterilising by autoclaving, were compared. The results suggested that microbial degradation was the process degrading TBT in

Table 5.7   *TBT in estuarine sediments*

Site	Depth (cm)	Concentration (ng Sn g^{-1} dry weight)	Reference
Suva Harbour, Fiji	Surface	<5–38 000	Stewart & de Mora, 1992
Arcachon Bay, France	0–50 cm	<5–5000	Astruc *et al.*, 1990
Kaipara Harbour, New Zealand	0–40 cm	<4–759	de Mora, King & Miller, 1989
San Diego Harbour, USA	0–20 cm	8–66	Stang & Seligman, 1986
San Diego Harbour, USA	Surface	143	Stang, Lee & Seligman, 1992
North Chesapeake Bay, USA	Surface	58–570	Matthias *et al.*, 1988
Poole Harbour, UK	Surface	8–213	Langston *et al.*, 1987
Sado Estuary, Portugal	Surface	6–213	Quevauviller *et al.*, 1988
Sydney Harbour, Australia	Surface	10–8000	Batley & Brockbank, unpublished results
Georges River, Sydney, Australia	0–40 cm	<0.2–20	Kilby & Batley, 1993
Great Bay Estuary, New Hampshire, USA	Surface	4–20	Weber *et al.*, 1988

sandy sediments, whereas in clayey sediments, abiotic processes appeared to dominate. The half-lives for TBT degradation were similar to those measured in solution alone. This study was complicated by the fact that adsorption/desorption and solution degradation reactions were occurring concurrently. The findings are relevant only to the behaviour of TBT on suspended particulate matter, where the sediments are aerobic and in low concentrations compared to the water volume.

The behaviour of TBT in deeper anoxic sediments is a quite different issue. Here the sediment to interstitial water ratio is high, and the diffusion of pore water especially in the clayey sediments is slow. Stang & Seligman (1986) examined the degradation of naturally contaminated San Diego Bay sediment (6 kg wet weight) in a polycarbonate aquarium (38 l) containing aerated sea water, and determined a half-life for the first-order degradation of 162 days. In addition, there was evidence that the degradation did not involve stepwise debutylation, but a direct conversion to MBT, although this apparent behaviour could also be explained by the rate of debutylation of TBT and DBT being similar.

Attempts have been made to calculate half-lives for TBT in sediments on the basis of measured depth profiles. Apparent degradation rates can be calculated assuming a constant sedimentation rate and a constant flux of TBT to the sediments, as well as a uniform sediment composition.

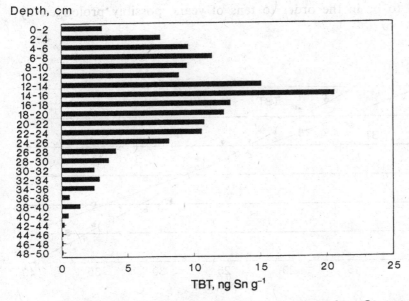

Figure 5.7 Depth profiles of TBT in a sediment from Georges River, NSW, Australia. (From Kilby & Batley, 1993.)

The latter is particularly important in controlling TBT adsorption, as has already been shown. Measurements on sediment cores from the Tamaki Estuary in Auckland, New Zealand, yielded a mean half-life of 1.85 years (de Mora *et al.*, 1989). The maximum TBT concentration in these sediments was 241 ng Sn g^{-1}. In sediment from the Georges River in Sydney, Australia, containing up to 21 ng Sn g^{-1}, two cores yielded a mean half-life of 3.8 years (Figure 5.7) (Kilby & Batley, 1993). Degradation of DBT was appreciably slower, with a half-life near 6 years (Figure 5.8). In both the New Zealand and Australian examples, good fits were obtained for first-order TBT decay plots ($r^2 = 0.92$–0.95). A second study in New Zealand found that the TBT degradation rate varied between 3.3 and 3.8 years for three cores form a marina with TBT concentrations ranging up to 350 ng Sn g^{-1} (de Mora, Stewart & Phillip, 1995). The differences in apparent half-lives in these studies may result from the assumption of a constant rate of deposition, but the values nevertheless illustrate that the persistence of TBT in deeper, anoxic sediments is considerably longer than in surface sediments. This may be associated with the greater stability of TBT-sulfide, as suggested in studies of Wulf & Byington (1976). Laboratory studies by Dowson *et al.* (1993) in both freshwater and marine sediments showed little difference in TBT degradation rates, with half-lives in surficial sediments ranging from 360–775 days. In anaerobic sediments the half-life appeared to be in the order to tens of years, possibly prolonged by

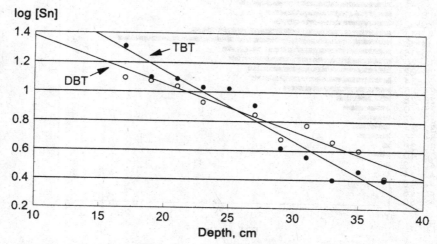

Figure 5.8 First-order decay plots for TBT and DBT in the sediment core shown in Figure 5.7. (From Kilby & Batley, 1993.)

limited biotic degradation activity. The findings have important implications for the environmental threat posed by TBT-containing dredged sediments.

Core profiles have been reported also for Arcachon Bay, France (Astruc *et al.*, 1990), and San Diego Bay (Stang & Seligman, 1986). In the former, TBT concentrations peaked at depths of around 10 cm, as was seen in the Australian example (Figure 5.7) (Kilby & Batley, 1993). Although this may be related to a decreased usage of TBT in recent years, it would also be anticipated on the basis of the more rapid degradation on oxic surface sediments described previously.

5.9 Conclusions

Measurements of TBT concentrations in waters from harbours and estuaries worldwide over the past ten years have shown that the extensive TBT contamination observed in the early 1980s near areas of high boating activity is now significantly reduced following legislation in most countries restricting or prohibiting its use in marine antifouling paints.

The persistence of TBT in waters is short, with degradation involving debutylation to DBT and MBT and inorganic tin, being dominated by biotic processes. In estuarine waters the typical half-life is 6–7 days at 28 °C, whereas in open ocean waters, low in nutrients, photolytic processes dominate and the half-life increases, at the same temperature, to 17 days. Evaporation is not a significant removal process.

The TBT molecule possesses both hydrophobic and polar characteristics, and both influence its partitioning in waters, particulates and the surface microlayer in marine and estuarine waters. Enrichment factors in the surface microlayer vary from 2 to > 40. Measurements of the sediment–water partitioning of TBT have yielded K_D values from 10^2 to 10^5. Adsorption decreases with increasing salinity.

Elevated TBT concentrations have also been detected in sediments near marinas or areas of high boating activity. Although in oxic surface sediments TBT degradation is microbially mediated and differs little from its half-life in waters, in deeper anoxic sediments, degradation is significantly slower and half-lives from 1.9 to 3.8 years have been estimated. This will have important consequences for biota should dredging result in exposure of these sediments to overlying waters.

5.10 References

Adelman, D., Hinga, K. R. and Pilson, M. E. Q., 1990. Biogeochemistry of butyltins in enclosed marine ecosystem. *Environ. Sci. Technol.*, **24**, 1027–32.

Alzieu, C., Sanjuan, J., Michel, P., Borel, M. and Dreno, J. P., 1989. Monitoring and assessment of butyltins in Atlantic coastal waters. *Mar. Pollut. Bull.*, **20**, 22–6.

Alzieu, C., Michel, P., Sanjuan, J. and Averty, B., 1990. Tributyltin levels in French Mediterranean coastal waters. *App. Organometall. Chem.*, 55–61.

Astruc, M., Lavigne, R., Pinel, R., Leguille, F., Desauziers, V., Quevauviller, P. and Donard, O., 1990. Speciation of tin in sediments of Arcachon Bay (France). In *Metal Speciation, Separation and Recovery*, Vol. 2, Patterson, J. W. and Passino, R. (eds), pp. 263–81.

Bacci, E. and Gaggi, C., 1989. Organotin compounds in harbour and marina waters from the Northern Tyrrhenian Sea. *Mar. Pollut. Bull.*, **20**, 290–2.

Barug, D., 1981. Microbial degradation of bis(tributyltin)oxide. *Chemosphere*, **10**, 1145–54.

Batley, G. E., Brockbank, C. I. and Scammell, M. S., 1992. The impact of banning of tributyltin-based antifouling paints on the Sydney rock oyster, *Saccostrea commercialis*. *Sci. Total Environ.*, **122**, 301–14.

Batley, G. E., Mann, K. J., Brockbank, C. I. and Maltz, A., 1989. Tributyltin in Sydney Harbour and Georges River waters. *Aust. J. Mar. Freshwater Res.*, **40**, 39–48.

Batley, G. E. and Scammell, M. S., 1990. Research on tributyltin in Australian estuaries. *App. Organometall. Chem.*, **5**, 99–105.

Bennett, R. F., 1983. Industrial development of organotin chemicals. *Ind. Chem. Bull.*, **2**, 171–6.

CEFIC (Conseil European des Federations de l'Industrie Chimique), 1994. *Use of organotin compounds in antifouling paints*. Paper presented at 35th Session of the Marine Environment Protection Committee, IMO, London, March 7–11.

Champ, M. A. and Pugh, W. L., 1987. Tributyltin antifouling paints, introduction and overview. In *Proceedings of the Organotin Symposium, Oceans 87*, Marine Technology Society, Washington, DC, Vol. 4, pp. 1296–308.

Clark, E. A., Sterritt, R. M. and Lester, J. N., 1988. The fate of tributyltin in the aquatic environment. *Environ. Sci. Technol.*, **22**, 600–4.

Clavell, C., Seligman, P. E. and Stang, P. M., 1986. Automated analysis of organotin compounds: a method for monitoring butyltins in the marine environment. In *Proceedings of the Organotin Symposium, Oceans 86*, Marine Technology Society, Washington, DC, Vol. 4, pp. 1152–4.

Cleary, J. J. and Stebbing, A. R. D., 1985. Organotin and total tin in coastal waters of southwest England. *Mar. Pollut. Bull.*, **16**, 350–5.

Cleary, J. J. and Stebbing, A. R. D., 1987a. Organotin in the surface microlayer and subsurface waters of southwest England. *Mar. Pollut. Bull.*, **18**, 238–46.

Cleary, J. J. and Stebbing, A. R. D., 1987b. Organotins in the water column – enhancement in the surface microlayer. In *Proceedings of the Organotin Symposium, Oceans 87*, Marine Technology Society, Washington, DC, Vol. 4, pp. 1405–10.

Connell, D. W., 1988. Bioaccumulation behaviour of persistent organic chemicals with aquatic organisms. *Rev. Environ. Contam. Toxicol.*, **101**, 117–54.

de Mora, S. J., King, N. G. and Miller, M. C., 1989. Tributyltin in marine sediments: Profiles and the apparent rate of degradation. *Environ. Technol. Lett.*, **10**, 901–8.

de Mora, S. J., Stewart, C. and Phillips, D., 1995. Sources and rates of degradation of tri(n-butyl)tin in marine sediments near Auckland, New Zealand. *Mar. Pollut. Bull.*, **30**, 50–7.

Dirkx, W., Lobinski, R., Ceulemans, M. and Adams, F., 1993. Determination of methyl- and butyltin compounds in waters of the Antwerp harbour. *Sci. Total Environ.*, **136**, 279–300.

Di Toro, D. M., Zarba, C. S., Hansen, D. J., Berry, W. J., Swartz, R. C., Cowan, C. C., Pavlou, S. P., Allen, H. E., Thomas, N. A. and Paquin, P. R., 1991. Technical basis for establishing sediment quality criteria for nonionic organic chemicals using equilibrium partitioning. *Environ. Toxicol. Chem.*, **10**, 1541–83.

Donard, O. F. X., Rapsomanikis, S. and Weber, J. H., 1986. Speciation of inorganic tin and alkyltin compounds by atomic absorption spectrometry using electrothermal quartz furnace hydride generation. *Anal. Chem.*, **58**, 772–7.

Donard, O. F. X., Short, F. T. and Weber, J. H., 1987. Regulation of tin and methyltin by the green alga Enteromorpha under simulated estuarine conditions. *Can. J. Fish. Aquatic Sci.*, **44**, 140–4.

Dowson, P. H., Bubb, J. M., Williams, T. P. and Lester, J. N., 1993. Degradation of tributyltin in freshwater and estuarine marina sediments. *Water Sci. Technol.*, **28**, 133–7.

Ebdon, L., Evans, K. and Hill, S., 1988. The variation of tributyltin levels with time in selected estuaries prior to the introduction of regulations governing the use of tributyltin-based antifouling paints. *Sci. Total Environ.*, **68**, 207–23.

Guard, H. E., Cobet, A. B. and Coileman, W. M., 1981. Methylation of trimethyltin compounds in estuarine sediments. *Science*, **213**, 770–1.

Hall, L. W., Lenkevich, M. J., Hall, W. S., Pinkney, A. E. and Bushong, S. J., 1987. Evaluation of butyltin compounds in Maryland waters of Chesapeake Bay. *Mar. Pollut. Bull.*, **18**, 78–83.

Hallas, L. E., Means, J. C. and Cooney, J. J., 1982. Microbial transformation of tin by estuarine organisms. *Science*, **213**, 1505–6.

Harris, J. R. W. and Cleary, J. J., 1987. Particle–water partitioning and organotin dispersal in an estuary. In *Proceedings of the Organotin Symposium, Oceans 87*, Marine Technology Society, Washington, DC, Vol. 4, pp. 1370–4.

Huggett, R. J., Unger, U. A., Seligman, P. F. and Valkirs, A. O., 1992. The marine biocide tributyltin: assessing and managing the environmental risks. *Environ. Sci. Technol.*, **26**, 232–7.

Jackson, J. A., Blair, W. R., Brinckman, F. E. and Iverson, W. P., 1982. Gas-chromatographic speciation of methylstannanes in the Chesapeake Bay using purge and trap sampling with a tin-selective detector. *Environ. Sci. Technol.*, **16**, 110–19.

Kilby, G. W. and Batley, G. E., 1993. Chemical indicators of sediment chronology. *Aust. J. Mar. Freshwater Res.*, **44**, 635–47.

King, N., Miller, M. and de Mora, S., 1989. Tributyltin levels in sea water, sediment and selected marine species in coastal Northland and Auckland, New Zealand. *N. Z. J. Mar. Freshwater Res.*, **23**, 287–94.

Langston, W. J., Burt, G. R. and Zhou Mingjiang, 1987. Tin and organotin in water, sediments, and benthic organisms of Poole Harbour. *Mar. Pollut. Bull.*, **18**, 634–9.

Lau, M. M., 1991. Tributyltin antifoulings: a threat to the Hong Kong marine environment. *Arch. Environ. Contam. Toxicol.*, **20**, 299–304.

Laughlin, R. B., Guard, H. E. and Coleman, W. M., 1986. Tributyltin in sea water:

164   G. Batley

speciation and octanol–water partition coefficient. *Environ. Sci. Technol.*, **20**, 201–4.

Laughlin, R. B. and Linden, O., 1985. Fate and effect of organotin compounds. *Ambio*, **14**, 88–94.

Lee, R. F., Valkirs, A. O. and Seligman, P. F., 1987. Fate of tributyltin in estuarine waters. In *Proceedings of the Organotin Symposium, Oceans 87*, Marine Technology Society, Washington, DC, Vol. 4, pp. 1411–15.

Lee, R. F., Valkirs, A. O. and Seligman, P. F., 1989. Importance of microalgae in the biodegradation of tributyltin in estuarine waters. *Environ. Sci. Technol.*, **23**, 1515–18.

Maguire, R. J., 1987. Environmental aspects of tributyltin. *App. Organometall. Chem.*, **1**, 475–98.

Maguire, R. J., Carey, J. H. and Hale, E. J., 1983. Degradation of tri-*n*-butyltin species in water. *J. Agric. Food Chem.*, **31**, 1060–5.

Maguire, R. J. and Tkacz, R. J., 1987. Concentration of tributyltin in the surface microlayer of natural waters. *Water Poll. Res. J. Canada*, **22**, 227–33.

Matthias, C. L., Bellama, J. M., Olson, G. J. and Brinckman, F. E., 1986. Comprehensive method for determination of aquatic butyltin and butylmethyltin species at ultratrace levels using simultaneous hydridization/extraction with gas chromatography-flame photometric detection. *Environ. Sci. Technol.*, **20**, 609–15.

Matthias, C. L., Bushong, S. J., Hall, L. W., Bellama, J. M. and Brinckman, F. E., 1988. Simultaneous butyltin determinations in the microlayer, water column and sediment of a northern Chesapeake Bay marina and receiving systems. *App. Organometall. Chem.*, **2**, 547–52.

Muller, M. D., Renberg, L. and Rippen, G., 1989. Tributyltin in the environment – sources, fate and determination: an assessment of present status and research needs. *Chemosphere*, **18**, 2015–42.

Olson, G. J. and Brinckman, F. E., 1986. Biodegradation of TBT by Chesapeake Bay microorganisms. In *Proceedings of the Organotin Symposium, Oceans 86*, Marine Technology Society, Washington, DC, Vol. 4, pp. 1196–201.

Quevauviller, P., Lavigne, R., Pinel, R. and Astruc, M., 1988. Organotin compounds in intertidal sediments of the Sado Estuary and mussels from the adjacent coastal area, Portugal. In *Heavy Metals in the Hydrological Cycle*, M. Astruc and J. N. Lester (eds), Selper, London, pp. 425–32.

Randall, L. and Weber, J. H., 1986. Adsorptive behaviour of butyltin compounds under simulated estuarine conditions. *Sci. Total Environ.*, **57**, 191–203.

Salomons, W. and Förstner, U., 1984. *Metals in the Hydrocycle*, Springer-Verlag, New York, p. 486.

Schatzberg, P., 1987. Organic antifouling hull paints and the U.S. Navy, a historical perspective. In *Proceedings of the Organotin Symposium, Oceans 87*, Marine Technology Society, Washington, DC, Vol. 4, pp. 1324–33.

Seligman, P. F., Grovhoug, J. G. and Richter, K. E., 1986a. Measurement of butyltins San Diego Bay, CA: a monitoring strategy. In *Proceedings of the Organotin Symposium, Oceans 86*, Marine Technology Society, Washington, DC, Vol. 4, pp. 1289–96.

Seligman, P. F., Valkirs, A. O. and Lee, R. F., 1986b. Degradation of tributyltin in San Diego Bay, California, waters. *Environ. Sci. Technol.*, **20**, 1229–35.

Seligman, P. F., Valkirs, A. O. and Lee, R. F., 1986c. Degradation of tributyltin in marine and estuarine waters. In *Proceedings of the Organotin Symposium, Oceans 86*, Marine Technology Society, Washington, DC, Vol. 4, pp. 1189–95.

Seligman, P. F., Adema, C. M., Stang, P. M., Valkirs, A. O. and Grouvhoug, J. G., 1987. Monitoring and prediction of tributyltin in the Elizabeth River and Hampton Roads, Virginia. In *Proceedings of the Organotin Symposium, Oceans 87*, Marine Technology Society, Washington, DC, Vol. 4, pp. 1357–63.

Seligman, P. F., Grovhoug, J. G., Valkirs, A. O., Stang, P. M., Fransham, R., Stallard, M. O., Davidson, B. and Lee, R. F., 1989. Distribution and fate of tributyltin in the United States marine environment. *App. Organometall. Chem.*, **3**, 31–47.

Stang, P. M. and Seligman, P. F. 1986. Distribution and fate of butyltin compounds in the sediment of San Diego Bay. In *Proceedings of the Organotin Symposium, Oceans 86*, Marine Technology Society, Washington, DC, Vol. 4, pp. 1256–61.

Stang, P. M. and Seligman, P. F., 1987. *In situ* adsorption and desorption of butyltin compounds from Pearl harbour, Hawaii sediment. In *Proceedings of the Organotin Symposium, Oceans 87*, Marine Technology Society, Washington, DC, Vol. 4, pp. 1386–91.

Stang, P. M., Lee, R. F. and Seligman, P. F., 1992. Evidence for rapid, nonbiological degradation of tributyltin compounds in autoclaved and heat-treated fine-grained sediments. *Environ. Sci. Technol.*, **26**, 1382–7.

Stewart, C. and de Mora, S. J., 1990. A review of the degradation of tri(*n*-butyl)tin in the marine environment. *Environ. Technol.*, **11**, 565–70.

Stewart, C. and de Mora, S. J., 1992. Elevated tri(*n*-butyl)tin concentrations in shellfish and sediments from Suva Harbour, Fiji. *App. Organometall. Chem.*, **6**, 507–12.

Takahashi, K., Yoshino, T. and Ohyagi, Y., 1987. Photodegradation of tributyltin chloride in sea water by ultraviolet light. *Chem. Soc. Japan*, 181–5.

Unger, M. A., MacIntyre, W. G. and Huggett, R. J., 1987. Equilibrium adsorption of butyltin chloride by Chesapeake Bay sediments. In *Proceedings of the Organotin Symposium, Oceans 87*, Marine Technology Society, Washington, DC, Vol. 4, pp. 1381–5.

Unger, M. A., MacIntyre, W. G. and Huggett, R. J., 1988. Sorption behaviour of tributyltin on estuarine and freshwater sediments. *Environ. Toxicol. Chem.*, **7**, 907–15.

Valkirs, A. O., Seligman, P. F. and Lee, R. F., 1986a. Butyltin partitioning in marine waters and sediments. In *Proceedings of the Organotin Symposium, Oceans 86*, Marine Technology Society, Washington, DC, Vol. 4, pp. 1165–70.

Valkirs, A. O., Seligman, P. F., Stang, P. M., Homer, V., Lieberman, S. H., Vafa, G. and Dooley, C. A. 1986b. Measurement of butyltin compounds in San Diego Bay. *Mar. Pollut. Bull.*, **17**, 319–24.

Valkirs, A. O., Stallard, M. O. and Seligman, P. F., 1987. Butyltin partitioning in marine waters. 1987. In *Proceedings of the Organotins Symposium, Oceans 87*, Marine Technology Society, Washington, DC, Vol. 4, pp. 1375–80.

Waldock, M. J., Waite, M. E. and Thain, J. E., 1988. Inputs of TBT to the marine environment from shipping activity in the UK. *Environ. Technol. Lett.*, **9**, 999–1010.

Watanabe, N., Sakai, S. and Takatsuki, H., 1992. Examination for degradation paths of butyltin compounds in natural waters. *Water Sci. Tech.*, **25**, 117–24.

Weber, J. H., Han, J. S. and François, R., 1988. Speciation of methyl- and butyltin compounds in compartments of the Great Bay Estuary. In *Heavy Metals in the*

Hydrological Cycle, M. Astruc and J. N. Lester (eds), Selper, London, pp. 395–400.

Wulf, R. G. and Byington, K. H., 1975. On the structure–activity relationships and mechanism of organotin induced, nonenergy dependent swelling of liver mitochondria. *Arch. Biochem. Biophys.*, **167**, 176–85.

6

○ ○ ○ ○ ○ ○ ○ ○ ○ ○ ○ ○ ○ ○ ○ ○ ○ ○ ○ ○

Biological effects of tributyltin on marine organisms

Claude Alzieu

A number of marine ecotoxicology studies devoted to TBT were initiated in the early 1980s following the harmful effects observed in oyster-growing areas, initially in France and later in Great Britain. This background implies that most of the data relate to mollusc shellfish, regarded as target species.

6.1 Bioaccumulation and metabolism

The presence of alkyl groups on the tin atom of TBT leads to an increase in its lipophilic properties, and thereby in an increased capacity to become bioaccumulated in living organisms. Theoretically, the octanol/water partition coefficient (K_{ow}) provides the possibility of calculating the bioconcentration factor (BCF).

In the case of TBTO, Laughlin, French & Guard (1986) have shown that for salinity values ranging from 0 to 25 psu, K_{ow} values vary between 7000 and 6000 respectively, while for higher salinity they range between 5000 and 6300. The above values are markedly higher than the value of 1550 found by Maguire, Carey & Hale (1983) in distilled water. Such disparity between K_{ow} values leads to considerable uncertainties regarding the bioconcentration factor (BCF) calculated on the basis of various correlation equations. Thus, the equation of Veith, Defoe & Bergstedt (1979) ($\log BCF = -0.70 + 0.85 \log K_{ow}$) would lead to a TBTO bioconcentration in freshwater fish ranging from approximately 100 ($K_{ow} = 1550$) to 370 ($K_{ow} = 7000$). Based on the equation of Mackay (1982), the BCF would range between 400 and 500.

The current state of the art fails to provide any possibility of assessing the potentials of butyltin bioconcentration on the basis of the K_{ow} coefficient and of the correlation equations commonly applied to organic pollutants. The bioconcentration factors thus calculated (100–500) seem low in comparison with certain high concentrations encountered in the environment and should be compared with experimental data.

6.1.1 Microorganisms

Blair *et al.* (1982) studied the TBTO accumulation in bacteria by isolates originating from sediments collected in contaminated areas and resistant to concentrations of $8 \, mg \, TBTO$-$Sn \, l^{-1}$. All isolates fixed considerable TBT quantities within less than one hour of initial contact. Bioconcentration factors for all isolates ranged between 356 and 855 in a glucose-free culture medium, and between 381 and 1039 in a glucose-enriched medium. In comparison, the *BCF*s for a *Pseudomonas* strain were 438 and 487, respectively, in glucose-free and glucose-enriched media. The accumulation mechanism would appear to be a process of chemical adsorption which does not induce any TBT metabolism. These results were later confirmed by Blair *et al.* (1988) who showed that estuarine microorganism biofilms could accumulate up to $120 \, ng \, TBT$-Sn per milligram of dry biomass, corresponding to a bioconcentration factor exceeding 7000. These data would seem to indicate that certain estuarine bacteria could constitute a tributyltin storage system likely to facilitate their integration into the food chains.

Data on bioconcentration by phytoplankton provide significant indications and would deserve further confirmation. According to unpublished data of Cobet, as cited by Laughlin *et al.* (1986), the *BCF* for the diatom *Isochrysis galbana* would be approximately 5500. This value appears notably lower than the value given by Maguire, Wong & Rhamey (1984) for the freshwater diatom *Ankistrodesmus falcatus*, i.e. approximately 30 000. Furthermore, according to these authors, this diatom would appear capable of debutylating TBT down to the elemental Sn stage within 14 days of exposure.

6.1.2 Invertebrates

The kinetics of TBT bioaccumulation are best known essentially in molluscs and crustaceans. Moore, Dillon & Suedel (1991) have shown that, following a 10 week exposure to concentrations ranging from 4 to $180 \, ng \, TBTCl$-$Sn \, l^{-1}$, the polychaete *Neanthes arenaceodentata* accumulates considerable TBT quantities of approximately $6 \, \mu g \, TBT$-$Sn \, g^{-1}$ dry weight. The polychaete tissues also contained significant quantities of dibutyltin ($1.5 \, \mu g \, Sn \, g^{-1}$) and monobutyltin ($2.6 \, \mu g \, Sn \, g^{-1}$) derived from TBT metabolism.

The kinetics of bioaccumulation and excretion were studied experimentally in several bivalve species: oysters, mussels, and scallops. Waldock, Thain & Miller (1983) studied TBTO bioaccumulation in the Pacific oyster *Crassostrea gigas* and in the flat oyster *Ostrea edulis* at two concentrations, i.e. 0.06 and $0.5\,\mu g\,Sn\,l^{-1}$. Bioaccumulation appears more significant in *C. gigas* than in *O. edulis*, and in both species the bioaccumulated quantities are greater at an exposure of $0.06\,\mu g\,Sn\,l^{-1}$. BCF values corresponding to low and high concentrations were as follows: 6000 and 2000 for *C. gigas*, 1500 and 1000 for *O. edulis*.

The uptake and excretion curves reveal that, in both species, the bioaccumulation kinetics are faster than in the excretion phase: accumulation reaches a plateau after one week, whereas 50% only of the bioaccumulated load is removed after 23 days of depuration. Lee (1986) has shown that TBT retention in oyster tissues is linked to the extremely low mixed functions oxygenase activity (MFO) in oyster, as compared with crustaceans or fish.

The above observations were confirmed by studies carried out on oysters *C. gigas* by Davies, Drinkwater & McKie (1988) and on scallops *Pecten maximus* by Davies, McKie & Paul (1986) placed in nets treated with a TBT-based antifoulant. Both bivalves quickly accumulated the TBT released. Once re-immersed in clean water for 31 weeks, depuration was greater in the oysters (90%) than in the scallops (20–30%).

Laughlin *et al.* (1986) studied short term (48 h) and long term (47 days) TBTO bioconcentrations in mussels *Mytilus edulis*. Accumulation occurs very rapidly during the first 90 min of exposure and it is twice as high in gills than in viscera. This difference between both organs decreases when the TBT uptake originates from contaminated phytoplankton. By the end of the long term test, branchiae and viscera presented comparable contaminations; however, the TBT content in water and in the tissues had not reached a state of equilibrium. Bioconcentration factors calculated on the basis of maximum content in tissues were approximately 5000 for the water and under 2 for the phytoplankton fed to the molluscs. Half-life of accumulated TBT was estimated at 14 days once mussels were re-immersed in an uncontaminated natural environment. Based on the above data on bivalves and TBTO, it appears that the TBT bioconcentration factor in molluscs is approximately 1000–6000 in relation to the water, i.e. values significantly higher than the predictive figures derived from the K_{ow} coefficients and from the correlation equations. Furthermore, the TBTO excretion kinetics are slow, particularly in oysters, due to the limited activity of the detoxification systems (MFO) in these species.

In the gastropod *Nucella lapillus*, Bryan *et al.* (1987) found a bioaccumulation factor of approximately 100 000 for TBTO concentrations in water ranging between 0.6 and 7.5 ng Sn l^{-1}. Bioconcentration is lower (30 000) at a concentration of 42.8 ng Sn l^{-1}, due to TBT metabolism into DBT in the tissues.

Available data on crustaceans were supplied by Evans & Laughlin (1984) who studied accumulation of ^{14}C-labelled TBTO in the crab *Rhithropanopeus harrisii* with exposure either from water or from the food supply. After 4 days of feeding with a contaminated diet, bioconcentration factors (*BCF*) were approximately 4400 in the hepatopancreas, and ranged between 500 and 1300 in other tissues. In the case of contamination via the water, the *BCF* values were 10–30 times lower, and the following tissues had accumulated TBTO in decreasing order: gills, exoskeleton and hepatopancreas. The authors suggested the possible existence of TBTO biomagnification processes in the case of contamination via food. However, it is essential to take into consideration the fact that, while a state of equilibrium was reached between concentrations in water and in tissues during the water exposure test, this was not the case when contamination occurred via the food. In addition, the ^{14}C-labelling technique used by Evans & Laughlin (1984) fails to take into account the possibility of TBTO metabolism by crustaceans. In this respect, Lee (1986) demonstrated that microsome preparations from the stomach of the crab *Callinectes sapidus* are capable of metabolising TBTO into (hydroxybutyl)dibutyltin, dibutyltin and monobutyltin. Further studies on bioconcentration and butyltin metabolism in crustaceans therefore appear necessary to determine whether or not there is a possibility of biomagnification of these substances in the food chains.

6.1.3 Vertebrates

Owing to their capacity to metabolise a large number of biocides, fish have been used as models in studies of accumulation/excretion kinetics and of TBT metabolic pathways in aquatic vertebrates.

Ward *et al.* (1981) subjected marine fish of the species *Cyprinodon variegatus* to a 58 days' exposure at a ^{14}C-labelled TBTO concentration of 0.8 μg Sn l^{-1}, followed with 28 days in uncontaminated sea water. The bioconcentration factors calculated for maximum tissue concentration on day 58 were 2600 for whole fish, 4580 in viscera, 1180 in muscles and 2120 in the remaining tissues. It should be noted that, by the end of the exposure period, no equilibrium was reached between water and tissue concentrations. In the depuration phase, 52% of the TBTO uptake was excreted after 7 days and 74% after 28 days. The presence of high proportions of DBT and MBT in the liver and viscera would suggest the possibility for this species to metabolise the accumulated TBT. Lee (1986) confirmed the ability of certain fish to metabolise TBTO in his studies on the perciform *Leistomus canthurus*.

In the mullet *Liza aurata*, Bressa *et al.* (1984) reported bioconcentration factors ranging from 3 in muscles to 19.6 in kidneys. These values, calculated at the point of equilibrium following a 60 days' TBTO exposure ($5 \mu g \, Sn \, l^{-1}$), are two orders of magnitude higher than those recorded by Ward *et al.* (1981).

The above values are notably lower than those found by Yamada & Takayanagi (1992) in three fish species, *Pagrus major*, *Mugil cephalus* and *Rudarius ercodes*, subjected to constant TBTO and TPTCl (triphenyltin chloride) concentrations for eight weeks in a flow-through system. *BCF* values ranged between 2400 and 11 000 depending on the species and on the concentrations tested. *Pagrus major* stands out from both other species with higher *BCF* values for TBTO, while the *BCF*s were of the same order of magnitude in the case of TPTCl: 3100 and 4100 respectively for *P. major* and for *R. ercodes*. The quantities accumulated in the organs were higher than in the muscles and no evidence of any correlation with lipid content could be found. TBTO excretion rates were lower in *P. major*, with a value of 0.024 day^{-1}.

TBT bioconcentration was studied *in situ* by Short & Thrower (1986) and Davies & McKie (1987) in salmonids raised in cages treated with TBT-base antifoulants. The latter two authors exposed young *Salmo salar* salmon to TBT concentrations of 40, 120 and 400 ng Sn l^{-1} for 26 days. The highest concentration was found in the liver ($BCF = 3.9$ at $40 \, ng \, l^{-1}$, and 1.62 at $400 \, ng \, l^{-1}$) and in the kidneys ($BCF = 0.9$ at $40 \, ng \, l^{-1}$, and 0.6 at $400 \, ng \, l^{-1}$). The state of equilibrium between contents in water and in tissues did not appear to be reached.

It has been shown that, in terrestrial mammals, TBT is metabolised by the hepatic microsomes in the presence of NADPH. The mechanism involves hydroxylation of the carbon–hydrogen bond located α or β to the tin atom. The hydroxylated α metabolite is unstable and rapidly transformed into dibutyltin and 1-butanol which in turn is transformed into butane. *In vitro* studies (Lee, 1986) conducted from hepatic microsomes confirmed the existence in marine fish of metabolic pathways comparable to those existing in mammals and implicating the hepatic cytochrome P450-dependent monoxygenase system. Fent & Stegeman (1991) recently showed *in vitro* that the accumulated TBT can interact with the monoxygenase system by destroying the native enzyme and inhibiting its activity.

The same authors (Fent & Stegeman, 1993) confirmed the above results *in vivo* by injecting high TBTCl doses (1.2–5.9 mg Sn kg^{-1}) to scuds *Stenotomus chrysops*. The hepatic microsomes analysed after 24 h of experimentation revealed, for all doses, a significant transformation of the cytochrome P450 into its degraded form as cytochrome P420, together with a decrease in the ethoxyresorufin O-deethylase (EROD) activity. Furthermore, the presence of larger quantities of the dibutyltin metabolite in the microsomes exposed to low doses confirms that the cytochrome P450-dependent system can be inactivated by TBT. The authors conclude that, considering the overall contamination of the environment by TBT, fish can lose their ability to metabolise other pollutants such as PAHs or PCBs due to the inhibition of the cytochrome P450-dependent system by TBT.

The above series of data demonstrate that, although the ability of fish to metabolise TBT is well established, large quantities of TBT can be accumulated in the liver. The considerable differences found experimentally in the resulting *BCF* values (1–4580) may be attributable either to specific features or to the inhibition of metabolic mechanisms, or to experimental conditions.

6.1.4 Conclusions

It may be concluded that, unlike inorganic tin which does not accumulate in living organisms, tributyltin is adsorbed or bioconcentrated by bacteria, phytoplankton, molluscs, crustaceans and fish. The bioconcentration factors shown in Table 6.1 present considerable variations (from < 1 to 30 000), and fail to provide the possibility of determining any range of values per group of species. A number of indications would tend to show that bioconcentration is greater in the crab when contamination occurs via the food, suggesting the possibility of biomagnification through the food chain. However, this assumption is in contradiction with the findings of the same researchers examining mussels for which the *BCF* values are lower when accumulation occurs through phytoplankton. When exposure takes place directly, the highest *BCF* values are found at the lowest TBT concentrations in the water.

It is well established that crustaceans and fish possess enzyme systems (MFO) capable of metabolising TBT into (hydroxybutyl)dibutyltin, dibutyltin and monobutyltin, although the fate of these metabolites still remains unknown at this time. However, in highly contaminated areas, TBT can inactivate the enzymatic mechanisms involved in the metabolism

Table 6.1 TBT bioconcentration factors (BCF) in tissues of marine organisms (concentrations in μg TBT-Sn l^{-1})

Organisms	Exposure	BCF	References
Microorganisms			
Estuarine bacteria	water	356–1039	Blair et al., 1982
Pseudomonas 244	water	438–487	
Phytoplankton			
Isochrysis galbana	water	5500	Laughlin et al., 1986
Ankistrodesmus falcata	water (8)	30000	Maguire et al., 1984
	water (16)	860	
Molluscs			
Crassostrea gigas	water (0.06)	6000	Waldock et al., 1983
Ostreas edulis	water (0.5)	2000	
	water (0.06)	1500	
	water (0.5)	1000	
Mytilus edulis	water (2)	5000	Laughlin et al., 1986
	phytoplankton	<2	
Crustaceans			
Rhithropanopeus harrisii	food	500–4400	Evans & Laughlin, 1984
Fish			
Cyprinodon variegatus	water (0.8)	118–4580	Ward et al., 1981
Liza aurata	water (5)	3–19.6	Bressa et al., 1984
Salmo salar	water (0.04)	0.9 (kidney) 3.9 (liver)	Davies & McKie, 1987
	water (0.4)	0.6 (kidney) 1.6 (liver)	
Pagrus major			
Rudarius ercodes	water	9400–11000	Yamada & Takayanagi, 1992
Mugil cephalus			

of TBT and of many xenobiotics. Comparatively, in molluscs, the absence of any cytochrome P450-dependent system may explain their ability to bioaccumulate TBT.

6.2   Contamination levels

The first data on TBT contamination levels in marine organisms were obtained in the early 1980s, based on overall analytical methods to derive the fraction of organic tin accumulated in the tissues. Alzieu *et al.* (1981) used this method to show that tin was preferentially accumulated by the oyster *Crassostrea gigas* in the digestive gland and in the gills. The highest concentrations were encountered in the Bay of Arcachon (France) in the vicinity of marinas and pleasure craft mooring areas. The highest content recorded at that time reached 7.03 and 17.37 mg Sn kg^{-1} dry weight in the digestive glands and in the gills respectively. Mean concentrations ranged from 0.9 to <0.15 mg Sn kg^{-1} dry weight between 1982 and 1985 following the ban on TBT use in antifouling paints as of January 1982. During the same period, Waldock *et al.* (1983) analysed oysters *Ostrea edulis* and *Crassostrea gigas* collected along the Essex coastline (UK). They showed that the TBT concentrations found in *C. gigas* were considerably higher, with values of 0.8 and 3.4 mg TBT-Sn kg^{-1} dry weight respectively. For *C. gigas* in New Zealand, King, Miller & de Mora (1989) reported a maximum concentration of 2.24 mg TBT-Sn kg^{-1} dry weight. Concentrations as high as 3.2 mg TBT-Sn kg^{-1} dry weight were found in mangrove oysters (*Crassostrea mordax*) collected in Suva Harbour, Fiji (Stewart & de Mora, 1992).

The data compiled over two years, between 1988 and 1990, by the Mussel Watch monitoring network along the west and east coasts of the United States (Uhler *et al.*, 1993) indicate that all samples of mussels *M. edulis*, *M. californianus* and oysters *Ostrea sandvincensis*, *Crassostrea gigas* analysed contained TBT. Overall figures for all molluscs show values ranging from 4 to 1600 µg TBT-Sn kg^{-1} dry weight on the west coast and from 4 to 2000 µg TBT-Sn kg^{-1} dry weight on the east coast. Approximately 40% of all samples from both coasts contained less than 0.2 mg TBT + DBT + MBT kg^{-1} dry weight (total butyltins). Approximately 7% of east coast samples and 12% of west coast samples presented total butyltins content exceeding 1 mg kg^{-1}. The most contaminated sites were encountered in Florida in the vicinity of marinas and mooring areas, in the harbours of San Diego (California) and of Honolulu (Hawaii). Dibutyltin concentrations varied between undetectable levels and

$800\,\mu g$ DBT-Sn kg^{-1} on the east coast and $1500\,\mu g$ DBT-Sn kg^{-1} on the west coast. In some cases, the DBT content represented up to 50% of total butyltins. Monobutyltin was only found occasionally and did not usually exceed 20% of the total butyltin load. Maximum concentrations reached 220 and $480\,\mu g$ TBT-Sn kg^{-1} on the east and west coast, respectively. A comparison with results from the same sites in 1987 reveals a slightly decreasing contamination in certain areas while other sites remained stable.

Batley, Scammell & Brockbank (1992) analysed oysters *Saccostrea commercialis* from George River on the Australian east coast and, in samples collected in 1988, they found concentrations ranging from 2 to $112\,\mu g$ TBT-Sn kg^{-1}. Contamination levels were found to have sharply decreased in 1991 with values ranging between <0.2 and $20\,\mu g$ TBT-Sn kg^{-1} following the ban on TBT use. Table 6.2 presents the distribution of TBT and its metabolites as measured throughout the various oyster organs by these authors. It indicates that the highest concentrations were found in the mantle for TBT, in the foot for DBT and in labial palpi for MBT. Total butyltin load (TBT + DBT + MBT) decreases in the following order: foot > mantle > labial palps > visceral mass > adductor muscle > gills.

Very high TBT burdens were reported in the gastropod *Nucella lapillus* (dogwhelk) living in the harbours and estuaries along the English coastline:

0.13–0.63 mg TBT-Sn kg^{-1} dry weight in enclosed waters of Plymouth and Torbay (Gibbs & Bryan, 1987),

0.37–0.79 mg TBT-Sn kg^{-1} dry weight in Fal Estuary located on the south-western coast of England (Bryan *et al.*, 1987).

Comparatively, TBT concentrations found in dogwhelks collected in Mull Island off the Scottish coast were below the detection threshold, i.e. 0.03 mg TBT-Sn kg^{-1}. In the Shetland Islands, dogwhelks and scallops collected in the vicinity of an oil terminal presented comparable

Table 6.2 *Distribution of butyltin species ($\mu g\,Sn\,kg^{-1}$) in oyster* Saccostrea commercialis *organs (from Batley* et al., *1992)*

Organ	TBT	DBT	MBT	ΣBT
Gills	190	167	5	362
Mantle	360	470	33	863
Adductor muscle	103	202	69	374
Foot	270	710	18	998
Visceral mass	116	430	96	642
Labial palps	128	500	127	755
Whole oyster	59	167	24	250

contamination levels, 0.16 and 0.23 mg TBT-Sn kg^{-1} respectively, while contamination was no longer detectable (<0.03 mg Sn kg^{-1}) outside the terminal waters (Bailey & Davies, 1988).

Monitoring programs set up along the Japanese coasts from 1985 to 1987 revealed contamination levels ranging from non-detectable to 0.11 mg TBT-Sn kg^{-1} wet weight in mussels *Mytilus edulis*, and from non-detectable to 0.68 mg TBT-Sn kg^{-1} wet weight in various fish species (EAJ, 1988).

Maguire *et al.* (1986), in their studies on fish from the Canadian coasts, reported levels of 0.24 mg TBT-Sn kg^{-1} in the herring *Clupea harengus* in the Bay of Vancouver. Humphrey & Hope (1987) found a maximum content of 4.4 mg TBT-Sn kg^{-1} in the finfish. Contamination can reach very high levels in salmon raised in cages treated with TBT-based antifoulants, as shown by Short & Thrower (1986) who reported TBT burdens ranging between 0.11 and 0.36 mg Sn kg^{-1} wet weight.

Table 6.3 summarises some data on TBT contamination levels in marine organisms. Data collected on molluscs, particularly in the context of the Mussel Watch program, show that contamination is localised in the vicinity of input sources: marinas, merchant ports, heavy maritime traffic areas and TBT-treated breeding cages. Outside of the above areas, the presence of TBT in marine organisms is most of the time undetectable. Furthermore, the ban regulations implemented to reduce TBT inputs (as early as 1982 in France) have led to significant decreases in the concerned areas. Generalised bans on the use of TBT-based antifouling paints for pleasure crafts should lead to a significant reduction of TBT contamination in molluscs and fish.

6.3   Biological effects

Organotin marine ecotoxicology is a recent concern linked to the dispersive use of biocides. Available data concern primarily trialkyl and triaryltins. Due to the specific aspects of its biological effects, each single trialkyltin constitutes a chemical compound of special interest from a toxicological standpoint. The TBT lethal and sublethal biological effects shall therefore be considered in this chapter in comparison with other trialkyltins.

6.3.1   Toxicity mechanisms

Aldridge & Street (1964) had shown that the action of trialkyltins at micromolar concentrations stimulates the adenosine triphosphate activity, inhibits oxidative phosphorylation, thereby causing malformations in the mitochondrial membranes. TBT is known to be a cytochrome P450 inhibitor and is capable of producing complexes by binding with the nitrogen or sulphur atoms of certain amino acids.

In the case of trialkyltins, the nature of the alkyl group determines the extent of their toxic effects on a given zoological phylum. Figure 6.1 (Evans & Smith, 1975) shows that tributyltin presents superior bactericidal and anti-cryptogamic properties in comparison with its methyl, ethyl and propyl homologues.

Due to the multiplicity of structures with co-ordination numbers from 0 to 4 and to the diversity of organic radicals fixed on the tin atom, organotin compounds present highly favourable conditions for the design and application of models simulating the correlation between molecular structure and biological activity, also called QSAR (Quantitative Structure Activity Relationship). Simple models merely take into account the lipophilic and hydrophilic characteristics. Thus, Wong et $al.$ (1982) discovered a relationship between log K_{ow} and growth inhibition in unicellular algae (IC_{50}) for certain trialkyltins. However, this relation is applicable only to highly lipophilic derivatives such as triphenyl-, tributyl- and tripropyltin.

Hansch's equation provides a better adequacy by correlating the biological effect (BE) with parameters describing the hydrophobicity (π), electron charge (σ) and steric factor (E_s) of organic substances, in accordance with the following law:

$$BE = K_1\pi + K_2\sigma + K_3E_s + K_4$$

Laughlin et $al.$ (1985) applied the above equation to the LC_{50} experimental values of 15 diorganotins (R_2SnX_2) and triorganotins (R_3SnX) obtained using crab $Rhithropanopeus$ $harrisii$ larvae. These authors show that toxicity is strongly correlated with parameter π and has no relation with parameter σ. This conclusion is confirmed by the existence of a linear relation between the sum of Hansch fragment constants (Fr) for groups R and the LC_{50} experimental values, as illustrated in Figure 6.2. The resulting relation, for which ($r^2 = 0.82$) can be expressed as:

$$LC_{50} = 0.908Fr + 10.62$$

It is possible with this model to calculate the toxicity on a scale of four orders of magnitude, taking into account the number and structure of the organic groups fixed on the tin atom. These results are in good agreement with those reported by Vighi & Calamari (1985) based on EC_{50} values

Table 6.3 Butyltin levels in marine organisms expressed as mg Sn kg^{-1} wet weight

Organisms	Concentrations		References
Algae			
Great Bay (USA)	nd–20 (TBT), nd (DBT), nd–2.4 (MBT)		Donard, Rapsomanikis & Weber, 1986
Oyster			
Burnham (UK)	<0.09–0.8 (TBT)	O. edulis	Waldock et al., 1983
Plagesham (UK)	<0.14–0.25 (TBT)	O. edulis	
Creeksea (UK)	1.4–3.45 (TBT)	C. gigas	
Arachon (F)	<0.15–0.9 (TOT)*	C. gigas	Alzieu et al., 1986
Tamaki (NZ)	<0.28–2.24 (TBT)*	C. gigas	King et al., 1989
Suva Harbour (Fiji)	<0.63–3.18 (TBT)*	C. mordax	Stewart & de Mora, 1992
Georges River (Aus)	0.002–0.11 (TBT) 1988	S. Commercialis	Batley et al., 1992
	<0.002–0.02 (TBT) 1991		
USA East Coast	0.004–1.61 (TBT)	C. virginica	Uhler et al., 1993
	0.005–0.76 (DBT)		
	nd–0.22 (MBT)		
USA West Coast	2.1 & 0.39 (TBT)	O. sandvincensis	
	1.5 & 0.41 (DBT)		
	0.48 & 0.04 (MBT)		
Mussels			
Sweden	<36–1480 (TBT)*	M. edulis	Linden et al., 1987
	40.5–496 (DBT)*		
	60.3–429 (MBT)*		
USA East Coast	0.004–0.48 (TBT)	M. edulis	Uhler et al., 1993
	0.004–0.29 (DBT)		
	nd–0.09 (MBT)		

Location	Species	Values	Reference
USA West Coast	*M. edulis*	0.004–0.55 (TBT) 0.004–0.37 (DBT) nd–0.21 (MBT)	
	M. californianus	0.004–0.19 (TBT) 0.005–0.05 (DBT) nd–0.01 (MBT)	
Tokyo Bay (Japan)	*M. edulis*	0.008–0.1 (TBT) 0.02–0.27 (DBT) 0.01–0.08 (MBT)	Higashiyama *et al.*, 1991
Dogwhelks Plymouth (UK) Fal estuary (UK)	*N. lapillus*	0.13–0.63 TBT* 0.37–0.79 (TBT)*	Gibbs & Bryan, 1987 Bryan *et al.*, 1987
Fish North Sea Fish farm	Herring Salmon	<20–236 (TBT)* 112–360 (TBT)	Linden *et al.*, 1987 Short & Thrower, 1986

TOT = total organotin; * = dry weight basis.

obtained in *Daphnia magna* for 16 structurally different organotins and on a calculation of the three parameters of the Hansch equation. The following relationship:

$$\log 1/EC_{50} = 0.207 \log K_{ow} + 0.513\, pKa + 0.206 + 0.824$$

where $pKa = 2\,pH - \log C$ represents molar conductivity, provides a prediction of the toxicity of diorganotin and triorganotin compounds. It is not applicable to tetraorganotins, thereby confirming the specific structure of its derivatives.

6.3.2   Lethal and sublethal toxicity

With respect principally to TBT, data on lethal and sublethal butyltin toxicity to aquatic organisms have recently been compiled by Hall & Pinkney (1985), Rexrode (1987), Waldock, Thain & Waite (1987) and

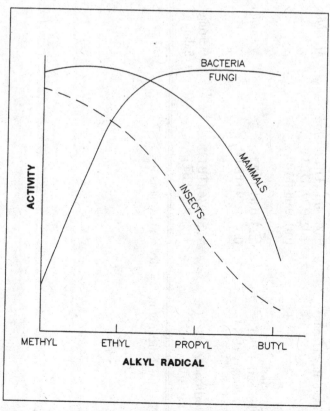

Figure 6.1 Toxicity of trialkyltin acetates according to nature of alkyl group (from Evans & Smith, 1975).

IPCS (1990). Table 6.4 summarises the most significant toxic effects assessed for the major groups of marine organisms.

6.3.2.1 Microorganisms

Butyltins are well known for their disinfecting and bactericidal properties against Gram-negative bacteria: TBT benzoate is widely used in hospitals for that purpose. It is therefore important to find out the thresholds beyond which the activity of aquatic microorganisms can be inhibited. The data presented in Table 6.4 show that bacteriostatic thresholds for TBT halides vary by two orders of magnitude, i.e. from $10\,\mu g$ TBT-Sn l^{-1}

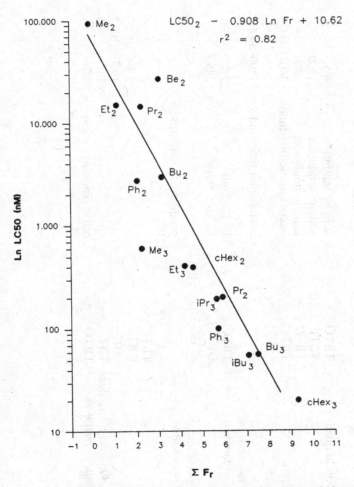

Figure 6.2 Correlation between of the sum of Hansch fragment constants (Σ *Fr*) and experimental LC_{50} values for crab *R. harrisii* larvae (from Laughlin *et al.*, 1985).

Table 6.4 *Lethal and sublethal effects of butyltin compounds for marine organisms (concentrations as μg Sn l⁻¹)*

Organisms	Butyltin	Effects	Concentration (μg Sn l^{-1})	References
Microorgamisms				
Activated sludges				
adapted	TBTO	no effect	380	Argaman, Hucks & Shelby, 1984
not adapted	TBTO	activity inhibition	9.5	Stein & Kuster *in* Argaman *et al.*, 1984
	TBTF	activity inhibition	1900	Argaman *et al.*, 1984
	TBTO	activity inhibition	3800	Polster & Halacka *in* Argaman *et al.*, 1984
Bacteria	TBTCl	bacteriostatic threshold	360–1800	
		bactericidal threshold	1800–3600	
Photobacterium phosphoreum	TBTCl	EC$_{50}$, 5 min	7.0	Dooley & Denis, 1987
	DBTCl$_2$	EC$_{50}$, 5 min	246.8	
Phytoplankton				
Ankistrodesmus falcatus		productivity inhibition IC$_{50}$		Wong *et al.*, 1982
	TBT		8	
	DBT		3.4×10^{-6}	
	MBT		16.7×10^{-6}	
Lac Ontario microalgae	TBT		1.2	
Dunaliella tertiolecta	TBTO	cell division reduction	3.8	Beaumont & Newman, 1986

Species	Compound	Endpoint	Value	Reference
Pavlova lutheri				
Skeletonema costatum	TBTO	death in 2 d	1.9	Walsh *et al.*, 1985
		Algistatic	0.04	
		LC_{50}, 48 h	5.7	Thain, 1983
		algistatic, 5 d	0.4–7	
Tetraselmis suesica	TBTO	algicide, 5 d	>7	
		algistatic, 5 d	224–400	
		algicide, 5 d	>400	
Zooplankton				
Acartia tonsa	TBTO	LC_{50}, 96 h	0.4 ± 0.08	U'Ren, 1983
		EC_{50}, 144 h	0.02	
		death from 6 d	0.01	
	TBTO	LC_{50}, 48 h	0.44	Bushong *et al.*, 1987
		18% spatfall reduction	0.004	Johansen & Mohlenberg, 1987
Brachionus plicatillis	TBT	LC_{50}, 24 h	30.4	Ambrogi, Saroglia & Scarano, 1982
Tishe furcata			34	
Artemia salina			50.8	
Nitocra spinipes	TBTO	LC(I), 96 h	0.8	Linden *et al.*, 1979
Eurytemora affinis	TBT	LC_{50}, 72 h	0.24	Bushong *et al.*, 1987
Daphnia magna	TBT Cl	phototaxis inversion, 96–144 h	1.8	Meador, 1986
		LC_{50}, 96–144 h	1.3–2.8	
	Alkyls and phenyls	EC_{50}, 24 h	4.7–49 000	Vighi & Calamari, 1985
Acanthomysis sculpta	TBT	LC_{50}, 96 h juveniles	0.17	Davidson, Valkirs & Seligman, 1986
		effects on growth	0.12	
		effects on reproduction	0.06	

Table 6.4 continued

Organisms	Butyltin	Effects	Concentration (μg Sn l^{-1})	References
Molluscs				
Crassostrea gigas				
Adults	TBTF paints	100% death in 1 month	0.76	Alzieu et al., 1980
	TBTF paints	30% death in 113 d	0.08	
	TBTO	50% death in 25 d	0.19	Héral et al., 1983
		LC$_{50}$, 48 h	684	Thain, 1983
	TBTO	LC$_{50}$, 96 h	110	
	TBTO	LC$_{50}$, 21 d	0.47	Waldock et al., 1983
	TBTO	LC$_{50}$, 48 d	0.09	Gendron, 1985
Juveniles	TBTO	growth stop	0.6	Thain & Waldock, 1983
		effect on:		
		O$_2$ consumption	> 0.02	Lawler & Aldrich, 1987
		feeding rate	> 0.007	
		hypoxie correction	> 0.004	
		calcification	< 0.004	
Larvae	TBTO	LC$_{50}$, 48 h	0.6	Thain, 1983
	TBT acetate	fecondity inhibition	35	His & Robert, 1983–5
		100% death in 8 d	0.17	
		no observable effect	0.007	
Crassostrea virginica				
Adults	TBT paints	LC$_{50}$, 30 d	1.0	Henderson, 1986
		condition index decrease	0.29	Valkirs, Davidson & Seligman, 1987
		no effect on condition index	0.02	
Juveniles	TBTO	effect on growth	0.76	Thain, 1986
		no growth effect	0.08	

Species / stage	Compound	Effect	Concentration	Reference
Ostreas edulis				
Adults	TBTO	LC_{50}, 48 h/96 h	>114/84	Thain, 1983
		no growth effect	0.1	Thain, 1986
		LC_{50}, 48 h	1.36	Thain & Waldock, 1983
Juveniles	TBTO	1 d larvae 50% reduction of growth in 20 d	0.02	
Mytilus edulis				
Adults	TBTO	LC_{50}, 48 h/96 h	114/14	Thain, 1983
		LC_{50}, 66 d	0.39	Valkirs et al., 1987
Juveniles	TBT	growth reduction in 7 d	>0.16	Stromgren & Bongard, 1987
Larvae	TBT	growth after 63 d	28	Salazar & Salazar, 1987
	TBTO	LC_{50}, 48 h	0.9	Thain, 1983
	TBTO	LC_{50}, 15 d	0.04	Beaumont & Budd, 1984
	TBTO	LC_{100}, 5 d/10 d	3.8/0.4	
	TBTO	no genotoxic effect		Dixon & Prosser, 1986
Mytilus galloprovincialis				
Larvae	TBT acetate	LC_{95}, 3 d	3.5	Robert & His, 1981
		LC_{70}, 5 d	1.7	
		LC_{25}, 9 d	1.0	
Mercenaria mercenaria				
Veligers	TBT	LC_{100}, 7 d/LC_{70}, 8 d	0.4/0.2	Laughlin, Pendoley & Gustafson, 1987
Post larvae		LC_{100}, 25 d/LC_{50}, 25 d	4.0/100	
Venerupis deccussata				
Spat	TBTO	no growth after 7 weeks	1.0	Thain & Waldock, 1986
		50% growth after 7 weeks	0.1	
Nucella lapillus	TBT	imposex	$<4 \times 10^{-4}$	Gibbs & Bryan, 1987

Table 6.4 *continued*

Organisms	Butyltin	Effects	Concentration (µg Sn l^{-1})	References
Fish				
Agonus cataphractus	TBTO	LC$_{50}$, 48 h/96 h	9.88/6.08	Thain, 1983
Brevoortia tyrannus	TBT	LC$_{50}$; 96 h, juveniles	1.7	Bushong *et al.*, 1987
	TBTO	avoiding reactions	2.1	Hall *et al.*, 1984
	TBT, Cl	no survival effects juveniles 28 d	0.2	Hall *et al.*, 1988
Cyprinodon variegatus	TBTO	LC$_{50}$, 21 d	0.36	Ward *et al.*, 1981
Dicentrarchus labrax	TBTO	LC$_{50}$, 96 h	26.3	Ambrogi *et al.*, 1982
Fundulus heteroclitus	TBTO	30% reduction in embryos survivance	1.1	Weis *et al.*, 1987a
Menidia beryllina (larvae)	TBT	LC$_{50}$, 72 h	1.8	Bushong *et al.*, 1987
	TBT, Cl	no survival effects larvae 28 d	0.2	Hall *et al.*, 1988
Morone saxitilis	TBTO	reactions avoiding	9.46	Hall *et al.*, 1984
Oncorhunchus mykiss	TBT	LC$_{50}$, 24 h/48 h	11.2/8.4	Alabaster, 1969
		no rheotaxis physiological effects	4.7	Chliamovitch & Kuhn, 1977
Solea solea	TBTO			
Adults		LC$_{50}$, 48 h/96 h	33.4/13.7	Thain, 1983
Larvae		LC$_{50}$, 48 h/96 h	3.2/0.8	
Crustaceans				
Carcinus maenas	TBTO			
Larvae		LC$_{50}$, 48 h/96 h	41.8/3.8	Thain, 1983

Species	Compound	Effect	Value	Reference
Crangon crangon				
Adults	TBTO	LC$_{50}$, 48 h/96 h	2.7/15.6	
Larvae	TBTO	LC$_{50}$, 48 h/96 h	2.5/0.6	
Gammarus oceanicus	TBTO/TBTF	survival reduction	0.1	Laughlin, Norlund & Leuder, 1984
Hemigrapsus nudus				
Larvae	TBTO	LC$_{100}$, 8 weeks	1.1	
(zoe)	TBTO	50% alive in 6.2 d	9.5	Laughlin & French, 1980
Homarus americanus (larvae)	TBTO	100% death in 24 h	7.6	
		growth decrease	3.8	Walsh, 1986
Palaemonetes pugio	TBTO	LC$_{50}$, 96 h	7.6	
Rhithropanopeus harrisii (zoe)	TBTO	survival effects/growth	>3.8/5.7	Laughlin, French & Guard, 1983
Uca pugilator	TBTS	survival effects/growth	>7.3/1.8	
	TBT	morphological anomalies	0.2	Weis *et al.*, 1987a,b

for activated sludge microorganisms to $4000\,\mu g\,TBT\text{-}Sn\,l^{-1}$ for similar sludges previously adapted to TBT degradation (Argaman *et al.*, 1984). Dooley & Denis (1987) studied the effects of 14 alkyltins, both saturated and unsaturated, on the bioluminescence of the Gram-negative bacterium *Photobacterium phosphoreum* (Microtox test). Based on concentrations leading to a 50% reduction in bioluminescence after a 5 min exposure (EC_{50}, 5 min), these authors showed that tributyltins are more toxic than tetraalkyltins, with EC_{50} values varying from 0.06 to $0.82\,\mu mol\,l^{-1}$ and from 0.9 to $41.1\,\mu mol\,l^{-1}$ respectively. TBTCl is the organotin presenting the highest toxicity while tetrapropyltin is the least toxic. Triphenyltin toxicity is comparable to that of tributyltins: EC_{50}, 5 min $= 0.36\,\mu mol\,l^{-1}$. These data are similar to those reported by Argaman *et al.* (1984) on activated sludge.

Certain marine bacteria may be TBT-resistant. Thus, in Japanese coastal waters, Suzuki, Fukagawa & Takama (1991) isolated nine resistant strains, including one highly resistant strain capable of developing in media containing up to $14.4\,mg\,TBTCl\text{-}Sn\,ml^{-1}$. These authors suggested that the resistance mechanisms may be likened to an efflux system induced by the chromosome DNA. Uchida (1993) reported inhibitory concentrations for estuarine bacteria ranged between 2 to $40\,\mu g\,TBTO\text{-}Sn\,l^{-1}$. Surprisingly, results showed that the inhibitory activity was on the following decreasing order: dibutyltin, tributyltin and tetrabutyltin.

Laurence, Cooney & Gadd (1989) tested the toxicity of nine organotins on an isolate from the marine yeast *Debaryomyces hansenii* collected in the waters of Chesapeake Bay (USA), by measuring the K^+ release at concentrations of $50\,\mu mol\,l^{-1}$. The highest K^+ losses were found to be caused by TBTCl, TPTCl and MBTCl. Comparatively, methyl- and ethyltins had no effect. Maximum rates of K^+ release were encountered for TBTCl, TPTCl and MBTCl with 2.58, 2.15 and $2.27\,n\,mol\,K^+\,min^{-1}$ respectively. Surprisingly, with a rate of only $0.21\,n\,mol\,K^+\,min^{-1}$, the DBT effect was significantly lower than that of MBT, whereas it usually proves to be more toxic than MBT. The authors imputed this result to the specific features of the strain studied presenting DBT-resistant characteristics. K^+ release is contingent upon the environmental conditions: maximum toxicity was found at pH 6.5 for TBT, MBT and TPT and at pH 5.0 for DBT. Toxicity is reduced at salinity values close to those of sea water. These results demonstrate that in highly polluted water ($> 100\,\mu g\,TBT\text{-}Sn\,l^{-1}$), TBT can lead to decreased cell viability of marine yeasts via K^+ losses.

6.3.2.2 Plankton

Wong *et al.* (1982) measured the inhibition of the primary productivity of freshwater plankton on a strain of *Ankistrodesmus falcatus* and on a set of microalgae collected in Lake Ontario. Concentrations leading to a 50% inhibition (IC_{50}) varied from a few micrograms of tin per litre for TBT and tripropyltin to 16 750 μg Sn l^{-1} for MBT. As a general rule, the microalgae collected from the natural environment proved more sensitive to alkyltins than the pure strain. As an example, the respective IC_{50} values for TBT and trimethyltin were 1.2 and 245 μg Sn l^{-1} for microalgae, 8 and 3850 μg Sn l^{-1} for *A. falcatus*.

As regards marine diatoms, algistatic TBTO concentrations were recorded at 4 μg Sn l^{-1} in *Skeletonema costatum* (Beaumont & Newman, 1986) and 400 μg Sn l^{-1} in *Tetraselmis suesica* (Thain, 1983). Reduced cell division rate was observed in *Dunaliella tertiolecta* and *Pavlova lutheri* at a concentration of 4 ng Sn l^{-1} (Beaumont & Newman, 1986). Puddu *et al.* (1989) found estimated EC_{50}, 48 h values of 1.4 μg TBT-Sn l^{-1} for the diatom *Dunaliella tertiolecta*, after correction of adsorption by suspended cells and by the walls of the culture vessel. Based on batch cultures, Liying, Xiankun & Bingyi (1990) found EC_{50}, 72 h values with a similar order of magnitude for TBTF (0.4 μg Sn l^{-1}), TBTO and TPTCl (0.3 μg Sn l^{-1}).

The above data provide evidence that undesirable effects on primary productivity may exist at concentration levels compatible with those of certain TBT-contaminated coastal areas. In the case of zooplankton, the acute toxicity data compiled in Table 6.4 indicate lethal 50% TBT concentrations, (expressed as μg TBT-Sn l^{-1}) ranging from 0.02 to 0.8 for the copepods *Acartia tonsa* (U'ren, 1983), *Nitocra spinipes* (Linden *et al.*, 1979) and *Eurytemora affinis* (Bushong *et al.*, 1987). Conversely, according to Ambrogi *et al.* (1982), *Tisbe furcata* would seem less sensitive with LC_{50}, 24 h = 34. Sublethal effects, such as reduced spawning, were observed in *A. tonsa* at TBTO concentrations as low as 0.004 (Linden *et al.*, 1979).

Vighi & Calamari (1985) studied the acute toxicity of 14 alkyl- and phenyltins in the Cladocera *Daphnia magna*. EC_{50}, 24 h values ranged from 4.7 for TBTCl to 49 000 for methyltin trichloride. In the case of trialkyltins, toxicity decreased in the following order: TBTCl (4.7), TBTO (5.6), tripropyl SnCl (15.5), triethyl SnBr (103), trimethyl SnCl (280). TPTCl (5.8) and DPTCl$_2$ (46.2) presented a toxicity comparable to trialkyltins. Among microcrustaceans, the Mysidacea *Acanthomysis sculpta* appeared to be highly sensitive to TBTO (Davidson *et al.*, 1986): effects on reproduction were observed at chronic exposures of $0.06 \, \mu g$ TBTO-Snl^{-1}. These authors set $0.04 \, \mu g$ TBTO-Snl^{-1} as the maximum level beyond which reproduction disturbances could occur.

An analysis of the above data reveals that plankton is particularly sensitive to alkyl- and phenyltins. Moreover, significant effects have been reported for phytoplankton and zooplankton of TBT concentrations lower than $1 \, \mu g \, l^{-1}$.

$0.04 \, \mu g$ TBT-Snl^{-1} = algistatic concentration for *Skeletonema costatum*

$0.04 \, \mu g$ TBT-Snl^{-1} = no effect level for the reproduction of the microcrustacean *A. sculpta*

$0.004 \, \mu g$ TBT-Snl^{-1} = effect on spawning in the copepod *A. Tonsa*.

Based on the three values above, it appears reasonable to estimate that maximum concentrations approximately equal to (or less than) $0.4 \, ng$ TBT-Snl^{-1} should not be exceeded in order to ensure plankton protection in the natural environment. Although no specific observations have yet been made in the natural environment, we should not disregard the fact that the current contamination of certain coastal waters may be harmful to the proper development of highly sensitive plankton organisms.

6.3.2.3 Molluscs

Data related to mollusc bivalves constitute the major part of information available on TBT toxicity relative to short term acute effects and long term chronic effects on growth, reproduction and calcification. This derives from the fact that TBT has been used for its molluscicidal action to fight certain disease vector carriers in freshwater environments. In addition, its use as an active agent in marine antifouling paints has resulted in severe disorders in shellfish farms.

6.3.2.3.1 LETHAL TOXICITY

Table 6.5 compiles data on acute toxicity for adult *C. gigas*, *C. virginica* and *M. edulis* individuals. The data show that LC_{50} values for the above species range between 180 and $720 \, \mu g$ TBTO-Snl^{-1} when exposure time does not exceed 48 h (Thain, 1983). Conversely, if exposure time exceeds 96 h, mortality rates remain high at concentration under $0.5 \, \mu g$ TBTO-Snl^{-1}. Thus Alzieu *et al.* (1980) exposed adult oysters *C. gigas* to water contaminated by TBT-based antifouling paint effluents and found a 30% mortality rate after 110 days, with an estimated TBT content in the water of $80 \, ng$ TBTF-Snl^{-1}. This observation is compatible with the results obtained by Henderson (1986) in *Crassostrea virginica* exposed to a constant concentration of $1 \, \mu g \, TBT-Sn \, l^{-1}$ originating from TBTO and TBT methacrylate-based paint effluents. Henderson reported mortality rates of 50% and 80% respectively after 30 and 57 days of exposure. Furthermore, during the same experiment, the oyster condition index was strongly affected at a TBT concentration of $40 \, ng \, Sn \, l^{-1}$, which would suggest that their long term survival was threatened. This condition index did not seem to be modified at a concentration of $16 \, ng \, TBT-Sn \, l^{-1}$ for 57 days.

The above results show that chronic exposure of adult bivalves at TBT concentrations of approximately $1 \, \mu g \, TBT \, l^{-1}$ leads to significant mortalities by the end of a 1–2 month period. In addition, concentrations as low as $0.2 \, \mu g \, l^{-1}$ $(80 \, ng \, TBT-Sn \, l^{-1})$ affect oyster survival beyond the second month.

Although no threshold leading to long term mortality has been determined experimentally, a number of indications on condition indices (Henderson, 1986) would suggest that it is close to $16 \, ng \, TBT-Sn \, l^{-1}$.

6.3.2.3.2 SUBLETHAL PHYSIOLOGICAL EFFECTS

Widdows & Page (1993) measured sublethal physiological responses, clearance rate, oxygen uptake, absorption efficiency and scope for growth (SFG) in the mussel *Mytilus edulis* exposed to TBTCl and DBTCl. Their results on the physiological parameters in relation to the concentrations accumulated within the tissues show the following:

Table 6.5 *TBT acute toxicity for* C. gigas, O. edulis *and adult* M. edulis

Concentration	Toxicity	Species	References
720	LC_{50}, 48 h	*C. gigas*	Thain, 1983
>120		*O. edulis*	
120		*M. edulis*	
116	LC_{50}, 96 h	*C. gigas*	
84		*O. edulis*	
15.2		*M. edulis*	
1	LC_{50}, 30 d	*C. virginica*	Henderson, 1986
0.8*	LC_{100}, 30 d	*C. gigas*	Alzieu *et al.*, 1980
0.5	LC_{50}, 21 d	*C. gigas*	Waldock *et al.*, 1983
0.39	LC_{50}, 66 d	*M. edulis*	Valkirs *et al.*, 1987
0.2*	LC_{50}, 25 d	*C. gigas*	Héral *et al.*, 1983
0.1*	LC_{50}, 48 d	*C. gigas*	Gendron, 1985
0.08*	LC_{50}, 113 d	*C. gigas*	Alzieu *et al.*, 1980

LC expressed as $\mu g \, TBT-Sn \, l^{-1}$; *estimated concentrations.

Respiration increases with TBT concentration up to a value of $4\,mg\,TBT\text{-}Sn\,kg^{-1}$ dry weight, then drops suddenly due to interrupted pumping. Clearance rate is maintained at rates not significantly different from the control until TBT levels reach tissue concentrations above $1.6\,mg\,TBT\text{-}Sn\,kg^{-1}$.

There were no significant effects on food absorption efficiency below $2.5\,mg\,TBT\text{-}Sn\,kg^{-1}$.

SFG is not significantly reduced until the TBT concentration in tissues increases above $1.6\,mg\,TBT\text{-}Sn\,kg^{-1}$.

In contrast, DBT experiments showed the underlying inhibitory effects of DBT (measured effect less estimation of effects due to TBT as DBT contaminant reagent) on clearance rate above $10\,mg\,DBT\text{-}Sn\,kg^{-1}$ and on respiration above $18\,mg\,DBT\text{-}Sn\,kg^{-1}$.

6.3.2.3.3 EFFECTS ON JUVENILE GROWTH

The influence of TBT on farmed bivalve juvenile growth was studied during periods ranging from one week to two months. Table 6.6 summarises the available experimental data and shows that thresholds approximately equal to or less than $50\,ng\,TBT\text{-}Sn\,l^{-1}$ have an influence on the growth of the six species considered (*C. gigas*, *O. edulis*, *C. virginica*, *M. edulis*, *V. decussata* and *V. semi decussata*), even for the shortest exposure times. These data obtained from relatively dissimilar experimental systems are nevertheless homogeneous overall for any single species. Based on these results, no-effect levels (expressed in $ng\,TBT\text{-}Sn\,l^{-1}$) for each species may be estimated approximately as follows.

S. commercialis, C. gigas	<0.2	Nell & Chvojka, 1992
M. edulis	<16 (66 days)	Valkirs et al., 1987
C. gigas	60 (56 days)	Thain & Waldock, 1983
V. decussatus	<100 (20 days)	Thain, 1986
V. semi decussatus, O. edulis	<100 (20 days)	Thain, 1986
C. virginica	120 (66 days)	Valkirs et al., 1987

Thus, the mussel *Mytilus edulis* would appear to be more sensitive than the other tested species. In oysters, the North American *Crassostrea virginica* and the flat European oyster *Ostrea edulis* proved more resistant than *Crassostrea gigas*. However, Nell & Chvojka (1992) found highly significant reduction in *Saccostrea commercialis* and *Crassostrea gigas* spat growth when exposed to concentrations of $0.2\,ng\,TBTO\text{-}Sn\,l^{-1}$ for 4 weeks. Weight reductions in comparison with the control were more significant for *C. gigas* (79%) than for *S. commercialis* (58%).

These experimental data remain yet to be corroborated or invalidated

Table 6.6 *TBT effects on juvenile growth of bivalves (concentration in µg TBT-Sn l^{-1})*

Species	Concentration (exposure days)	Growth	References
C. gigas	0.1 (20)	strong decrease	Thain, 1986
	0.6 (56)	stop	Thain & Waldock, 1983
	0.06 (56)	50% decrease	
C. virginica	0.8 (20)	decrease	Thain, 1986
	0.08 (20)	no effect	
	0.3 (66)	weight decrease	Valkirs et al., 1987
	0.016–0.124 (66)	no effect	
O. edulis	0.1 (20)	no effect	Thain, 1986
M. edulis	1.04 (20)	stop	Salazar & Salazar, 1987
	0.03 (63)	weight length decrease	Thain, 1986
	0.1 (20)	strong decrease	Stromgren & Bongard, 1987
	>0.16 (7)	length decrease	Valkirs et al., 1987
	0.016 (66)	no weight effect length decrease	
V. decussata	0.1 (20)	strong decrease	Thain, 1986
V. semi decussata	0.1 (20)	no effect	Thain, 1986
	1.04 (20)	strong decrease	

by studies in the natural environment, due in particular to the difficulties encountered in finding sites where TBT concentrations are relatively constant over prolonged time periods. This has led a number of authors, including for instance Salazar & Salazar (1987), to consider that TBT effects on mollusc growth are overestimated due to the stress induced by experimental conditions; important parameters being temperature, turbidity, and food. Thain & Waldock (1983) have shown, however, that water turbidity ($30\,mg\,l^{-1}$) had no significant influence on the growth of C. gigas spat in comparison with waters free from any suspended matter and containing equivalent TBTO levels. Furthermore, Stephenson et al. (1986) have shown a trend of reduced growth in the mussels M. edulis and M. californianus grown at four different stations located in the Bay of San Diego (California) following the gradient of TBT contamination. The TBT concentrations recorded in bay waters at various periods indicated levels of approximately $4–12\,ng\,TBT$-$Sn\,l^{-1}$ in the vicinity of the control station, and $90–370\,ng\,TBT$-$Sn\,l^{-1}$ close to the most polluted station. In addition, high mortality rates were observed in M. edulis juveniles at all four stations.

Thus it would appear, both from experimental data and field observations, that the growth of farmed bivalve juveniles is affected at TBT concentrations well below $1\,\mu g\,TBT$-$Sn\,l^{-1}$. The mussel M. edulis, with a no-effect level below $16\,ng\,TBT$-$Sn\,l^{-1}$, seems to be the most sensitive species, while among oysters C. gigas proves least resistant. The above thresholds, corresponding to levels measured in the natural waters in the vicinity of TBT input areas, may provide a reasonable explanation for certain disturbances observed for individuals transferred into a contaminated environment.

6.3.2.3.4   EFFECTS ON REPRODUCTION

(a) Sexuality   The influence of TBT on gastropod sexuality is a well known effect, described in Chapter 7. It leads to the development of male sexual characters in the females, a phenomenon known as imposex.

Thain (1986) observed in bivalves a predominance of males in O. edulis individuals exposed to concentrations of $0.01\,\mu g\,TBTO$-$Sn\,l^{-1}$ for 75 days as compared with a control population. Concurrently, individuals exposed to a higher concentration (1.0) presented little or no sexual differentiation. Histological examinations suggested that the absence of any larval production by individuals exposed to $0.01\,\mu g\,TBT$-$Sn\,l^{-1}$ may be due to a delay in the sexual changeover phase from male to female during the gametogenesis cycle. At the highest concentration, gonad development is fully inhibited. These results may explain the limited reproduction capability of *O. edulis* along the south-western coast of England observed since the considerable stock decrease which occurred in 1962. By contrast, Roberts *et al.* (1987) observed no modification in the fertility of oysters *C. virginica* exposed to TBT concentrations ranging from 0.02 and to $0.4 \mu g \, Sn \, l^{-1}$ for 8 weeks. Sex ratios were not significantly different from the controls and the formed gametes were mature and fertilisable. This would tend to corroborate the observations of His & Robert (1983–5) on *Crassostrea gigas* spawners collected in 1981 from the highly contaminated waters of the Bay of Arcachon (France) which, whether matured in the Bay or not, produced perfectly viable larvae under laboratory conditions.

Although fragmentary, these indications would suggest that TBT effects on bivalve sexuality are far less significant than in gastropods. However, *Ostrea edulis* reproduction may be affected via a delayed sexual maturity. Conversely, the gametogenesis of *Crassostrea virginica* and *Crassostrea gigas* does not appear disturbed, even at relatively high concentrations.

(b) *Larval development* Table 6.3 presents the various data available on acute TBT toxicity at different larval stages in molluscs. Short term lethal concentrations (at least one week) for bivalves range between 0.2 (LC_{70}, 8 d, *M. mercenaria* veligers, Laughlin *et al.*, 1987) and $4 \mu g TBT\text{-}Sn \, l^{-1}$ (LC_{100}, 25 d, *M. mercenaria* postlarvae, Laughlin *et al.*, 1987; 95% mortality in 3 d, *M. galloprovincialis*, Robert & His, 1981). However, data on *C. gigas* constitute the most comprehensive information available on TBT effects. His & Robert (1983–5) proposed a relationship scale between TBT acetate concentrations (expressed in $\mu g \, AcTBT\text{-}Sn \, l^{-1}$) and its effects on embryogenesis and *C. gigas* larval development:

35	inhibition of fecundation,
17.5	inhibition of segmentation,
8.7	partial inhibition of segmentation (40%),
3.5	absence of trocophore formation,
1 & 1.75	no veligers, monstrous trocophores,
0.35	abnormal veligers, total mortality within 6 days,
0.17	numerous abnormal larvae, total mortality in 8 days, disturbances in trophic regime intensified from day 4 to day 8, extremely reduced growth,
0.07	lower percentage of abnormal D larvae, disturbance in trophic regime as early as day 4, progressive then total mortality on day 12, reduced growth,

0.03	majority of normal D larvae, notable disturbances in trophic regime as early as day 6, reduced growth up to day 6, total mortality on day 12,
0.017	normal D larvae, notable disturbance in trophic regime on day 8, significant mortality as of day 10, reduced growth,
0.007	normal D larvae, limited mortality, adequate growth. No-effect level.

It should be noted that the no-effect level for TBT is approximately 1000 times lower than the threshold proposed by the same authors for copper chloride ($CuCl_2$), i.e. $25 \mu g l^{-1}$, thereby demonstrating the extreme susceptibility of C. gigas larval development to TBT.

Because no concentration/effect correlation has yet been published for other bivalves, it is impossible at this time to know whether C. gigas embryogenesis and larval development are particularly sensitive to TBT, or whether the 0.007 no-effect level is applicable as well to other bivalves. Nevertheless, the scale defined by His & Robert (1983–5) has been validated by observations on water quality in the Bay of Arcachon. Following the reduction in TBT inputs, C. gigas was able to reproduce normally, whereas its reproduction had previously been down to a deficit or zero level. Monitoring studies on the contamination of bay waters (1985–7) revealed that concentrations had fallen well below 0.007 μg TBT-$Sn l^{-1}$ in growing areas (Alzieu et al., 1989), thereby confirming the no-effect level proposed by His & Robert (1983–5).

6.3.2.3.5 EFFECTS ON CALCIFICATION MECHANISMS

The correlation between TBT contamination of the environment and shell calcification anomalies in the oyster C. gigas was established by Alzieu et al. (1980). These anomalies (Figure 6.3) consist of a wafer-like shell structure with formation of an interlamellar gel. They were first observed in 1974 in the Bay of Arcachon on the French Atlantic coast, before spreading to various extents to most oyster-growing sites along the Atlantic coastline.

The bay of Arcachon, located along the French Atlantic coast between the Gironde estuary and the Spanish border, is a triangle-shaped enclosed bay (about 20 km along each side), opening onto the Atlantic ocean through a narrow channel. The 1000-ha wide Crassostrea gigas oyster beds occupy the central portion of the bay and, on an average year, produce 10–15 000 tons, i.e. approximately 10% of the entire French

(a)

(b)

(c)

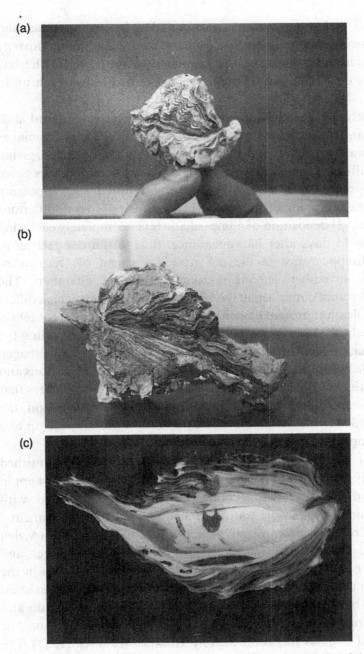

Figure 6.3 TBT effects on calcification mechanisms of the oyster
Crassostreas gigas. (a) Acute malformation showing on 'ball shaped' shell
sampled in the bay of Cadiz (Spain); (b) shell from Arcachon Bay (France)
grown during TBT pollution period (left side) and after the ban on
antifouling paints (right side); (c) cut of the same shell as (b) showing
calcification anomalies (chambers in the external part right side) and normal
calcification (inner part of the shell).

production. The harvest areas are surrounded by ten marinas and numerous seasonal moorings, representing a total accommodation capacity of 7800 pleasure crafts. During the summer season, traffic in the bay waters is very dense and the number of pleasure boats can reach up to 10–15 000.

Concurrently with spatfall failure, cases of stunted growth and shell calcification anomalies were observed as early as 1974. These anomalies consisted of wafered 'chambers' filled with a gelatinous substance, giving the shell a ball-like shape. The chambering process occurred over three successive phases (Héral et al., 1981): 1) hypersecretion of a gel appearing suddenly at the beginning of July and synchronously across populations of various ages; 2) deposition of a fine calcitic layer completely enclosing the gel within 15 days after its appearance, thus forming a gelatinous pocket; 3) disappearance of the gel around the end of October or beginning of November, leaving a cavity called the 'chamber'. The gelatinous substance contained in the 'chambers' is a protein which differs from the calcification protein (conchyolin) in that its threonine content is higher and it shows a lower proportion of the amino acids providing the bond with calcium: aspartic acid, glycine and serine (Krampitz, Drolshagen & Deltreil, 1983). Observations have shown that shell malformations and stunted growth occurrences were all the more significant for oysters that grew closer to the marinas. Thus, in the Bay of Marennes-Oléron, the influence of a 300 boat marina on shell calcification was notable up to a distance of approximately 1.5 km (Alzieu et al., 1980, 1981).

The role of TBT in the anomalies described above was established experimentally, from the study of batches of anomaly-free oysters kept in a marina and in experimental tanks in which plates coated with TBT-based paint were placed. Control batches were placed in experimental tanks, and on oyster beds known to be free of any malformations (Alzieu et al., 1981). Observations showed that the formation of gel and subsequently of gelatinous pockets developed concurrently both in the marina and in the TBT-contaminated tanks, whereas it was non-existent in the control batches placed in uncontaminated experimental tanks and in the TBT-free natural environment. The influence of TBT on the calcification of C. gigas was confirmed by Thain & Waldock (1983). The mechanism of TBT action on calcification remains unknown today. However, recent indications would suggest that TBT may inhibit calcification of C. gigas at concentrations below $0.8\,ng\,TBT\text{-}Sn\,l^{-1}$ (the lowest concentration tested by Chagot et al., 1990).

In addition, Okoshi, Mori & Nomura (1987) have shown that

sensitivity to chambering was not identical for two stocks of *C. gigas* grown in Japan. This behaviour could be explained by differences in genetic characteristics. This would tend to corroborate the observations made on stocks grown in the Bay of Arcachon and on the behaviour of hybrids derived from different morphotypes.

6.3.2.3.6 CONCLUSIONS

An analysis of ecotoxicological data on marine molluscs reveals that these populations are highly sensitive to the presence of extremely low TBT concentrations. No-effect levels for sensitive species, whether measured or estimated from experimental data, are among the lowest known to date.

16 ng TBT-Sn l^{-1}	long term survival of adult bivalves and no effect on juvenile growth (2 months)
7 ng TBT-Sn l^{-1}	no effect on embryogenesis and larval growth of *C. gigas*
0.7 ng TBT-Sn l^{-1}	no-effect level for calcification anomalies in *C. gigas*

The above thresholds are comparable to the contamination levels encountered in TBT input areas or in their vicinity. They may explain certain imbalances and disturbances observed in bivalve farms and in the wild marine fauna.

6.3.2.4 Fish

6.3.2.4.1 LETHAL TOXICITY

The LC_{50}, 48 and 96 h values for TBT in adult fish presented in Table 6.3 range between 6.4 µg TBT-Sn l^{-1} LC_{50}, 96 h for *Agonus cataphractus* and 2.2 µg TBT-Sn l^{-1} LC_{50}, 48 h for *Solea solea* (Thain, 1983). On the longer term (21 d), Ward *et al.* (1981) found an LC_{50} value of approximately 0.4 for *Cyprinodon variegatus*, i.e. lower by a factor of 10–90 than the above mentioned LC_{50}, 48 and 96 h. In larvae or juveniles of the various species, the published LC_{50}, 96 h values are generally below 2 and over 0.8 µg TBT-Sn l^{-1} (Bushong *et al.*, 1987; Thain, 1983).

6.3.2.4.2 SUBLETHAL EFFECTS

Little information has been published to date on sublethal TBT effect in fish. Hall *et al.* (1984) found behavioural changes (avoidance responses) to TBTO contaminated environment for the species *Morone saxatilis* and *Brevoortia tyrannus*. Concentrations leading to a flight response, expressed in micrograms of total organotin per litre, were equal to 24.9 and 5.5 respectively. Pinkney *et al.* (1985) report that concentrations of approximately $0.5-1.5\,\mu g\,Sn\,l^{-1}$ induce a similar reaction in *Fundulus heteroclitus*. In the rainbow trout *Salmo gairdneri* and *Tilapia rendalli*, Chliamovitch & Kuhn (1977) observed a loss of positive rheotaxis reached after 3 min at a concentration of $47\,\mu g\,TBT\text{-}Sn\,l^{-1}$, accompanied by increased pigmentation of the body.

The same authors observed histological lesions such as separation of the branchial epithelium from the basal membrane ($2.34\,mg\,Sn\,l^{-1}$ for 10 min), lesions of biliary cells, activation of mucus secretion, corneal lesions. Similar lesions were observed by Weis *et al.* (1987a) in *Fundulus heteroclitus* embryos. At the highest TBT concentration ($12\,\mu g\,TBT\text{-}Sn\,l^{-1}$), teratogenic effects included mono- and bilateral anophtalmia, microphtalmia and heart malformations. However, the extent of these effects was highly variable from one experimental batch to another. These authors reported nevertheless that the teratogenic effects observed in the laboratory were similar to those encountered for that species in areas known to be contaminated by TBT.

6.3.2.5 Crustaceans

6.3.2.5.1 LETHAL TOXICITY

Information on lethal TBT toxicity to crustaceans is restricted to the shrimp *Crangon crangon*, with LC_{50}, 48 and 96 h values of 29.2 and $16.4\,\mu g\,TBT\text{-}Sn\,l^{-1}$ respectively according to Thain (1983), and to *Palaemonetes pugio*: LC_{50}, 96 h $= 8\,\mu g\,TBT\text{-}Sn\,l^{-1}$ according to Walsh (1986). As regards larvae survival, Laughlin *et al.* (1984) found that a TBT concentration of $12\,\mu g\,Sn\,l^{-1}$ limited the growth of *Gammarus oceanicus* larvae after 8 weeks of exposure. At a concentration of $120\,ng\,TBT\text{-}Sn\,l^{-1}$, survival was zero. By contrast, lobster larvae seemed less sensitive, with total mortality occurring within 24 h at a TBT concentration of $8\,\mu g\,Sn\,l^{-1}$, and stunted growth at $0.4\,\mu g\,TBT\text{-}Sn\,l^{-1}$ (Laughlin & French, 1980). These authors tested three organotins on zoea of the crab *Hemigrapsus nudus*. Zoea survival did not exceed 2 days at concentrations of 200 and $400\,\mu g\,TBTO\text{-}Sn\,l^{-1}$, while all individuals had died by day 10 at the lowest concentration ($10\,\mu g\,TBTO\text{-}Sn\,l^{-1}$). Toxicity decreased in the following order: TBTO > trimethyl SnO > tripropyl SnO > triethyl SnO.

6.3.2.5.2 SUBLETHAL TOXICITY

Laughlin *et al.* (1983) studied the influence of TBTO and its presumed derivative (TBTS) in the environment on zoea larvae of the crab *Rhithropanopeus harrisii* exposed to concentrations of 0.2–10 and 0.18–$18\,\mu g\,Sn\,l^{-1}$, respectively. Growth was significantly stunted for TBTO and TBTS concentrations above $0.36\,\mu g\,Sn\,l^{-1}$ and the development rate was reduced even at the lowest concentrations. The average metamorphosis time of 11.9 days in the controls, increased by 2 days at $10\,\mu g$ TBTO-Sn$\,l^{-1}$ and by 6 days at $18\,\mu g$ TBTS-Sn$\,l^{-1}$. Conversely, at low concentrations, larval growth would seem to be slightly stimulated.

Weis *et al.* (1987b) studied the leg regeneration capacity in the crab *Uca pugilator* exposed to TBT concentration ranging between 0.2 and $10\,\mu g$ TBTO-Sn$\,l^{-1}$. At the lowest concentration, delays were observed in regeneration and exuviation, along with morphological anomalies in the regenerated parts. Malformation rates varied from 16.7% in controls to 66.7% at a concentration of $10\,\mu g$ TBTO-Sn$\,l^{-1}$.

The scarce information available on lethal and sublethal TBT toxicity in crustaceans concern a very small number of species. Nevertheless, they indicate that reproduction and exuviation may be affected at concentrations of $0.2\,\mu g$ TBT-Sn$\,l^{-1}$. No TBT no-effect level on crustacean reproduction and growth has been established to date.

6.3.2.6 Effects on ecosystems

A comparison between TBT no-effect levels on various species and the contamination levels encountered in the natural environment suggests that functioning of coastal ecosystems, whether commercially exploited or not, may be disturbed by butyltin contamination. A number of studies have been carried out in this field either in microcosms or from observations of farm-grown populations.

6.3.2.6.1 MICROCOSM STUDIES

Henderson (1986) used a so-called 'Portable Environmental Test System' (PETS) with a continuous natural sea water supply (temperature = 25–28 °C, salinity = 33–36 psu) to assess the long term TBT effects on biofouling component organisms. Tested species were derived from natural fouling on Plexiglass panels exposed for 14 weeks to colonising in natural waters and 5 weeks for adaptation purposes in the PETS. The TBT input was supplied via leaching of a TBTO- and TBT methacrylate-based paint. Concentrations recorded in the experimental tanks were 16, 40, 216, 708 and 1000 ng TBT-Sn$\,l^{-1}$. The number and diversity of species were observed to decrease rapidly from the onset of exposure at concentrations of $216\,\mu g\,l^{-1}$ and above. At the end of exposure (2 months), the average abundance of species exposed to concentrations equal to or above $200\,ng\,l^{-1}$ was significantly lower than at other concentrations. Out of the 20 epifaunal taxa present in all controls, only one species was absent in the presence of $16\,ng\,l^{-1}$ and two at a concentration of $40\,ng\,l^{-1}$. Five species, including one anemone, one ascidium, two worms and one gastropod survived at the highest concentration. On completion of the recovery period (2 months), the species abundance was identical to the controls at all tested concentrations. These authors suggest that the high TBT content may reduce the abundance of the biofouling component species and promote the dominance of resistant species such as tunicates, annelids and gastropods (vermetids).

Salazar & Salazar (1987) used the same experimental system and found that TBT concentrations of 26–77 ng TBT-Sn l^{-1} did not disturb the abundance nor the species distribution of biofouling organisms after a 7 months' exposure.

6.3.2.6.2   EFFECT ON BENTHIC COMMUNITIES

It has been proved difficult to identify the effect of a contaminant in the natural environment on a benthic population owing to the hydraulic and climatic conditions and to the presence of other pollutants liable to interfere. Such is the case for TBT, for which the highest concentrations are found in harbour areas subject additionally to multiple inputs of pollutants. Thus, in the Bay of San Diego (California), Lenihan, Oliver & Stephenson (1990) studied hard bottom communities and showed that hydrographic variations in embayments in some harbours were not sufficient to explain the observed modifications. Back-bay areas with a high boat density presented less community cover with a lower biomass and contained a smaller number of species than comparable areas with lower boat density. TBT-polluted areas were characterised by the presence of serpulid polychaetes, fibrous seaweeds and a single tunicate, i.e. *Ciona intestinalis*. In areas with low mooring density, these groups disappeared to the benefit of other types of sessile invertebrates, such as mussels, sponges, bryozoans and tunicates.

6.3.2.6.3   EFFECTS ON HARVESTED ECOSYSTEMS

Alzieu *et al.* (1980) were the first authors to report the harmful effects on the growth and calcitic development of oysters *C. gigas* related to the use of TBT-based antifouling paints. His & Robert (1983–5) later showed that in the Bay of Arcachon the total absence of any oyster spatfall from 1977 to 1981 could be attributed to TBT contamination of the bay waters. Alzieu *et al.* (1986) have since confirmed that the January 1982 ban on the use of TBT-based paints for boats under 25 m long resulted in a decrease of total Sn content in oyster tissues and in a correlated improvement of growth and shell quality. The reduction of TBT inputs also had an influence on the annual oyster production of the Bay of Arcachon, which recovered its normal level only two years after implementation of the regulations on antifouling paints (Table 6.7). The harmful consequences of TBT-related water contamination on spatfall were reported as well by Thain (1986) for farmed oysters *Ostrea edulis* in England and by Minchin, Duggan & King (1987) for scallops in Ireland. The calcification malformations observed in the oyster *Saccostrea commercialis* in New South Wales (Australia) disappeared within two years after the ban of TBT-based antifouling paints for boats under 25 m was implemented (Batley *et al.*, 1992).

In intensive fish aquaculture farms, the use of nets and cages coated with TBT-based paints for biofouling protection has proved harmful both to the farmed fish and to the surrounding biotope. Short & Thrower (1986) observed significant mortalities in Pacific salmon juveniles placed in cages treated with TBT-based paints. Davies *et al.* (1987) have shown that TBT-treated cages could have an impact on the chambering of *C. gigas* shells located up to 1 km away. The TBT concentrations recorded in Loch Sween waters ranged from 2 to 24 ng TBT-Sn l^{-1} and were correlated with the gradient of imposex ratio in *N. lapillus*.

6.4 Conclusions

The ecotoxicology of organotins is a recent concern linked to the dispersive use of biocides, and information available on the subject is essentially related to triorganotins (TBT, TPT). Trialkyltins are oxidative phosphorylation inhibitors, and the nature of the alkyl group determines the extent of toxic effects on the different zoological branches. Thus, tributyl derivatives are especially toxic for molluscan species, fungi and bacteria, while trimethyl derivatives are noxious for mammals and insects.

Table 6.7 *Production variations of oysters Crassostrea gigas in Arcachon Bay 1978–85 (Alzieu, 1991)*

Period	Production (tons)
1978–9	10 000
1979–80	6 000
1980–1	3 000
1981–2	5 000
1982–3	8 000
1983–4	12 000
1984–5	12 000

An analysis of the data available for aquatic life reveals that significant effects on the growth and reproduction of phytoplankton and zooplankton have been observed at TBT concentrations under $0.4\,\mu g\,TBT\text{-}Sn\,l^{-1}$. However, molluscs are the species exhibiting the greatest sensitivity to TBT-related water contamination. Chronic exposure of adult bivalves to TBT concentrations in the order of $0.4\,\mu g\,TBT\text{-}Sn\,l^{-1}$ causes significant mortalities within a 1–2 months time period. Although the threshold ensuring long term survival of adult bivalves has not yet been determined experimentally, a number of indicators would suggest that it may be close to $16\,ng\,TBT\text{-}Sn\,l^{-1}$.

On the basis both of experimental data and field observations, it would appear that the growth of juvenile bivalves in marine farms is impaired at TBT concentration rates of 1–2 orders of magnitude below $0.4\,\mu g\,TBT\text{-}Sn\,l^{-1}$. The mussel *Mytilus edulis* seems to be the most sensitive species, while in oysters, *C. gigas* is the least resistant. The above thresholds correspond to levels determined in natural waters in the vicinity of TBT input areas, and could explain certain disturbances observed in individuals transferred into a contaminated environment.

Results from laboratory experiments, corroborated by observations in the natural environment, provide evidence that TBT concentrations below $0.4\,ng\,TBT\text{-}Sn\,l^{-1}$ can significantly modify the sexual characters of marine gastropods. The effects reported include the development of male genital organs in females, a phenomenon known as imposex. In the final stages, the fertility and reproduction capabilities of the species are affected: the decline of certain populations may be explained by the proximity of TBT input sources. As regards bivalves, fragmentary information would suggest that TBT could impair the reproduction of the oyster *Ostrea edulis* by delaying sexual maturity. By contrast, the gametogenesis of oysters *Crassostrea virginica* and *Crassostrea gigas* does not appear to be impaired even at relatively high TBT concentrations $(0.4\,\mu g\,TBT\text{-}Sn\,l^{-1})$. However, larval development of *C. gigas* is affected once concentration exceeds $8\,ng\,TBT\text{-}Sn\,l^{-1}$.

The influence of TBT on shell calcification in *C. gigas* has been identified both *in situ* and in the laboratory. It leads to the formation of chambers filled with a jelly-like substance induced by concentrations in solution as low as $0.8\,ng\,TBT\text{-}Sn\,l^{-1}$. Differences in sensitivity have been detected in two species of *C. gigas*, implying some role of genetic phenomena in the mechanisms of appearance and development of the chambers.

An analysis of ecotoxicological data relevant to the effects of butyltins on aquatic marine organisms highlights the scarcity of information and the necessity of further evaluation for a large number of species. Similarly, there is a paucity of data for organotin derivatives other than TBT. Nevertheless, available data provide, in certain cases, the possibility of estimating the probable thresholds ensuring survival and reproduction of the organisms. Contamination levels in excess of the following values (expressed as TBT-Sn l^{-1}) are likely to induce disturbances in aquatic organisms:

0.4 ng   no-effect level on phytoplankton and zooplankton,

<0.8 ng   no-effect level on calcification anomalies in oysters *C. gigas*,

8 ng   no-effect level on reproduction of *C. gigas*,

0.4–4 µg   effects on fish reproduction,

0.4–40 µg   modification of fish behaviour (avoidance response, rheotaxis),

>200 µg   effects on exuviation of crustacean species.

6.5   References

Alabaster, J. S. (1969). Survival of fish in 164 herbicides, insecticides, fungicides wetting agents and miscellaneous substances. *International. Pest. Control.*, **11**, 287–97.

Aldridge, W. N. & Street, B. W. (1964). Oxidative phosphorylation: biochemical effects and properties of trialkyltins. *Biochemistry Journal*, **91**, 287–97.

Alzieu, Cl. (1991). Environmental problems caused by TBT in France: assessment, regulations, prospects. *Marine Environmental Research*, **32**, 7–17.

Alzieu, Cl., Héral, M., Thibaud, Y., Dardignac, M. J. & Feuillet, M. (1981). Influence des peintures antisalissures à base d'organostanniques sur la calcification de la coquille de l'huître *Crassostrea gigas*. *Revue des Travaux des Pêches maritimes*, **45**, 2, 101–16.

Alzieu, Cl., Sanjuan, J., Deltreil, J. P. & Borel, M. (1986). Tin contamination in Arcachon Bay: effects on oyster shell anomalies. *Marine Pollution Bulletin*, **17**, 11, 494–8.

Alzieu, Cl., Sanjuan, J., Michel, P., Borel, N. & Dréno, J. P. (1989). Monitoring and assessment of butyltins in Atlantic coastal waters. *Marine Pollution Bulletin*, **20**, 1, 22–6.

Alzieu, Cl., Thibaud, Y., Héral, M. & Boutier, B. (1980). Evaluation des risques dus à l'emploi des peintures antisalissures dans les zones conchylicoles. *Revue des Travaux des Pêches maritimes*, **44**, 4, 301–48.

Ambrogi, R., Saroglia, M. G. & Scarano, G. (1982). Water treatment for fouling prevention residual toxicity of an organotin compounds towards marine organisms. In *IX Annual Aquatic Toxicity Workshop*, Edmonton, Alberta, Canada, Nov. 2–5, 1982.

Argaman, Y., Hucks, C. E. & Shelby, S. E. (1984). The effects on the activated sludge process. *Water Research*, **18**, 5, 535–42.

Bailey, S. K. & Davies, I. M. (1988). Tributyltin contamination around an oil terminal in Sullom Voe (Shetland). *Environmental Pollution*, **55**, 161–72.

Batley, G. E., Scammell, M. S. & Brockbank, C. I. (1992). The impact of the banning of tributyltin-based antifouling paints on the Sydney rock oyster, *Saccostrea commercialis*. *The Science of the Total Environment*, **122**, 301–14.

Beaumont, A. R. & Budd, (1984). High mortality of the larvae of the common mussel at low concentrations of tributyltin. *Marine Pollution Bulletin*, **15**, 11, 402–5.

Beaumont, A. R. & Newman, P. B. (1986). Low levels of tributyltin reduce growth of marine micro-algae. *Marine Pollution Bulletin*, **17**, 10, 457–61.

Blair, W. R., Olson, G. J., Brinckman, F. E. & Iverson, W. P. (1982). Accumulation and fate of tri-n-butyltin cation in estuarine bacteria. *Microbiology and Ecology*, **8** 241–51.

Blair, W. R., Olson, G. J., Trout, T. K., Jewett, K. L. & Brinckman, F. E. (1988). Accumulation and fate of tributyltin species in microbial biofilms. In *Proceedings of the Organotin Symposium, Oceans '88 Conference*, Baltimore, Maryland, Oct. 31–Nov. 2, 1988, **4**, 1668–72.

Bressa, G., Cima, L., Canova, F. & Caravello, G. U. (1984). Bioaccumulation of tin in the fish tissues *Liza aurata*. In *Environmental Contamination, International Conference*, London, July 1984, UNEP IRPTC, pp. 812–15.

Bryan, G. W., Gibbs, P. E., Burt, G. R. & Hummerstone, L. G. (1987). The effects of tributyltin TBT accumulation on adult dog-whelks, *Nucella lapillus*: long term field and laboratory experiments. *Journal of the Marine Biological Association of the United Kingdom*, **67**, 525–44.

Bushong, S. J., Hall, W. S., Johnson, W. E. & Hall, L. W. (1987). Toxicity of tributyltin to selected Chesapeake Bay biota. In *Proceedings of the Organotin Symposium, Oceans '87 Conference*, Halifax, Nova Scotia, Canada, Sept. 28–Oct. 1, 1987, **4**, 1499–503.

Chagot, D., Alzieu, Cl., Sanjuan, J. & Grizel, H. (1990). Sublethal and histopathological effects of trace levels of tributyltin fluoride on adult oysters *Crassostrea gigas*. *Aquatic Living Resources*, **3**, 121–30.

Chliamovitch, Y. P. & Kuhn, C. (1977). Behavioural, hematological and histological studies on acute toxicity of bis, tri-n-butyltin oxide on *Salmo gairdneri* Richardson and *Tilapia rendalli* Boulenger. *Journal of Fisheries Biology*, **10**, 575–85.

Davidson, B. M., Valkirs, A. O. & Seligman, P. F. (1986). Acute and chronic effects of tributyltin on the mysid *Acanthomysis sculpta*, crustacea, Mysidacea. In *Proceedings of the Organotin Symposium, Oceans '86 Conference*, Washington, Sept. 23–25, 1986, **4**, 1219–25.

Davies, I. M., Drinkwater, J. & McKie, J. C. (1988). Effects of tributyltin compounds from antifoulants on Pacific oysters, *Crassostrea gigas* in Scottish Sea Lochs. *Aquaculture*, **74**, 307–17.

Davies, I. M., Drinkwater, J., McKie, J. C. & Balls, P. (1987). Effects of the use of tributyltin antifoulants in mariculture. In *Proceedings of the Organotin Symposium, Oceans '87 Conference*, Halifax, Nova Scotia, Canada, Sept. 28–Oct. 1, 1987, **4**, 1477–81.

Davies, I. M. & McKie, J. C. (1987). Accumulation of total tin and tributyltin in muscle tissue of farmed atlantic salmon. *Marine Pollution Bulletin*, **18**, 7, 405–7.

Davies, I. M., McKie, J. C. & Paul, J. D. (1986). Accumulation of tin and tributyltin from antifouling paint by cultivated scallops *Pecten maximus* and pacific oysters *Crassostreas gigas*. *ICES, C. M.* 1986/f 11.

Dixon, D. R. & Prosser, H. (1986). An investigation of the genotoxic effects of an organotin antifouling compound, bistributyltin oxide on the chromosomes of the edible mussel, *Mytilus edulis. Aquatic Toxicology*, **8**, 185–94.

Donard, O., Rapsomanikis, S. & Weber, J. H. (1986). Speciation of inorganic tin and alkyltin compounds by atomic absorption spectrometry using electrothermal quartz furnace after hydride generation. *Analytical Chemistry*, **58**, 772–7.

Dooley, C. A. & Denis, P. (1987). Response of bioluminescent bacteria to alkyltin compounds. In *Proceedings of the Organotin Symposium, Oceans '87 Conference*, Halifax, Nova Scotia, Canada, Sept. 28–Oct. 1, 1987, **4**, 1517–24.

EAJ, (1988). *Outline of TBT compounds monitored in Japan*, Tokyo, Environment Agency of Japan (OECD Clearing House Project on Organotins).

Evans, D. W. & Laughlin, R. B. (1984). Accumulation of bis tributyltin oxide by the mud crab, *Rhithropanopeus harrisii. Chemosphere*, **13**, 1, 213–19.

Evans, J. C. & Smith, P. J. (1975). Organotin based antifouling systems. *Journal of the Oil and Colour Chemists' Association*, **58**, 160–8.

Fent, K. & Stegeman, J. J. (1991). Effects of tributyltin chloride *in vitro* on the hepatic microsomal monooxygenase system in the fish *Stenotomus Chrysops. Aquatic Toxicology*, **20**, 159–68.

Fent, K. & Stegeman, J. J. (1993). Effects of tributyltin *in vivo* on hepatic cytochrome P450 forms in marine fish. *Aquatic Toxicology*, **24**, 219–40.

Gendron, F. (1985). Recherches sur la toxicité des peintures à base d'organostanniques et de l'oxyde de tributylétain vis-à-vis de l'huître *Crassostrea gigas. Thèse doctorat es-Sciences*, Université Aix-Marseille, France.

Gibbs, P. E. & Bryan, G. W. (1987). TBT paints and the demise of the dogwhelk *Nucella lapillus*, Gastropoda. In *Proceedings of the Organotin Symposium, Oceans '87 Conference*, Halifax, Nova Scotia, Canada, Sept. 28–Oct. 1, 1987, **4**, 1482–87.

Hall, L. W., Bushong, S. J., Ziegenfuss, M. C., Johnson, W. E., Herman, R. L. & Wright, D. A. (1988). Chronic toxicity of tributyltin to Chesapeake Bay biota. *Water, Air and Soil Pollution*, **39**, 365–76.

Hall, L. W. & Pinkney, A. E. (1985). Acute and sublethal effects of organotin compounds on aquatic biota: an interpretative literature evaluation. *CRC Critical Reviews in Toxicology*, **14**, 159–209.

Hall, L. W., Pinkney, A. E., Zeger, S., Burton, D. T. & Lenkevich, M. J. (1984). Behavioral responses to two estuarine fish species subjected to bis, tri-n-butyltin oxide. *Water Resources Bulletin*, **20**, 2, 235–9.

Henderson, R. S. (1986). Effects of organotin antifouling paint leachates on Pearl Harbour organisms: a site specific flowthrough bioassay. In *Proceedings of the Organotin Symposium, Oceans '86 Conference*, Washington, Sept. 23–25, 1986, **4**, 1226–37.

Héral, M., Alzieu, Cl., Caux, O., Razet, D. & Garnier, J. (1983). Effets des sels organiques de tributylétain sur l'huître adulte *Crassostrea gigas*. In *18th European Marine Biology Symposium*, Oslo, poster 13 p.

Héral, M., Berthomé, J. P., Polanco-Torrès, E., Alzieu, Cl., Deslou-Paoli, J. M., Razet, D. & Garnier, J. (1981). Anomalies de croissance de la coquille de *Crassostrea gigas* dans le bassin de Marennes-Oléron. Bilan de trois années d'observations. *CIEM, C. M.* 1981/K: 31.

Higashiyama, T., Shiraishi, H., Otsuki, A. & Hashimoto, S. (1991). Concentrations of organotin compounds in blue mussels from the wharves of Tokyo bay. *Marine Pollution Bulletin*, **22**, 12, 585–7.

His, E. & Robert, R. (1983–5). Développement des veligères de *Crassostrea gigas* dans le bassin d'Arcachon. Etudes sur les mortalités larvaires. *Revue des Travaux de l'Institut des Pêches maritimes*, **47**, 1–2, 63–88.

Humphrey, B. & Hope, D. (1987). Analysis of water, sediments and biota for organotin compounds. In *Proceedings of the Organotin Symposium, Oceans '87 Conference*, Halifax, Nova Scotia, Canada, Sept. 28–Oct. 1, 1987, **4**, 1348–51.

IPCS, (1990). *Environmental Health Criteria 116 – Tributyltin compounds*. WHO Geneva, 273 p.

Johansen, K. & Mohlenberg, F. (1987). Impairment of egg production in *Acartia tonsa* exposed to tributyltin oxide. *Ophelia*, **27**, 2, 137–41.

King, N., Miller, M. & de Mora, S. J. (1989). Tributyl tin levels for sea water, sediment, and selected marine species in coastal Northland and Auckland, New Zealand. *New Zealand Journal of Marine and Freshwater Research*, **23**, 287–94.

Krampitz, G., Drolshagen, H. & Deltreil, J. P. (1983). Soluble matrix compounds in malformed oyster shells. *Experientia*, **39**, 1105–6.

Laughlin, R. B. & French, W. J. (1980). Comparative study of the acute toxicity of a homologous series of trialkyltins to larval shore crabs, *Hemigrapsus nudus*, and lobster, *Homarus americanus*. *Bulletin of Environmental Contamination and Toxicology*, **25**, 802–9.

Laughlin, R. B., French, W. J. & Guard, H. E. (1983). Acute and sublethal toxicity of tributyltin oxide, TBTO and its putative environmental product, tributyltin sulfide, TBTS to zoeal mud crabs *Rhithropanopeus harrissii*. *Water, Air and Soil Pollution*, **20**, 69–79.

Laughlin, R. B., French, W. J. & Guard, H. E. (1986). Accumulation of bis tributyltin oxide by the marine mussel *Mytilus edulis*. *Environmental Science and Technology*, **20**, 9, 884–90.

Laughlin, R. B., Johannesen, R. B., French, W., Guard, H. & Brinckman, F. E. (1985). Structure activity relationships for organotin compounds. *Environmental Toxicology and Chemistry*, **4**, 343–51.

Laughlin, R. B., Nordlund, K. & Linden, O. (1984). Longterm effects of tributyltin compounds on the baltic amphipod, *Grammarus oceanicus*. *Marine Environmental Research*, **12**, 243–71.

Laughlin, R. B., Pendoley, P. & Gustafson, R. G. (1987). Sublethal effects of tributyltin on the hard shell clam, *Mercenaria mercenaria*. In *Proceedings of the Organotin Symposium, Oceans '87 Conference*, Halifax, Nova Scotia, Canada, Sept. 28–Oct. 1, 1987, **4**, 1494–8.

Laurence, O. S., Cooney, J. J. & Gadd, G. M. (1989). Toxicity of organotins towards the marine yeast *Debaryomyces hansenii*. *Microbiology and Ecology*, **17**, 275–85.

Lawler, I. F. & Aldrich, J. C. (1987). Sublethal effects of bis, tri-n-butyltin oxide on *Crassostrea gigas* spat. *Marine Pollution Bulletin*, **18**, 6, 274–8.

Lee, R. F. (1986). Metabolism of bis tributyltin oxide by estuarine animals. In *Proceedings of the Organotin Symposium, Oceans '86 Conference*, Washington, Sept. 23–25, 1986, **4**, 1182–8.

Lenihan, H. S., Oliver, J. S. & Stephenson, M. (1990). *Marine Ecology Progress Series*, **60**, 147–59.

Linden, E., Bengtsson, B. E., Svanberg, O. & Sunstrom, G. (1979). The acute toxicity of 78 chemical and pesticide formulations against two brackish water organisms, the bleak, *Albumus albunus* and the harpacticoïd, *Nitocra spinipes*. *Chemosphere*, **11**, 843–851.

Liying, Z., Xiankun, L. & Bingyi, S. (1990). Toxic effects of organotin on marine diatoms. *Journal of Oceanography of the University of Quingdao*, **20**, 125–31.

Mackay, D. (1982). Correlation of bioconcentration factors. *Environmental Science and*

Technology, **16**, 274–8.

Maguire, R. J., Carey, J. H. & Hale, E. J. (1983). Degradation of the tri-n-butyltin. In *Proceedings of the Organotin Symposium, Oceans '86 Conference*, Washington, Sept. 23–25, 1986, **4**, 1177–81.

Maguire, R. J., Tkacz, R. J., Chau, Y. K., Bengert, G. A. & Wong, P. T. S. (1986). Occurrence of organotin compounds in water and sediment in Canada. *Chemosphere*, **15**, 253–74.

Maguire, R. J., Wong, P. T. S. & Rhamey, J. S. (1984). Accumulation and metabolism of tri-n-butyltin cation by a green algae, *Ankistrodesmus falcatus*. *Canadian Journal of Fisheries and Aquatic Sciences*, **41**, 537–40.

Meador, J. P. (1986). An analysis of photobehaviour of *Daphnia magna* exposed to tributyltin. In *Proceedings of the Organotin Symposium, Oceans '86 Conference*, Washington, Sept. 23–25, 1986, **4**, 1213–18.

Minchin, D., Duggan, C. B. & King, W. (1987). Possible effects on organotins on scallop recruitment. *Marine Pollution Bulletin*, **18**, 11, 604–8.

Moore, D. W., Dillon, T. M. & Suedel, B. C. (1991). Chronic toxicity of tributyltin to the marine polychaete worm, *Neanthes arenaceodentata*. *Aquatic Toxicology*, **21**, 181–98.

Nell, J. A. & Chvojka, R. (1992). The effect of bis-tributyltin oxide (TBTO) and copper on the growth of juvenile rock oysters *Saccostrea commercialis* (Iredale and Roughley) and Pacific oyster *Crassostrea gigas* Thunberg. *Science of the Total Environment*, **125**, 193–201.

Okoshi, K., Mori, K. & Nomura, T. (1987). Characteristics of shell chamber formation between the two local races in the japanese oyster, *Crassostrea gigas*. *Aquaculture*, **67**, 313–20.

Pinkney, A. E., Hall, L. W., Lenkevitch, M. J. & Burton, D. T. (1985). Comparison of avoidance response of an estuarine fish *Fundulus heteroclitus* and crustacean, *Palemonetes pugio*, to bis, tri-n-butyltin oxide. *Water, Air and Soil Pollution*, **25**, 33–40.

Puddu, A., Pettine, M., La Noce, T. & Pagnotta, R. (1989). Toxic effects of organotin compounds on marine phytoplankton. In *International Conference Heavy Metals in the Environment*, Geneva, September 1989, Vernet, J.-P. (ed.), **2**, 166–9.

Rexrode, M. (1987). Ecotoxicity of tributyltin. In *Proceedings of the Organotin Symposium, Oceans '87 Conference*, Halifax, Nova Scotia, Canada, Sept. 28–Oct. 1, 1987, **4**, 1443–55.

Robert, R. & His, E. (1981). Action de l'acétate de tributylétain sur les oeufs et larves D de deux mollusques d'intérêt commercial: *Crassostrea gigas*, Thunberg et *Mytilus galloprovincialis*, Lmk. *ICES, C. M.* 1981/F: 42.

Roberts, M. H., Bender, M. E., de Lisle, P. F., Sutton, H. C. & Williams, R. L. (1987). Sex ratio and gamete production in american oysters exposed to tributyltin in the laboratory. In *Proceedings of the Organotin Symposium, Oceans '87 Conference*, Halifax, Nova Scotia, Canada, Sept. 28–Oct. 1, 1987, **4**, 1471–6.

Salazar, M. H. & Salazar, S. M. (1987). Tributyltin effects on juvenile mussel growth. In *Proceedings of the Organotin Symposium, Oceans '87 Conference*, Halifax, Nova Scotia, Canada, Sept. 28–Oct. 1, 1987, **4**, 1504–10.

Short, J. W. & Thrower, F. P. (1986). Tri-n-butyltin caused mortality of chinook salmon, *Oncorhyncus tshawytscha* on transfer to a TBT-treated marine net pen. In *Proceedings of the Organotin Symposium, Oceans '86 Conference*, Washington, Sept. 23–25, 1986, **4**, 1202–5.

Stephenson, M. D., Smith, D. R., Goetzl, J., Ichikawa, G. & Martin, M. (1986). Growth abnormalities in mussels and oysters from areas with high levels of tributyltin in San Diego Bay. In *Proceedings of the Organotin Symposium, Oceans '86 Conference*, Washington, Sept. 23–25, 1986, **4**, 1246–51.

Stewart, C. & de Mora, S. J. (1992). Elevated tri(n-butyl)tin concentrations in shellfish and sediments from Suva Harbour, Fiji. *Applied Organometallic Chemistry*, **6**, 507–12.

Stromgren, T. & Bongard, T. (1987). The effect of tributyltin oxide on growth of *Mytilus edulis*. *Marine Pollution Bulletin*, **18**, 1, 30–1.

Suzuki, S., Fukagawa, T. & Takama, K. (1991). Isolation and characterization of tributyltin resistant marine Vibrio sp. In *II International Marine Biotechnology Conference*, Baltimore, MD (USA), 13–16 Oct. 1991, p. 88.

Thain, J. E. (1983). The acute toxicity of bis, tributyltin oxide to the adults and larvae of some marine organisms. *ICES, C. M.* 1983/E: 13.

Thain, J. E. (1986). Toxicity of TBT to bivalves: effects on reproduction, growth and survival. In *Proceedings of the Organotin Symposium, Oceans '86 Conference*, Washington, Sept. 23–25, 1986, **4**, 1306–13.

Thain, J. E. & Waldock, M. J. (1983). The effect of suspended sediment and bis, tributyltin oxide on the growth of *Crassostrea gigas* spat. *ICES, C. M.* 1983/E: 10.

Thain, J. E. & Waldock, M. J. (1986). The impact of tributyltin, TBT antifouling paints on molluscan fisheries. *Water Science and Technology*, **18**, 193–202.

Uchida, M. (1993). Inhibitory activity of organotin compounds against colony formation of estuarine bacteria. *Bulletin of Japanese Society of Science Fisheries*, **59**, 2037–42.

Uhler, A. D., Durell, G. S., Steinhauer, W. G. & Spellacy, A. M. (1993). Tributyltin levels in bivalve mollusks from the east and west coast of the United States: results from the 1988–1990 national status and trends mussel watch project. *Environmental Technology and Chemistry*, **12**, 139–53.

U'ren, S. C. (1983). Acute toxicity of bis, tributyltin oxide to a marine copepod. *Marine Pollution Bulletin*, **14**, 8, 303–6.

Valkirs, A. O., Davidson, B. M. & Seligman, P. F. (1987). Sublethal growth effects and mortality to marine bivalves from long term exposure to tributyltin. *Chemosphere*, **16**, 1, 201–20.

Veith, G. D., Defoe, D. L. & Bergstedt, B. V. (1979). Measuring and estimating the bioconcentration factor of chemicals in fish. *Journal of the Fisheries Research Board of Canada*, **36**, 9, 1040–8.

Vighi, M. & Calamari, D. (1985). QSARS for organotin compounds on *Daphnia magna*. *Chemosphere*, **14**, 11/12, 1925–32.

Waldock, M. J., Thain, J. E. & Miller, D. (1983). The accumulation and depuration of bistributyltin oxide in oysters: a comparison between the Pacific oyster, *Crassostrea gigas* and the European flat oyster, *Ostrea edulis. ICES, C. M.* 1983/E: 59.

Waldock, M. J., Thain, J. E. & Waite, M. E. (1987). The distribution and potential toxic effects of TBT in UK estuaries during 1986. *Applied Organometallic Chemistry*, **1**, 287–301.

Walsh, G. E. (1986). Organotin toxicities studies conducted with selected marine organisms at EPA'S environmental research laboratory, Gulf Breeze, Florida. In *Proceedings of the Organotin Symposium, Oceans '86 Conference*, Washington, Sept. 23–25, 1986, **4**, 1210–12.

Walsh, G. E., McLaughlan, L., Lores, E. M., Louie, M. K. & Deans, C. H. (1985). Effects of organotins on growth and survival of two marine diatoms *Skeletonema costatum* and *Thalassiosira pseudonana*. *Chemosphere*, **14**, 383–92.

Ward, G. S., Gramm, G. C., Parrish, P. R., Trachman, H. & Slesinger, A. (1981). Bioaccumulation and chronic toxicity of bistributyltin oxide TBTO: tests with a saltwater fish. In *Aquatic Toxicology and Hazard Assessment: Fourth Conference*, Branson & Dickson (eds), ASTM, STP 737, Philadelphia, pp. 183–200.

Weis, J. S., Weis, P. & Wang, F. (1987a). Developmental effects of tributyltin on the fiddler crab, *Uca pugilator* and the Killifish, *Fundulus heteroclitus*. In *Proceedings of the Organotin Symposium, Oceans '87 Conference*, Halifax, Nova Scotia, Canada, Sept. 28–Oct. 1, 1987, **4**, 1456–60.

Weis, J. S., Gottlieb, J. & Kwiatkowski, J. (1987b). Tributyltin retards regeneration and produces deformities of limbs in the fiddler crab, *Uca pugilator*. *Archives of Environmental Contamination and Toxicology*, **16**, 321–6.

Widdows, J. & Page, D. S. (1993). Effects of tributyltin and dibutyltin on the physiological energetics of the mussel *Mytilus edulis*. *Marine Environmental Research*, **35**, 233–49.

Wong, P. T. S., Chau, Y. K., Kramar, O. & Bengert, G. A. (1982). Structure toxicity relationship of tin compounds on algae. *Canadian Journal of Fisheries and Aquatic Sciences*, **39**, 483–8.

Yamada, H. & Takayanagi, K. (1992). Bioconcentration and elimination of bis(tributyltin) oxide (TBTO) and triphenyltin chloride (TPTC) in several marine fish species. *Water Research*, **26**, 12, 1589–95.

7

○ ○ ○ ○ ○ ○ ○ ○ ○ ○ ○ ○ ○ ○ ○ ○ ○ ○ ○ ○

TBT-induced imposex in neogastropod snails: masculinization to mass extinction

Peter E. Gibbs and Geoffrey W. Bryan[†]

7.1 Introduction

Prosobranch gastropods exhibit all types of sexuality but in most the sexes are separate and unchanged throughout the life of the individual (Fretter & Graham, 1962). Before the late 1960s, distinguishing males from females was a routine matter involving a simple examination to determine whether or not a specimen possessed a penis. The situation was radically changed around 1970 when surveys of populations of four neogastropod species in different parts of the world (both coasts of the United States and Atlantic Europe) revealed that some females were now penis-bearing (see Gibbs, Pascoe & Bryan, 1991c). In his short note on one of these species, the nassariid *Ilyanassa obsoleta*, Smith (1971) coined the term 'imposex' to describe 'a superimposition of male characters onto unparasitized and parasitized females'; his subsequent work (Smith, 1980, 1981a–d) established the primary cause of female masculinization (development of a penis and vas deferens (sperm duct)) to be exposure to the leachates of marine antifouling preparations containing tributyltin (TBT) compounds. In the case of *I. obsoleta*, no major effects on population ecology and reproduction were detectable, but later studies of other species demonstrated that imposex resulting from exposure to remarkably low TBT concentrations can, in fact, profoundly disrupt breeding activity and lead to population extinctions on a broad scale.

Marine antifouling paints containing tributyltin (TBT) compounds were introduced in the mid-1960s and rapidly gained popularity because of both their effectiveness and ease of maintenance. It was not until the late 1970s that the possible effect of leachates on non-target organisms became a matter of concern. Widespread disturbance in oyster production

[†]Geoffrey Bryan died 17 September 1993.

in French waters was attributed to usage of TBT paints and this prompted legislation in 1982 restricting their application to larger vessels (>25 m length) (see Alzieu, 1991). Similar regulations were introduced for UK waters in 1987 and subsequently for most European and North American regions. This action has resulted in TBT pollution declining in coastal waters but the problem still remains acute with the continued use of TBT paints on large vessels and, perhaps more significantly, the release of TBT persisting in estuarine and coastal sediments. TBT compounds are now known to produce a variety of pathological conditions at relatively low concentrations (see Rexrode, 1987; Bryan & Gibbs, 1991) but to date none has been demonstrated to rival the sensitivity of the imposex response of neogastropods.

'Imposex' is a term now widely used to describe the condition of female gastropods that exhibit male characteristics. However, some caution in the application of the term would seem appropriate. The definition implies the active intervention of an external agent causing an unnatural degree of masculinization; thus the presence of a vestigial penis on a female should not be identified as imposex *per se* unless corroborative evidence links the condition to TBT exposure. Examples of females with vestigial penes collected before the advent of TBT paints, such as those for *Urosalpinx cinerea*, should be regarded as having a separate status to those described after 1960 (see Gibbs & Bryan, 1994). Good evidence that imposex is a modern phenomenon has been provided by comparisons of historical and present-day samples from the same populations, as in the case of the European dog-whelk, *Nucella lapillus* (see Bryan *et al.*, 1986; Bailey & Davies, 1988). Nevertheless, even when dealing with a common and well-studied species such as *Nucella lapillus*, suitable historical collections are often surprisingly scarce.

Whilst females of some mesogastropods exhibit male characters, imposex studies have focused largely on neogastropods, the most advanced prosobranchs. Overall, the neogastropods are a fairly uniform taxonomic group in terms of their shell characters, external features and internal anatomy. They are distinguished from mesogastropods by the extension of the mantle edge and shell to form the siphon and siphonal canal respectively, in the structure of the radula and in the mode of proboscis retraction. All are carnivorous, often with a narrow choice of prey; this specialization in food is reflected in the wide variation in gut structure, but otherwise different species remain similar in organization and are often sympatric (Fretter & Graham, 1984). Males always have a penis, fertilization is internal and eggs are encapsulated within the oviduct before expulsion and attachment to the substratum via the ventral pedal gland, a pit-like structure on the sole of the female foot (see Fretter, 1941, 1946; Kool, 1993). Development within the capsule may be of relatively short duration, hatching taking place at the veliger stage to be followed by a planktonic phase of variable length; commonly, intracapsular development is prolonged, the planktonic phase suppressed and the young resemble miniature adults on hatching. To a large extent, the mode of larval development dictates the dispersive powers of the species and is thus an important factor in the assessment of the effects of imposex at the population level.

Females exhibiting signs of masculinization have been recorded for many species, genera and families within the suborder Neogastropoda. Although the coverage is wide, much of the basic information on imposex in relation to TBT pollution is centred on relatively few species (Table 7.1), notably the temperate forms *Nucella lapillus*, *Ocenebra erinacea*, *Ilyanassa obsoleta*, *Nassarius* (*Hinia*) *reticulatus* and *Lepsiella scobina*.

Table 7.1 *List of neogastropod species for which imposex and TBT pollution have been linked and which are mentioned in text. Geographical area of investigation is indicated. Species marked with an asterisk are those known to be sterilized by advanced imposex*

Family and species	Geographical area
Muricidae	
Haustrum haustorium (Gmelin)*	New Zealand
Lepsiella scobina (Quoy & Gaimard)*	New Zealand
Lepsiella vinosa (Lamarck)	SE Australia
Morula marginalba (Blainville)	SE Australia
Nucella emarginata (Deshayes)	NW North America
Nucella lamellosa (Gmelin)*	NW North America
Nucella lapillus (L.)*	Europe
Nucella lima (Martyn)*	NW North America
Ocenebra erinacea (L.)*	Europe
Thais haemastoma (L.)*	Europe
Thais orbita (Gmelin)*	SE Australia
Urosalpinx cinerea (Say)*	Europe
Nassariidae	
Ilyanassa obsoleta (Say)	NE North America
Nassarius (*Hinia*) *reticulatus* (L.)	Europe
Buccinidae	
Buccinum undatum (L.)	Europe
Conidae	
Conus spp.	W Australia

On a worldwide scale, there is a requirement to extend the scope of studies to include more tropical forms such as those included in the survey of Ellis & Pattisina (1990); conids appear to have potential (see Kohn & Almasi, 1993). The universality of the imposex response is impressive. In fact, where searched for, penis-bearing females seem to have been detected in virtually all neogastropods at all sites of TBT contamination: the only exception on record appears to be the buccinid *Cominella glandiformis* (see Smith & McVeagh, 1991).

The literature on imposex has expanded considerably since 1990 and a complete review of the subject is beyond the scope of this chapter. Several detailed accounts of the morphological changes involved during its development in the key species *N. lapillus* have been published (see for example Gibbs *et al.*, 1987; Gibbs, Pascoe & Burt, 1988; Oehlmann, Stroben & Fioroni, 1991) and events are only briefly summarized here. The same applies to a consideration of its employment as a biomonitoring tool for the detection of TBT pollution (see Gibbs & Bryan, 1994).

7.2 The imposex scenario as exemplified by *Nucella lapillus*

In male *Nucella lapillus*, as in all neogastropods, the penis is situated behind the right tentacle and sperm are carried to it across the mantle cavity through a subsurface duct (vas deferens) formed by an infolding and fusion of the epithelium. The posterior section of the vas deferens is enlarged to form the glandular 'prostate'. When *Nucella lapillus* females are exposed to TBT, masculinization proceeds in a dose-responsive fashion (Table 7.2). The degree of vas deferens and penis formation depends on the ambient TBT concentration: the initial stages (proximal sperm duct/penis primordium) appear at a TBT concentration below the level of chemical detection (<0.5 ng TBT-Sn l^{-1}), at $1-2$ ng TBT-Sn l^{-1} the vas deferens extends from the penis to the anterior oviduct. Imposex development to this stage does not inhibit breeding since the capsules constructed within the capsule gland enclosing the fertilized eggs can still be expelled through the oviduct opening (vulva). But when TBT levels exceed 2 ng TBT-Sn l^{-1} the vas deferens extends to overgrow the genital papilla so as to block the vulva; in advanced cases, hyperplasic tumours are formed and/or prostatic tissue replaces the bursa copulatrix (a pouched structure at the anterior end of the oviduct that receives the sperm during copulation). Closure of the vulva prevents the passage of capsules from the oviduct to the exterior and thus affected females are effectively sterilized. This blockage does not, however, inhibit capsule formation and these accumulate within the oviduct, frequently massing to such an extent that the distended oviduct wall ruptures. The capsule masses released may then fuse to the interior of the shell (see Gibbs & Bryan, 1986, Plate 1). Whilst females appear to survive for some time after incurring this injury, it is thought that such a major trauma must hasten death and account, at least in part, for the low proportion of females in declining populations.

Nucella lapillus has a 'direct' development, the young escaping from the capsule as 'crawlaways' into the parental habitat. The lack of a dispersive, pelagic phase renders the species particularly vulnerable to TBT pollution since there is no recruitment to sterilized populations (apart from occasional, chance strandings).

The complete sequence of events in the demise of Nucella lapillus populations from initial exposure to TBT pollution to eventual extinction required a considerable period of time. Blaber (1970) first observed penis-bearing females in populations on the inner shores of Plymouth Sound in 1969–70; these same populations were still extant in the mid- to late-1980s but composed solely of old individuals, most of which were male, the few females all being sterile. Thus, about two decades elapsed between the introduction of TBT paints and final extinction of some of the first-exposed populations. To a great extent, the length of this period

Table 7.2 *Summary of effects of TBT exposure on the reproductive system of female* Nucella lapillus, *based on laboratory and field observations (after Gibbs et al., 1988)*

TBT-Sn in water ($ng\,l^{-1}$)	Morphological modifications of genital tract
<0.5	Breeding normal. Development of penis and vas deferens. RPSI <5. VDSI <4.
1–2	Breeding capacity retained by some females; others sterilized by blockage of oviduct as indicated by presence of aborted capsule masses. RPSI 40+. VDSI >4.
3–5	Virtually all females sterilized. Oogenesis apparently normal. RPSI 40+. VDSI 5–6.
10+	Oogenesis suppressed. Spermatogenesis initiated.
20	Testis developed to variable extent. Vesicula seminalis with ripe sperm in most-affected animals.
100	Sperm-ingesting gland undeveloped in some individuals.

reflects the longevity of the species which in such a sheltered locality probably exceeds ten years. The sterilization of the inner Sound populations may well have been complete by the mid-1970s but remained undetected until 1984 when the current series of investigations began.

7.3 Characterization of imposex intensity

Given the appropriate TBT exposure, the masculinization of females can be so extensive as to effect a complete sex reversal. Females in which oogenesis is suppressed and spermatogenesis promoted to the stage of sperm production have been observed in several species including *Nucella lapillus* (Gibbs *et al.*, 1988), *Ocenebra erinacea* (Gibbs *et al.*, 1990) and *Lepsiella scobina* (Stewart *et al.*, 1992): this feature is often associated with an imposed sperm duct extending from penis to 'testis'. Whether or not these putative males ever become functional is questionable: it would seem to be highly unlikely since all females in the population would be similarly suffering from advanced imposex and therefore unable to copulate.

Relatively few individuals are encountered in which masculinization has proceeded to the point where the sex of the specimen is in question. For the most part, the female character of the oviduct is retained in that the capsule, albumen and sperm-ingesting glands (see Fretter, 1941) remain defined. Gauging the intensity of imposex thus centres on devising meaningful indices to measure the relative development of both penis and vas deferens. In terms of impact, the presence of a vas deferens is of greater significance because it is this tissue that causes oviduct blockage in some species. The presence of a penis does not appear to affect the functioning of the female but, in being the most conspicuous structure, penis expression has proved to be the most commonly used marker.

Comparing the size of the female penis with that of the male is a seemingly straightforward operation. However, the morphology of the penis is highly variable in different species: many taper to a point whilst others are blunt-ended; most are curved to some degree, and various structural elaborations in the form of side lobes are present in the more complex types. Examples are illustrated in Figure 7.1. Obtaining a reasonably accurate, quickly executed measure of size can thus pose difficulties and a variety of methods have been employed. The penis of *Nucella lapillus* is roughly columnar in form and the weight (or volume) is related to the cube of its length and thus the ratio (length of the female penis^3/length of male penis^3) × 100 provides a good approximation of the relative mass of the female penis (Bryan *et al.*, 1986). Mean values for females and males within a population (Relative Penis Size Index (RPSI): Gibbs *et al.*, 1987) when plotted over geographic areas clearly indicate point sources of TBT pollution. The RPSI is based on measurements of non-narcotized specimens; obviously, the volume to length relationship (and cubing) cannot be applied to narcotized specimens (cf. Oehlmann, Stroben & Fioroni, 1992). In the latter case, an RPSI based on uncubed lengths is appropriate, as reported by Stroben, Oehlmann & Fioroni (1992b). Subsequent authors have introduced variations of the index to suit penis morphology. Thus, Stewart *et al.* (1992) found an index based on relative penis lengths (RPLI) gave a better description of the more complex penis of *Lepsiella scobina*; similarly, for *Thais* species, RPSIs based on penis area (Wilson, Ahsanallah & Thompson, 1993) and ratio of penis weight to body weight (Foale, 1993) have been used. Regardless of methodology, female penis expression has proven to be a reliable indicator of TBT pollution and subject species include at least one exhibiting an annual cycle of genital development (*Ilyanassa obsoleta*: see Smith, 1980; Bryan *et al.*, 1989a; Curtis & Barse, 1990). It is particularly useful in areas of trace contamination (few ng TBT-Sn l^{-1}): at higher TBT levels, information on vas deferens formation and oviduct malformation becomes more relevant.

Vas deferens development has been studied in relatively few species. In young male *N. lapillus*, the development of this subsurface duct commences

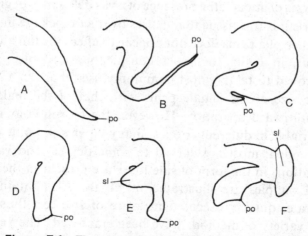

Figure 7.1 The penes of muricid species: a selection to illustrate diversity of form. (A) *Muricanthus fulvescens* (Sowerby); (B) *Nassa serta* (Bruguière); (C) *Thais nodosa* (L.); (D) *Cymia tecta* (Wood); (E) *Cronia amygdala* (Kiener); (F) *Morula uva* (Röding). Abbreviations: po, penial opening; sl, side lobe. (after Kool, 1993.)

at two separate sites, proximally close to the anterior prostate and distally as an extension of the penial duct; the two infoldings of the epithelium migrate towards each other across the floor of the mantle cavity and eventually join about halfway to complete the duct. The vas deferens imposed on females follows the same course of development and various stages can be used to construct an index. The basis of the Vas Deferens Sequence (VDS) index suggested by Gibbs *et al.* (1987) can be summarized briefly thus: the initiation of imposex is marked by the appearance of the proximal infolding (stage 1), followed by the appearance of the distal infolding (stage 2), extension of the two infolding sections (stage 3) and their subsequent fusion (stage 4). Vas deferens imposition to this stage does not hinder breeding activity but further tissue may overgrow the genital papilla causing the blockage described above (stage 5) and consequent retention of capsules (stage 6). Inevitably, a degree of variability is encountered, both between individuals and populations: stage 1 may not be apparent in some individuals/populations and occasionally penis formation is not initiated. The significance of these deviations may lie in the timing and dosage of TBT exposure or could have a genetic basis (see below).

The general principle of the VDS scheme has been applied to a number of species, suitably modified to suit specific variations, e.g. *Nucella lima* (Short *et al.*, 1989), *Thais haemastoma* (Spence, Hawkins & Santos, 1990) and *Lepsiella scobina* (Stewart *et al.*, 1992). An extended version of the VDS scheme that includes variations in the sequence of penis and vas deferens formation, has been utilized by Fioroni and co-workers (see, for example, Stroben, Oehlmann & Fioroni, 1992a). Further species need to be investigated to ascertain the full range of modifications to the female anatomy but it is evident that the structure of the male tract provides important clues as to the net effect of imposex when it is promoted to an advanced stage. From the limited information available it would appear the Muricidae, which includes all the sterilized species mentioned above (Table 7.1), are prone to sterilization because the male tract follows the same path within the mantle cavity as that of the female tract and thus its imposition disrupts the female organization; disruption may be compounded by the fact that in these species the large prostate is at the same level as the capsule gland section of the oviduct. The anatomy of the male tract in other groups differs significantly in that the course followed by the posterior section of the pallial vas deferens is parallel to the oviduct, not overlapping and, additionally, no prostate enlargement is evident. In these cases the vas deferens contacts the oviduct some distance behind the genital papilla and does not cause any major anatomical modification which interferes with copulation or capsule release. Examples of this type are found in at least the Nassariidae, e.g. *Ilyanassa obsoleta* (Smith, 1980), *Nassarius reticulatus* (Bryan & Gibbs, 1991; Bryan *et al.*, 1993b; Stroben *et al.*, 1992a) and Buccinidae, e.g. *Buccinum undatum* (Gibbs, unpublished) (Figure 7.2).

7.4   Mechanisms of sterilization

Sterilization results from the masculinization of the oviduct, especially during its development in the maturing subadult stage. The underlying causes preventing normal breeding activity have been studied in relatively few species and to date two mechanisms have been described but, given the many subtle variations of the basic structure of the neogastropod reproductive system that have evolved (see e.g. Kool, 1993), there can be little doubt other mechanisms remain to be discovered.

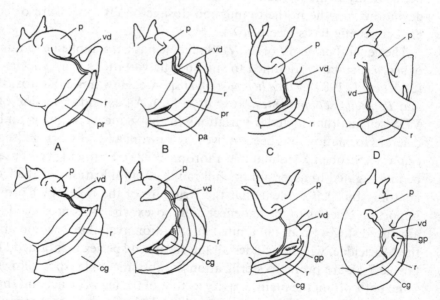

Figure 7.2 Comparison of the effect of imposex in species having male and female tracts following the same path and possessing a glandular prostate (A, *Nucella lapillus*; B, *Ocenebra erinacea*: both sterilized), and species in which the two tracts follow separate paths and no prostate is differentiated (C, *Nassarius reticulatus*; D, *Buccinum undatum*: breeding apparently unaffected in both). Upper line: males; lower line: females. Not drawn to scale. All specimens from the vicinity of Plymouth. Abbreviations: cg, capsule gland; gp, genital papilla; p, penis; pa, prostate aperture; pr, prostate; r, rectum; vd, vas deferens.

One mechanism is the simple blockage of the oviduct by vas deferens tissue (as described above for *N. lapillus*), a phenomenon readily identified by the accumulation of capsules within the oviduct; the other is the presence of a longitudinal split in the capsule gland of the adult, an abnormality noted in *Ocenebra erinacea* (Gibbs *et al.*, 1990; Oehlmann *et al.*, 1992) and *Urosalpinx cinerea* (Gibbs, Spencer & Pascoe, 1991d) which probably inhibits capsule formation (Figure 7.2) (see Gibbs, 1996).

In the case of *N. lapillus*, the blockage may be a simple overgrowth of the genital papilla causing occlusion of the vulva, but the process can involve a more profound reorganization of the anterior oviduct with prostate tissue replacing the bursa copulatrix: the latter certainly occurs in females reared experimentally (field and laboratory) in waters containing elevated TBT (Gibbs *et al.*, 1988). Oviduct blockage in *Nucella lamellosa* resembles that of *N. lapillus* (Bright & Ellis, 1990) whereas in *Lepsiella scobina* it occurs without any obvious overgrowth of the genital papilla (Stewart *et al.*, 1992). Masses of aborted capsules have also been detected in *Thais orbita* and *Haustrum haustorium* (Stewart *et al.*, 1992) but details have yet to be published.

The split oviduct condition is present in *Ocenebra erinacea* females of all ages in populations near to TBT sources and obviously the underlying cause lies in the early ontogeny of the female reproductive tract. It can be interpreted as a masculine character (Gibbs *et al.*, 1991c). There is general agreement that the closed condition of the neogastropod gonoduct arose by fusion of the edges of an open groove of the type found in primitive genera. In male *O. erinacea* the posterior section of the prostate remains open, the lumen connecting the mantle cavity via a slit-like aperture. This character is transferred to the oviduct as part of the masculinizing process that occurs during the development of the female tract (Figure 7.3). The same hypothesis can be applied to *Urosalpinx cinerea* but its wider application remains to be tested with the discovery of further examples of this type of mechanism.

7.5 Bioaccumulation of TBT and the mechanism/specificity of its effect

Uptake of TBT both from solution and from food have been studied experimentally in *Nucella lapillus* (Gibbs *et al.*, 1988; Bryan *et al.*, 1987, 1989b; Stroben *et al.*, 1992b), *Nucella lima* (Stickle *et al.*, 1990) and *Nassarius reticulatus* (Stroben *et al.*, 1992b). Results for these species are in general agreement. Long-term exposure of *Nucella lapillus*, for example,

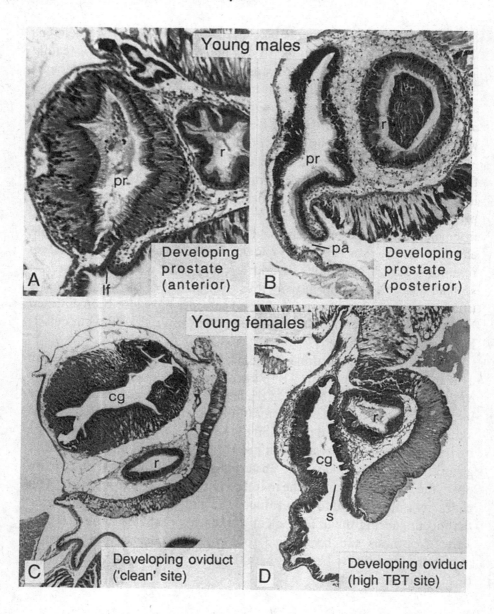

Figure 7.3 *Ocenebra erinacea*: masculinization of the oviduct. The structure (TS) of the developing prostate – anterior (A) and posterior (B) sections – compared to that of the oviduct (capsule gland) of young females from sites of trace ('clean') (C) and high TBT (D) pollution. Note similarity of (B) and (D). Male and female (C) from Bude, north Cornwall; (D) from mouth of Fal Estuary, south Cornwall. Abbreviations: cg, capsule gland; lf, line of gonoduct fusion; pa, prostate aperture; pr, prostate; r, rectum; s, split in capsule gland.

to water TBT concentrations below $20 \, \text{ng} \, \text{TBT-Sn} \, l^{-1}$ resulted in bioaccumulation factors of about 100 000 (dry tissue/water) after 6–9 months but at the highest concentration tested ($107 \, \text{ng} \, \text{TBT-Sn} \, l^{-1}$) a steady state of 30 000 was reached after about 3 months. When this species was fed on ^{14}C-TBT-labelled mussels (*Mytilus edulis*) in water containing about $8 \, \text{ng} \, \text{TBT-Sn} \, l^{-1}$ it was found that fed animals absorbed TBT two to three times more rapidly than controls exposed to the water only. After 49 days the diet accounted for about one half of the body burden and bioaccumulation factors for labelled TBT were 60 000 and 30 000 for fed and unfed animals respectively. Laboratory and field (transplantation) experiments have demonstrated that imposex in adult females is promoted with increased TBT exposure and consequent tissue accumulation. No general remission has been observed with a lowering of ambient TBT and a decline in tissue TBT through loss and depuration (half-times range from 50 to 100 days depending on conditions) although data on field populations of *Nassarius reticulatus* extending over six years suggest female penis lengths do decrease slowly (see Bryan *et al.*, 1993b).

The physiological and biochemical mechanisms underlying imposex are poorly understood but some evidence has been gained of TBT affecting both the neuroendocrine system and steroid metabolism. Using *in vitro* culture of *Ocenebra erinacea* tissues, Féral & Le Gall (1982, 1983) demonstrated that penis growth results from the release of neural factors from the pedal ganglia through the influence of TBT on the cerebropleural ganglia. Radioimmunoassays of steroid concentrations in *Nucella lapillus* (Spooner *et al.*, 1991) revealed an increase in the level of testosterone in females exposed to TBT but the levels of progesterone and oestradiol 17β were unchanged; injection of testosterone in the absence of TBT was found to promote female penis development. These observations suggest imposex is caused by the accumulation of testosterone which, in turn, is caused by an inhibition of the cytochrome P-450 dependent aromatase responsible for the conversion of testosterone to oestradiol 17β. Further studies are required to determine whether aromatase activity is affected directly by TBT exposure or whether such an inhibition is the result of an influence of TBT on the neuroendocrine system controlling steroid metabolism (Spooner *et al.*, 1991). With respect to the latter, the observed affinity of nervous tissue for butyltins (Bryan *et al.*, 1993a) may be highly significant.

The fundamental question of the specificity of the imposex response has received surprisingly little attention. Bryan, Gibbs & Burt (1988) investigated this aspect by exposure of *Nucella lapillus* to both dissolved and injected organotins, namely tetrabutyltin (TTBT), tri-*n*-butyltin (TBT), dibutyltin (DBT), monobutyltin (MBT), tri-*n*-propyltin (TPrT) and triphenyltin (TPT). Injection was employed to obviate the problem of differential absorption through the body wall. DBT, MBT and TPT produced no measurable effect in terms of promoting penis development; a response was produced by TTBT and TPrT but TBT was by far the most effective promoter. The response invoked by TTBT may have been an effect of TBT as impurity but the degradation of TTBT to TBT in the tissues is a more likely explanation. It was concluded that the capacity of TPrT and TTBT to promote imposex does not detract from the use of imposex as a TBT indicator since TPrT is a little-used compound and the effect of TTBT is probably dependent on its degradation to TBT.

Imposex has been used successfully as an indicator of TBT pollution in many parts of the world, and its use validated by numerous field and laboratory experiments. However, doubts as to the TBT specificity of the imposex response in *Lepsiella vinosa* have been expressed by Nias, McKillup & Edyvane (1993): their experimental results suggest it could be induced additionally by the paint matrix, copper or even environmental stress. With regard to copper as a promotor of imposex, negative results were obtained by Bryan *et al.* (1987) in field experiments wherein the shells of *Nucella lapillus* were coated with a copper-based paint.

7.6   Recovery from imposex following amelioration of TBT pollution

Studies of *N. lapillus* (see above) have shown that the intensity of imposex development is dose-dependent, irreversible and most pronounced in the late juvenile stage when the formation of the reproductive tract is proceeding. These characteristics facilitate the measurement, by means of indices, of any intensification of imposex shown by populations during periods of increasing pollution since not only do juveniles exhibit a higher level of response but that of adults increases also (as demonstrated by transplantation experiments: see Gibbs & Bryan, 1994). However, because of its irreversibility, the use of imposex during periods of declining pollution becomes a less sensitive monitoring method because a decline in intensity within a population only becomes apparent with the replacement of more-affected females with less-affected. The rate at which this turnover occurs depends largely on the life-history characteristics of the population/ species under consideration, particularly age at maturity and longevity. In the case of *N. lapillus*, for example, females mature towards the end of their second or third year (Moore, 1938; Feare, 1970; Gibbs *et al.*, 1988) and thus, in situations of declining TBT, less-affected individuals will enter the adult population only after an interval of two or three years; given that individuals survive at least six or seven years (Feare, 1970), probably much longer in sheltered habitats, a significant drop in imposex intensity of the overall population will be noticeable only after appreciable mortality of the older, more affected females. Thus, a period of perhaps more than five years can be expected to elapse before an appreciable decline in imposex intensity will reflect an amelioration in pollution levels.

Recovery can be considered at three levels of effect: (i) a lessening of the intensity of imposex in those populations exhibiting early to intermediate imposex (VDS stages 1–4) in which breeding is not thought to have been compromised; (ii) an increase in the breeding capacity of those populations in which sterilization by imposex has caused a reduction in fecundity; and (iii) recolonization of those areas denuded of individuals. In non-sterilized species (e.g. *Nassarius reticulatus*) only (i) will apply. All three levels have been studied in *Nucella lapillus* around southern England before and after the 1987 UK legislation.

In an area where TBT levels are below the limit of reliable detection by chemical means (less than $0.5 \, \text{ng} \, \text{TBT-Sn} \, l^{-1}$) female penis length has been used to gauge the effectiveness of the paint restrictions. In the relatively clean waters of the north Cornwall coast all adult *N. lapillus* females exhibit imposex but no sterilized individual has been discovered. Mean penis lengths for adults and subadults (defined below) in the years 1985–93 are shown in Figure 7.4. In adults, penes reached maximal development in the 1987–8 winter, some six months after the legislation (no doubt boat painting was intense so as to consume existing paint stocks); subsequently, the overall trend has been towards a decrease in penis length but with increased sampling variation. This latter is attributable to the recruitment to the adult population of individuals that have suffered less exposure. Subadults sampled during the winters of 1989–90 and 1990–1 showed minimal penis development and these less-affected cohorts would be represented in the samples of adults taken in 1992 and 1993.

Greater sensitivity in monitoring can be achieved by restricting observations to young animals. In populations of *N. lapillus* that exhibit continuous breeding (capsules present throughout the year) it is possible to select subadult females (using size and shell characters) that have yet to reach maturity (i.e. oviduct and ovary still developing) and aged about one year. This stage is thought to be the most sensitive period in terms of susceptibility to TBT-induced abnormality. The degree of imposex in

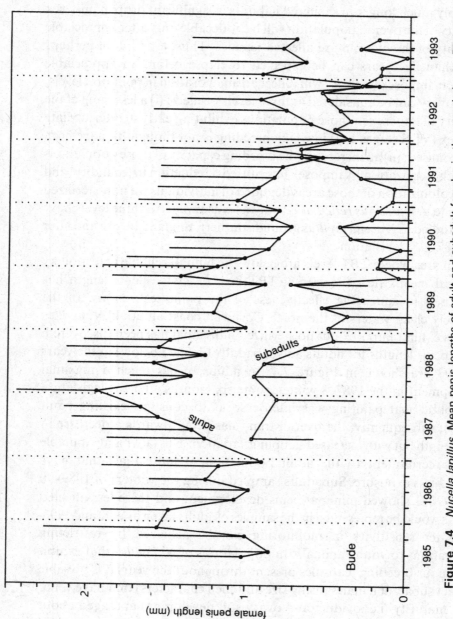

Figure 7.4 *Nucella lapillus.* Mean penis lengths of adult and subadult females in the population at Bude, north Cornwall, 1985–93. Vertical bars indicate standard deviations: values for adults based on samples of 10–20 individuals, those for subadults on 20–30.

such animals can be used to gauge contamination levels during the year preceding sampling, imposex development being expressed in terms of, for example, mean penis length as described above, or simply as frequency of penis development (VDS stages 3 and 4). The latter is illustrated in Figure 7.5: overall, penis development has declined since the 1987 legislation but an underlying seasonal fluctuation is detectable (relatively fewer in winter–spring) and this may reflect subtle changes in contamination levels associated with the intensity of boating activity. Overall, the data indicate that even in areas far removed from a major source of TBT, imposex remains readily detectable some six years after legislation restricting paint usage (see also Evans *et al.*, 1994).

In an area of relatively high TBT pollution, recovery of the breeding capacity in a population exhibiting a condition of advanced imposex is illustrated by the data for the *N. lapillus* population at Renney Rocks, close to Plymouth, which has been surveyed over the eight-year period 1986–93 (Figure 7.6). In the years 1986–90 some 70–80% of the females were sterile (VDS stages 5 and 6). The first sign of recovery did not appear

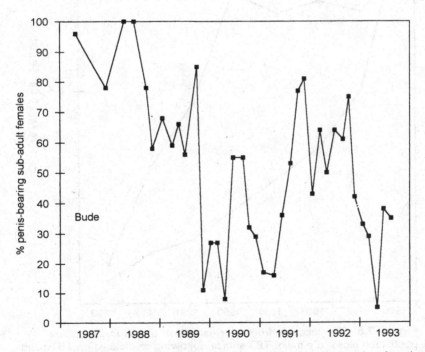

Figure 7.5 *Nucella lapillus.* Percentages of penis-bearing females in the subadult population at Bude, north Cornwall, 1987–93. Based on samples of 20–30 individuals.

until 1991 when this percentage dropped to below 50% and this trend continued so that in 1993 less than 10% were found to be sterile. Levels of TBT in the water at this site have shown a general reduction: apart from one or two occasions when peaks have appeared (naval operations?) the background is now below the critical level of $2\,\mathrm{ng\,TBT\text{-}Sn\,l^{-1}}$. TBT concentrations in female body tissues have declined markedly and hence the degree of masculinization has decreased. Young females are no longer being sterilized before they reach maturity and, with the death of the older, sterilized females, the proportion of fertile individuals has increased

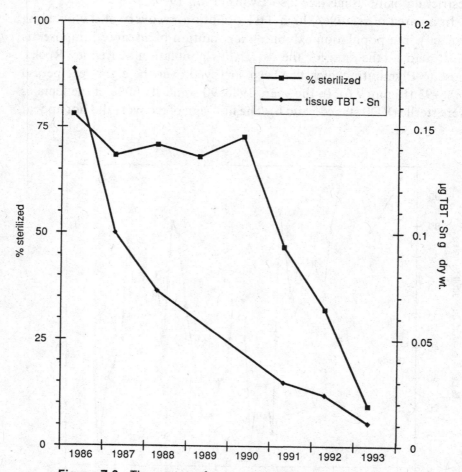

Figure 7.6 The recovery from near-extinction of a *Nucella lapillus* population close to a major TBT source, following restrictions on TBT paint usage in 1987, as indicated by percentages of females sterilized. Tissue TBT-Sn concentrations are also shown. Data relate to the population at Renney Rocks, at the entrance to Plymouth Sound.

to the point where the future of the population, which was formerly threatened, now seems secure.

In Plymouth Sound virtually all of the *Nucella lapillus* population(s) inhabiting the inner shores were exterminated during the 1980s. These shores have been kept under surveillance but no recolonisation has been detected despite declining water TBT concentrations; this is attributable to the poor powers of dispersal of the species. On a broader geographic scale, similar exterminations took place close to the major ports along the south coast of England (Figure 7.7). With the continued use of TBT paints on large vessels there seems little prospect of recovery at any level in areas close to major shipping lanes; for example, recolonization of the north coast of the Isle of Wight bordering Southampton Water (see Langston *et al.*, 1990) is unlikely in the foreseeable future.

The life-history patterns of species within the genus *Nucella* are diverse, and consequently differences will also be found in the response to imposex. The population effects observed for *N. lapillus* may not be typical for all species. The life histories of the Pacific coast forms *N. emarginata* and *N. lamellosa* have been extensively studied. *N. emarginata* matures, and typically dies, within one year and breeding is continuous throughout the year. *N. lamellosa* remains juvenile for nearly four years, many (40–60%) surviving beyond the first year, and breeding is restricted to the spring months (Spight, 1979). Because of the differences in maturation periods and population turnover rates for the two species the levels of imposex exhibited by adults will reflect ambient TBT on different

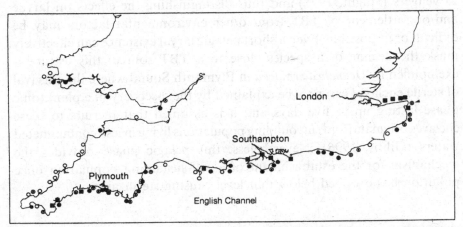

Figure 7.7 *Nucella lapillus.* Incidence of female sterility around southern Britain between 1986 and 1989: (○) none sterilized; (⊙) less than 50%; (●) more than 50%; (■) site where population recorded previously but now absent. Site of enclave exhibiting Dumpton Syndrome is arrowed.

time scales, i.e. about 6–12 months in *N. emarginata* but 2–3 years in *N. lamellosa*. The continuous breeding of *N. emarginata* may be advantageous in that at least some juveniles should avoid a coincidence of the most sensitive period of development with the highest TBT levels during the summer pleasure-boating season. Interestingly, Bright & Ellis (1990) found this upper-shore species remained abundant and without any sign of sterilization whilst the lower shore *N. lamellosa* was sterilized in the same fashion as *N. lapillus* and numbers were declining.

Few species have been studied with respect to population turnover rates but it is worth noting that some muricids are very long-lived. For example, longevity has been estimated to exceed ten years in *Shaskyus festivus* (Hinds) (19.6 y: Fotheringham, 1971), *Ocenebra poulsoni* Carpenter (14.8 y: Fotheringham, 1971), *Dicathais orbita* (Gmelin) (19 y: Philips & Campbell, 1974) and *Urosalpinx cinerea* (Say) (14 y: Cole, 1942). The inclusion of females of such extreme age in populations will inevitably cause some distortion of any assessment of present-day pollution using imposex. On the other hand, there is some evidence from sex ratios in *N. lapillus* populations showing signs of terminal decline that males tend to outlive females (see Bryan *et al.*, 1986), possibly because the effects of advanced imposex may abbreviate the normal female life-span.

Another factor to be considered in assessing the condition of populations in contaminated areas is recruitment. For intertidal muricids, most temperate forms hatch as miniature adults and thus the success of breeding can be readily gauged; however, in the tropics most forms hatch as veligers (Spight, 1977) and thus distinguishing the effects on larvae and/or settlement of TBT from other environmental factors may be difficult or impossible. Even a short pelagic larval existence can effectively mask the decline of a species close to a TBT source: this feature is exemplified by *Ocenebra erinacea* in Plymouth Sound where the survival of sterile enclaves can now be explained by the discovery of a planktonic phase lasting up to five days and it is assumed that recruits to these enclaves originate from surrounding populations living in less-contaminated waters (Gibbs, 1996). Nevertheless, this pelagic phase provides the mechanism for the establishment of a self-sustaining population once pollution has dropped below that level causing sterilization.

7.7   Genetic aspects of imposex

Reference has been made above to the variability that can be observed in the expression of imposex, especially in *Nucella lapillus*. The focus on this species can be explained not only by the fact that the species is one of the best-known of neogastropods but also because its restricted powers of dispersal (limited adult movement; development without a planktonic phase) has favoured the evolution of local forms or enclaves. In fact, variation is a basic feature of this widespread form, not only in its shell morphology (see, for example, Crothers, 1985; Kitching, 1985) but also in its chromosomal (Robertsonian) polymorphism, manifest in the 2n chromosome number ranging between 26 and 36 (see Dixon *et al.*, 1994). Oehlmann *et al.* (1991) consider that different chromosomal complements give rise to different paths of imposex development although this idea remains to be substantiated.

Given the multitude of discrete enclaves of *Nucella lapillus* that exist, it would perhaps be surprising not to find at least some enclaves that show signs of resistance to the masculinizing effect of TBT. Certainly, imposex expression in certain *Nucella lapillus* enclaves shows significant variation, most commonly in the degree of penis development, exceptionally in both penis and vas deferens development. One simple example was discovered (in 1989) on the south Brittany coast at Pte du Cabellou, close to the port of Concarneau: at this site about half the females had large penes and were mostly sterile (the typical state for such a polluted locality), but the remainder had small penes and their breeding capacity was seemingly unimpaired (see Gibbs, Bryan & Pascoe, 1991a). A more widespread variant involves development of the vas deferens but not the penis, a possible indication that formation of the two organs may be separately controlled. Female 'aphally' is generally not common but has been noted in populations around south-west England (Gibbs, unpublished) and Brittany (Oehlmann *et al.*, 1991, as stage 3b). Since penis differentiation appears to be under hormonal control (see above), reduced or suppressed development could arise through environmental influence(s) but an underlying genetic factor may be operating.

A third, and rare, anomaly is an apparent absence of imposex in some females in a population otherwise showing advanced imposex. Before the amelioration of TBT pollution, it was usual to find all females within any population exhibited a similar level of imposex, i.e. the VDS stages occur in clusters – all at stages 0–2, or 3–4, or 4–6, or in heavily polluted sites, only 5–6 – a feature indicative of steady-state or increasing pollution. Populations containing females at all stages of imposex (VDS stages 0–6) are atypical and warrant close inspection. A striking example of this was discovered on the North Foreland coast of SE England (Kent) in a region where TBT pollution has eliminated virtually all neogastropods (see

Gibbs, Bryan & Spence, 1991b; Gibbs, 1993). The survival of an enclave centred on Dumpton Gap, close to the ferry port of Ramsgate (Figure 7.7), obviously deserved attention, especially when it was noted that not only were many females aphallic, but a proportion of the males lacked penes also. The importance of this feature was realised when a literature search failed to uncover any previous report of a neogastropod male lacking a penis. In fact, the genital deficiency in the Dumpton males involved the whole gonoduct in that not only was the penis undeveloped or underdeveloped but the vas deferens was incompletely closed throughout its length, with the prostate remaining small and open to the mantle cavity. The same characters (collectively termed 'Dumpton Syndrome') appeared in laboratory-bred animals; hence the deficiency is thought to be genetically based. Given the axiom of imposex being a faithful transfer of specific male characters to the female, it follows that underdevelopment in the male will result in the same abnormality being manifest when imposed on the female. Hence, the sterilizing influence of imposex is lessened in carrier females which are able to continue breeding with non-carrier males. Dumpton Syndrome appears to be one of those rare examples of an otherwise deleterious mutation conferring an advantage in a particular environment. Whether the disorder was present before the advent of TBT paints or whether it has arisen since the mid-1960s is unknown, but whatever its origin it seems to have favoured the survival of an isolated enclave in an area of high TBT pollution. No doubt selection has increased its frequency over the last two decades concomitant with the rise in pollution in the Thames Estuary and Dover Strait: whether a state of equilibrium has now been reached should be evident from the results of ongoing studies.

7.8   Summary

Studies of imposex have now been carried out on a variety of species. Sufficient evidence has been accumulated to define the character and time-course of the response:

(1) it is induced by exposure to TBT at water concentration about or below $0.5 \, \text{ng} \, \text{TBT-Sn} \, l^{-1}$;

(2) its degree of development is dose-related; indices of intensity provide accurate bioindicators of exposure;

(3) it is irreversible, no major remission being evident with a decline in the TBT body burden; hence its intensity is an indicator of past exposure, particularly during juvenile life;

(4) the specific characters of the male tract are transferred to the female;

(5) depending on (4), sterilization may or may not result when development is advanced;

(6) the effects of sterilization at the population level may be masked in those species having a dispersive phase in their life history;

(7) terminally declining populations of sterilized species with non-planktonic development are composed solely of old individuals, often predominantly male; the rate of extinction can be expected to be related to the longevity of the species.

7.9 References

Alzieu, C., 1991. Environmental problems caused by TBT in France: assessment, regulations, prospects. *Mar. envir. Res.* **32**: 7–17.

Bailey, S. K. & I. M. Davies, 1988. Tributyltin contamination in the Firth of Forth (1975–87). *Sci. total Envir.* **76**: 185–92.

Blaber, S. J. M., 1970. The occurrence of a penis-like outgrowth behind the right tentacle in spent females of *Nucella lapillus* (L.). *Proc. malac. Soc. Lond.* **39**: 231–3.

Bright, D. A. & D. V. Ellis, 1990. A comparative survey of imposex in northeast Pacific neogastropods (Prosobranchia) related to tributyltin contamination, and choice of a suitable bioindicator. *Can. J. Zool.* **68**: 1915–24.

Bryan, G. W. & P. E. Gibbs, 1991. Impact of low concentrations of tributyltin (TBT) on marine organisms. In *Ecotoxicology of Metals: Current Concepts and Application*, Newman, M. C. & A. W. McIntosh (eds), Lewis Publishers Inc., Ann Arbor, Boca Raton, Boston, pp. 323–61.

Bryan, G. W., D. A. Bright, L. G. Hummerstone & G. R. Burt, 1993a. Uptake, tissue distribution and metabolism of ^{14}C-labelled tributyltin (TBT) in the dog-whelk, *Nucella lapillus*. *J. mar. biol. Ass. U.K.* **73**: 889–912.

Bryan, G. W., G. R. Burt, P. E. Gibbs & P. L. Pascoe, 1993b. *Nassarius reticulatus* (Nassariidae: Gastropoda) as an indicator of tributyltin pollution before and after TBT restrictions. *J. mar. biol. Ass. U.K.* **73**: 913–29.

Bryan, G. W., P. E. Gibbs & G. R. Burt, 1988. A comparison of the effectiveness of tri-n-butyltin chloride and five other organotin compounds in promoting the development of imposex in the dog-whelk, *Nucella lapillus*. *J. mar. biol. Ass. U.K.* **68**: 733–44.

Bryan, G. W., P. E. Gibbs, G. R. Burt & L. G. Hummerstone, 1987. The effects of tributyltin (TBT) accumulation on adult dog-whelks, *Nucella lapillus*: long-term field and laboratory experiments. *J. mar. biol. Ass. U.K.* **67**: 525–44.

Bryan, G. W., P. E. Gibbs, R. J. Huggett, L. A. Curtis, D. S. Bailey & D. M. Dauer, 1989a. Effects of tributyltin pollution on the mud snail, *Ilyanassa obsoleta*, from the York River and Sarah Creek, Chesapeake Bay. *Mar. Pollut. Bull.* **20**: 458–62.

Bryan, G. W., P. E. Gibbs, L. G. Hummerstone & G. R. Burt, 1986. The decline of the gastropod *Nucella lapillus* around south-west England: evidence for the effect of tributyltin from antifouling paints. *J. mar. biol. Ass. U.K.* **66**: 611–40.

Bryan, G. W., P. E. Gibbs, L. G. Hummerstone & G. R. Burt, 1989b. Uptake and transformation of ^{14}C-labelled tributyltin chloride by the dog-whelk, *Nucella*

234 **P. E. Gibbs & G. W. Bryan**

lapillus: importance of absorption from the diet. *Mar. envir. Res.* **28**: 241–5.

Cole, H. A., 1942. The American whelk tingle, *Urosalpinx cinerea* (Say), on British oyster beds. *J. mar. biol. Ass. U.K.* **25**: 477–508.

Crothers, J. H., 1985. Dog-whelks: an introduction to the biology of *Nucella lapillus* (L.). *Field Stud.* **6**: 291–360.

Curtis, L. A. & A. M. Barse, 1990. Sexual anomalies in the estuarine snail *Ilyanassa obsoleta*: imposex in females and associated phenomena in males. *Oecologia*, **84**: 371–5.

Dixon, D. R., P. L. Pascoe, P. E. Gibbs & J. J. Pasantes, 1994. The nature of Robertsonian chromosomal polymorphism in *Nucella lapillus*: a re-examination. In *Genetics and Evolution of Aquatic Organisms*, A. R. Beaumonnt, (ed.), Chapman & Hall, London pp. 389–99.

Ellis, D. V. & L. A. Pattisina, 1990. Widespread neogastropod imposex: a biological indicator of global TBT contamination. *Mar. Pollut. Bull.* **21**: 248–53.

Evans, S. M., S. T. Hawkins, J. Porter & A. M. Samosir, 1994. Recovery of dogwhelk populations on the Isle of Cumbrae, Scotland following legislation limiting the use of TBT as an antifoulant. *Mar. Pollut. Bull.* **28**: 15–17.

Feare, C. J., 1970. Aspects of the ecology of an exposed shore population of dogwhelks *Nucella lapillus* (L.). *Oecologia (Berlin)*, **5**: 1–18.

Féral, C. & S. Le Gall, 1982. Induction expérimentale par un polluant marin (le tributylétain), de l'activité neuroendocrine contrôlant la morphogenése du pénis chez les femelles d'*Ocenebra erinacea* (Mollusque Prosobranche gonochorique). *C. r. hebd. séanc. Acad. Sci.* **295**: 627–30.

Féral, C. & S. Le Gall, 1983. The influence of a pollutant factor (tributyltin) on the neuroendocrine mechanism responsible for the occurrence of a penis in the females of *Ocenebra erinacea*. In *Molluscan Neuro-endocrinology*. Proceedings of the International Minisymposium on Molluscan Endocrinology, 1982, Lever, J. & H. H. Boer (eds), North Holland Publishing Company, Amsterdam, pp. 173–5.

Foale, S., 1993. An evaluation of the potential of gastropod imposex as a bioindicator of tributyltin pollution in Port Philip Bay, Victoria. *Mar. Pollut. Bull.* **26**: 546–52.

Fotheringham, N., 1971. Life history patterns of the littoral gastropods *Shaskyrus festivus* (Hinds) and *Ocenebra poulsoni* Carpenter (Prosobranchia: Muricidae). *Ecology*, **52**: 742–57.

Fretter, V., 1941. The genital ducts of some British stenoglossan prosobranchs. *J. mar. biol. Ass. U.K.* **25**: 173–211.

Fretter, V., 1946. The pedal sucker and anal gland of some British Stenoglossa. *Proc. malac. Soc. Lond.* **27**: 126–30.

Fretter, V. & A. Graham, 1962. *British Prosobranch Molluscs: Their Functional Anatomy and Ecology.* London, Ray Society.

Fretter, V. & A. Graham, 1984. The prosobranch molluscs of Britain and Denmark. Part 8. *J. mollusc. Stud. Suppl.* **15**: 435–556.

Gibbs, P. E., 1993. A male genital defect in the dog-whelk, *Nucella lapillus*, (Neogastropoda), favouring survival in a TBT-polluted area. *J. mar. biol. Ass. U.K.* **73**: 667–78.

Gibbs, P. E., 1996. Oviduct malformation as a sterilising effect of tributyltin-(TBT)-induced imposex in *Ocenebra erinacea* (Muricidae: Gastropoda). *J. mollusc. Stud.* (in press).

Gibbs, P. E. & G. W. Bryan, 1986. Reproductive failure in populations of the dog-whelk,

Nucella lapillus, caused by imposex induced by tributyltin from antifouling paints. *J. mar. biol. Ass. U.K.* **66**: 767–77.

Gibbs, P. E. & G. W. Bryan, 1994. Biomonitoring of tributyltin (TBT) pollution using the imposex response of neogastropod molluscs. In *Biomonitoring of Coastal Waters and Estuaries*, K. J. M. Kramer (ed.), CRC Press Inc., Boca Raton, Florida, pp. 205–26.

Gibbs, P. E., G. W. Bryan, P. L. Pascoe & G. R. Burt, 1987. The use of the dog-whelk, *Nucella lapillus*, as an indicator of tributyltin (TBT) contamination. *J. mar. biol. Ass. U.K.* **67**: 507–23.

Gibbs, P. E., P. L. Pascoe & G. R. Burt, 1988. Sex change in the female dog-whelk, *Nucella lapillus*, induced by tributyltin from antifouling paints. *J. mar. biol. Ass. U.K.* **68**: 715–31.

Gibbs, P. E., G. W. Bryan, P. L. Pascoe & G. R. Burt, 1990. Reproductive abnormalities in female *Ocenebra erinacea* (Gastropoda) resulting from tributyltin-induced imposex. *J. mar. biol. Ass. U.K.* **70**: 639–56.

Gibbs, P. E., G. W. Bryan & P. L. Pascoe, 1991a. TBT-induced imposex in the dog-whelk, *Nucella lapillus*: geographical uniformity of the response and effects. *Mar. envir. Res.* **32**: 79–87.

Gibbs, P. E., G. W. Bryan & S. K. Spence, 1991b. An assessment of the impact of TBT pollution on the populations of the dog-whelk, *Nucella lapillus*, around the coastline of south-east England (Sussex and Kent). *Oceanologica Acta*, Spec. Vol. **11**: 257–61.

Gibbs, P. E., P. L. Pascoe & G. W. Bryan, 1991c. Tributyltin-induced imposex in stenoglossan gastropods: pathological effects on the female reproductive system. *Comp. Biochem. Physiol.* **100C**: 231–5.

Gibbs, P. E., B. E. Spencer & P. L. Pascoe, 1991d. The American oyster drill, *Urosalpinx cinerea* (Gastropoda): evidence of decline in an imposex-affected population (River Blackwater, Essex). *J. mar. biol. Ass. U.K.* **71**: 827–38.

Kitching, J. A., 1985. The ecological significance and control of shell variability in dogwhelks from temperate rocky shores. In *The Ecology of Rocky Coasts*, P. G. Moore & R. Seed (eds), Hodder & Stoughton, London, pp. 234–48.

Kohn, A. J. & K. N. Almasi, 1993. Imposex in Australian *Conus*. *J. mar. biol. Ass. U.K.* **73**: 241–4.

Kool, S. P., 1993. Phylogenetic analysis of the Rapaninae (Neogastropoda: Muricidae). *Malacologia*, **35**: 155–259.

Langston, W. J., G. W. Bryan, G. R. Burt & P. E. Gibbs, 1990. Assessing the impact of tin and TBT in estuaries and coastal regions. *Funct. Ecol.* **4**: 433–43.

Moore, H. B., 1938. The biology of *Purpura lapillus*. Part II. Growth. *J. mar. biol. Ass. U.K.* **23**: 57–66.

Nias, D. J., S. C. McKillup & K. S. Edyvane, 1993. Imposex in *Lepsiella vinosa* from southern Australia. *Mar. Pollut. Bull.* **26**: 380–4.

Oehlmann, J., E. Stroben & P. Fioroni, 1991. The morphological expression of imposex in *Nucella lapillus* (Linnaeus)(Gastropoda: Muricidae). *J. mollusc. Stud.* **57**: 375–90.

Oehlmann, J., E. Stroben & P. Fioroni, 1992. The rough tingle *Ocenebra erinacea* (Neogastropoda: Muricidae): an exhibitor of imposex in comparison to *Nucella lapillus*. *Helgolander Meeresuntes.* **46**: 311–28.

Philips, B. F. & N. A. Campbell, 1974. Mortality and longevity in the whelk *Dicathais orbita* (Gmelin). *Aust. J. mar. Freshwat. Res.* **25**: 25–33.

Rexrode, M., 1987. Ecotoxicology of tributyltin. In *Proceedings Oceans '87 Conference*,

Halifax, Nova Scotia, September 28–October 1, 1987. Vol. 4. International Organotin Symposium. Institute of Electrical and Electronics Engineers, New Jersey, pp. 1443–55.

Short, J. W., S. D. Rice, C. C. Brodersen & W. B. Stickle, 1989. Occurrence of tri-*n*-butyltin-caused imposex in the North Pacific marine snail *Nucella lima* in Auke Bay, Alaska. *Mar. Biol.* **102**: 291–7.

Smith, B. S., 1971. Sexuality in the American mud-snail *Nassarius obsoletus* Say. *Proc. malac. Soc. Lond.* **39**: 377–8.

Smith, B. S., 1980. The estuarine mud snail, *Nassarius obsoletus*: abnormalities in the reproductive system. *J. mollusc. Stud.* **46**: 247–56.

Smith, B. S., 1981a. Reproductive anomalies in stenoglossan snails related to pollution from marinas. *J. appl. Toxic.* **1**: 15–21.

Smith, B. S., 1981b. Male characteristics on female mud snails caused by antifouling bottom paints. *J. appl. Toxic.* **1**: 22–5.

Smith, B. S., 1981c. Tributyltin compounds induce male characteristics on female mud snails *Nassarius obsoletus* = *Ilyanassa obsoleta*. *J. appl. Toxic.* **1**: 141–4.

Smith, B. S., 1981d. Male characteristics in female *Nassarius obsoletus*: variations related to locality, season and year. *Veliger*, **23**: 212–16.

Smith, P. J. & M. McVeagh, 1991. Widespread organotin pollution in New Zealand coastal waters as indicated by imposex in dogwhelks. *Mar. Pollut. Bull.* **22**: 409–13.

Spence, S. K., S. J. Hawkins & R. S. Santos, 1990. The mollusc *Thais haemastoma* – an exhibitor of 'imposex' and potential indicator of tributyltin pollution. *Mar. Ecol.* **11**: 147–56.

Spight, T. M., 1977. Latitude, habitat, and hatching type for muricacean gastropods. *The Nautilus*, **91**: 67–71.

Spight, T. M., 1979. Environment and life history: the case of two marine snails. In *Reproductive Ecology of Marine Invertebrates*, S. E. Stancyk (ed.), Columbia: Univ. of South Carolina Press, pp. 135–43. (Belle W. Baruch Library in Marine Science, 9.)

Spooner, N., P. E. Gibbs, G. W. Bryan & L. J. Goad, 1991. The effect of tributyltin upon steroid titres in the female dogwhelk, *Nucella lapillus*, and the development of imposex. *Mar. envir. Res.* **23**: 37–49.

Stewart, C., S. J. de Mora, M. R. L. Jones & M. C. Miller, 1992. Imposex in New Zealand neogastropods. *Mar. Pollut. Bull.* **24**: 204–9.

Stickle, W. B., J. L. Sharp-Dahl, S. D. Rice & J. W. Short, 1990. Imposex induction in *Nucella lima* (Gmelin) via mode of exposure to tributyltin. *J. exp. mar. Biol. Ecol.* **143**: 165–80.

Stroben, E., J. Oehlmann & P. Fioroni, 1992a. The morphological expression of imposex in *Hinia reticulata* (Gastropoda: Buccinidae): a potential indicator of tributyltin pollution. *Mar. Biol.* **113**: 625–36.

Stroben, E., J. Oehlmann & P. Fioroni, 1992b. *Hinia reticulata* and *Nucella lapillus*. Comparison of two gastropod tributyltin bioindicators. *Mar. Biol.* **114**: 289–96.

Wilson, S. P., M. Ahsanallah & G. B. Thompson, 1993. Imposex in neogastropods: an indicator of tributyltin contamination in eastern Australia. *Mar. Pollut. Bull.* **26**: 44–8.

8

○ ○ ○ ○ ○ ○ ○ ○ ○ ○ ○ ○ ○ ○ ○ ○ ○ ○ ○ ○

Environmental law and tributyltin in the environment

Klaus Bosselmann

8.1 Introduction

The pesticide tributyltin (TBT) is part of a family of organotin compounds, the biocidal properties of which were first recognised in the early 1950s.[1] TBT is extremely harmful to marine organisms. This character has been exploited with TBT used as the active ingredient in many formulations of marine antifouling paints. These paints are applied to boat hulls and other submerged structures, such as wharves, buoys, and fishpens to prevent fouling, that is, the attachment and growth of marine organisms such as barnacles, mussels, and algae.[2] These fouling organisms contribute to corrosion and to the floating weight of vessels. They also produce turbulent flow and increase drag across the hulls of ships, thus decreasing speed and increasing fuel consumption.[3] The antifouling paints work in the following way. Once the biocide is released from the paint film, a thin envelope of highly concentrated TBT is formed around the vessel hull. The larvae of fouling or nuisance organisms are killed or repelled when they encounter this layer, thus protecting the vessel.[4]

Because of their extreme toxicity, and their longer useful lifespan – seven years as compared to the two years of copper alternatives – TBT compounds are the most effective antifouling agents so far devised.[5] Their

---

[1] Goldberg, TBT: An Environmental Dilemma, 1986, *Environment* **28**(8), 18.

[2] Callow, Ship Fouling: Problems and Solutions, 1990, *Chemistry and Industry*, p. 123.

[3] Arthur, TBT: Why Scientists are looking for a Conviction, 1987, *Water and Waste Treatment* **30**(10), 36.

[4] Huggett, Unger, Seligman and Valkirs, The Marine Biocide Tributyltin; Unger, Seligman and Valkirs, The Marine Biocide Tributyltin: Assessing and Managing the Environmental Risks, 1992, *Environ. Sci. Technol.* **26**(2), 232.

application can save vessel operators hundreds of millions of dollars annually in fuel and maintenance costs. For example, the Environmental Protection Agency (EPA) of the United States has found that the United States Navy, by using TBT copolymer antifoulants, could save an estimated US$130 million a year over the standard copper formula.[6] As a result, TBT has been used extensively throughout the world in antifouling paint.

Ubiquitous use of these paints has created a significant environmental problem. The spatial distribution of TBT concentrations clearly implicates TBT based antifoulants as the main cause of this TBT pollution. The highest concentrations occur in areas with extensive shipping and boating activities: marinas, harbours, and shipping channels.[7] The scientific community became aware of the TBT invasion of the environment in the late 1970s. Although essentially a localised contaminant, case studies throughout the world have established that TBT is indeed a global pollutant.[8]

This is of serious concern given the considerable persistence of TBT in the marine environment. Because TBT is rapidly adsorbed into suspended particles, it accumulates in the sediments where degradation rates are very slow. TBT can therefore persist in the sediment for decades. In New Zealand, TBT has been detected in sediments deposited nearly 20 years ago, and it exhibits a half-life of about 2.5 years. As TBT is not completely adsorbed to particulate matter, slow release into overlying waters may occur for a long time, facilitated by storms or dredging.[9]

The potency, concentrations, and persistence of TBT within the marine environment has resulted in a wide range of deleterious biological effects on non-target organisms in coastal ecosystems. The environmental toxicology is well documented in scientific literature. TBT causes mortality and other deleterious effects in marine species at very low concentrations, for example, 0.1 parts per billion. Adverse biological effects even occur at concentrations where TBT is undetectable in the surrounding waters. Thus, a 'no effect' concentration of TBT in water has yet to be

[5] *Supra* Note 1, p. 18.
[6] Weis and Cole, Tributyltin and Public Policy, 1989, *Environ. Impact Assess. Rev.* **9**, 40.
[7] Stewart and de Mora, A Review of the Degradation of Tri(n-butyl)tin in Marine Sediment, 1990, *Environ. Technol.* **11**, 565, 566.
[8] de Mora, Stewart and Phillips, Sources and Rate of Degradation of Tri(n-butyl)tin in Marine Sediments near Auckland, New Zealand, 1995, *Marine Pollution Bulletin* **30**, 50.
[9] King, Miller and de Mora, Tributyl Tin Levels for Sea water, Sediment, and Selected Marine Species in Coastal Northland and Auckland, New Zealand, 1989, *New Zealand Journal of Marine and Freshwater Research* **23**, 293.

demonstrated.[10] Moreover, the larvae of marine organisms are much more susceptible to TBT than adult forms, so while certain concentrations may not have significant effects on adult forms, it may stop regeneration of marine life.[11]

Specific sublethal effects include poor growth rates and reduced recruitment, leading to population decline. The most obvious manifestation of TBT contamination is shell deformation (thickening) in Pacific oysters.[12] Shell thickness was directly related to the body burden of TBT.[13] Another sublethal effect reported to occur at even lower aqueous concentrations of TBT is the induction of imposex (the development of male sex organs in females) in marine neogastropods. This can effectively cause sterility and population decline. A positive relationship has been found between imposex intensity and the TBT burden in both gastropods and their environment.[14]

In light of these environmental impacts, Goldberg has drawn a strong parallel between the impact of TBT on non-target organisms and that of DDT in the 1960s. Like many, he has called for a total ban.[15] However, TBT differs from DDT in one crucial respect – its impact on human beings. TBT compounds can exhibit a strongly irritant effect on skin and mucous membranes when improperly used, however

> Experience in humans indicates that the depuration of TBT compounds from the body occurs relatively rapidly (ie within some days). Although TBT compounds have been in use for more than 25 years and there is extensive experience from the medical surveillance of employees in organotin production, there are no reports on cases of systematic poisoning or long-term adverse effects in humans. Available data on exposure in shipyards does not indicate particular health hazards.[16]

Thus the dispersion of TBT into the marine environment from antifoulants does not appear to threaten human health, the primary concern of much legislation.

[10] *Supra* Note 6, p. 38.

[11] British laboratory experiments have shown that concentrations of TBT encountered at marinas were sufficient to kill oyster and mussel larvae in 48 hours. See Wade, Garcia-Romero and Brooks, Tributyltin Contamination in Bivalves from United States Coastal Estuaries, 1988, *Environ. Sci. Technol.* **22**(12), 1488.

[12] *Supra* Note 1, p. 19.

[13] *Supra* Note 9, p. 287.

[14] Stewart, de Mora, Jones and Miller, Imposex in New Zealand Neogastropods, 1992, *Marine Pollution Bulletin* **24**, 207.

[15] *Supra* Note 3, p. 43.

[16] See CEFIC submission to the IMO, *TBT Copolymer Anti-Fouling Paints: The Facts*, 33rd session of the MEPC, October 1992, p. 9.

TBT pollution does, however, impact upon another area of human concern – the economy. For example, the decimation of a flourishing Pacific oyster industry in Arcachon Bay, France, was attributed to TBT.[17] Therefore, recognising that a problem related to the use of TBT antifoulants exists, and that there is some measure of risk and uncertainty associated with the problem, how can the use of TBT antifoulants be regulated?

There are a number of possible regulatory strategies and combinations of strategies that could be employed in order to effectively regulate the use of TBT antifoulants. These are as follows.

1. Do nothing.
2. Cover under existing general chemicals control legislation.
3. Place warning labels on containers.
4. Impose regulations on use near sensitive areas.
5. Regulate use to approved boat maintenance yards that include cut-off drains and collection sumps.
6. Regulate the availability to types of craft, structures, apparatus, and/or specific container sizes.
7. Regulate manufacture to particular types of formulations.
8. Combinations of 2 to 7.
9. Introduce a progressive total ban, for example, a warning of intent followed by consecutive bans on free association and copolymer formulations.
10. Combinations of 8 and 9.
11. Immediate total ban on the import, manufacture, sale, and use of all antifoulants containing organotins.

To do nothing is not a strategy recommended for the effective regulation of TBT based antifoulants. The evidence mentioned above identifies a very real need to have concern for the environmental impacts incurred through the use of antifoulants containing organotins. The steps being taken by nations worldwide to regulate the use of such antifoulants is indicative of the scientific and public concerns for their continued use.

Labels on antifoulant containers warning of the hazards of use, and providing advice on how to apply the antifoulants safely and dispose of waste materials, may be a means to alleviate some of the problems incurred. However, only limited information can be communicated by

[17] *Supra* Note 1, p. 19.

this means. Too much information may act as a disincentive to reading the label. Moreover, it is impossible to police and/or monitor user activity by this means. Consequently, warning labels and safety information alone also cannot be relied upon as a strategy for effective regulation of TBT based antifoulants.[18]

However, the use of TBT based antifoulants could be restricted with respect to their manufacture, application, availability, and use. For example, the use of TBT antifoulants could be restricted to localised areas, approved boatyards, types and/or size of craft or structures, types of formulations, or made available only in particular container sizes. Each of these regulatory strategies could be expected to make some contribution towards effectively regulating the environmental impacts of TBT within aquatic environments. Another possibility is the restriction of vessels coated with TBT based antifouling paint to discrete moorings.[19]

Partial bans on the use of TBT antifoulants could be put in place that include combinations of any or all of the above restrictions, including warning labels. For example, it may be appropriate to restrict a ban to contact leaching and soluble matrix/ablative organotin antifoulants. The copolymer could then be made available only in 20 l containers, which in turn, are available only to approved boatyards. Such boatyards could in turn apply the copolymer only to vessels that are of a given length and/or weight, or of aluminium construction.[20]

On the other hand, a total ban could be put into effect over a period of two to three or even four years. Such a ban would give manufacturers and users a longer period of time to adjust to a new market regime. A warning could be given one boating season in advance, for a total ban on contact leaching and soluble matrix/ablative organotin antifoulants. A ban on inland waters and use by smaller craft of the copolymer antifoulants could be put into effect the following season. All the factors of a partial ban described previously could be put into effect at this stage. A total ban on the use of the copolymer organotin antifoulants could follow a further one to two seasons later.[21]

Finally, an immediate total ban on the import, manufacture, sale, and use of all antifoulants containing TBT could be put into effect. However, the implications of such a ban would mean that paint manufacturers and users would have limited time to adjust to the new regulations. Specific provision for disposal would have to be made, as manufacturers and users

---

[18] Ministry for the Environment, *Report of the Working Party Reviewing the Use of Antifoulants Containing Organotins in New Zealand*, 1988, p. 34.

[19] *Ibid*, 34. [20] *Supra* Note 18, p. 35. [21] *Ibid*, 35.

would have existing stock. Furthermore, there may be a short period of time before alternative antifoulants commercially available could replenish depleted stocks of organotin antifoulants. Thus, these problems would need to be alleviated, and could be if the total ban were to take effect a few months after notification.[22]

As we can see, there are a number of possible regulatory strategies that could be employed in order to effectively regulate the use of TBT compounds.

8.2   The International Maritime Organisation

During the first half of the twentieth century, several international conventions on shipping were adopted. These conventions were formulated either by intergovernmental or private bodies. Over the years, the number of organisations engaged in the preparation of rules on shipping substantially increased. It became apparent that there was a need for a body which would be able to co-ordinate and establish further measures on a permanent basis.[23]

In order to meet such a need, the United Nations, on 6 March 1948, adopted a Convention which established the Intergovernmental Maritime Consultative Organisation (IMCO) as a body devoted exclusively to shipping matters. IMCO is a specialised agency of the United Nations. From 22 May 1982, the word 'Consultative' was removed from its name and it became the International Maritime Organisation (IMO).[24]

> The IMO's role is primarily to develop and adopt treaties and other regulations which are designed to improve the safety of international shipping and to prevent pollution of the world's oceans. Virtually every maritime nation in the world is a member of IMO and normally conventions are adopted unanimously.[25]

On 17–20 April 1990, participants at the Third International Organotin Symposium held in Monaco, recognised that the IMO was the appropriate body to regulate the use of organotin compounds internationally, and considered possible regulatory strategies with respect to TBT based

---

[22] *Supra* Note 18, p. 36.
[23] Mankabady, *The International Maritime Organisation*, 1986, Billing & Sons Limited, Worcester, p. 2.
[24] *Ibid*, 6.
[25] O'Neil, Secretary General, IMO, speaking on World Maritime Day, 1993, see *IMO Publication* 253/93.

antifoulants. Ten regulatory options and five regulatory requirements were introduced for consideration by the assembled group.[26] These were as follows.

Regulatory options

1. Total ban on the use of TBT in antifoulant paints.
2. Regulate the use of TBT in antifoulant paints by length of vessels. Prohibition less than 25 m. Ban all non-commercial vessel use.
3. Limit the amount of TBT (on a percentage basis) by paint formulation and no direct sale of TBT.
4. Limit the leach rate of TBT from antifouling paints to the adjacent water column.
5. Regulate the application/removal of antifouling paints which utilise TBT to trained and certified applicators.
6. Regulate the containment, clean-up, and disposal of removed TBT based antifouling paints in boatyards and drydock facilities.
7. Regulate the discharge rates of wastewaters from boatyards and drydocks by standard practices and clean-up and disposal procedures.
8. Regulate the dockage and port time of vessels that utilise TBT based antifouling paints by dockage time and specific time periods with no excess at anchor time in harbours or waterways, with exception for marinas and permanent moorings.
9. Foreign vessels painted with TBT based antifouling paints are required to pay an environmental degradation fee of US$50/hour for anchoring time for estuaries or ports.
10. Self-regulatory public information strategies for small boat owners – who previously (within past 5 years) have painted their boat with TBT based antifouling paints.[27]

Regulatory requirements

1. No or low cost regulatory options.
2. Biocide producer pays all registration fees.
3. Benefit user pays user fees as an environmental degradation fee.
4. Create an environmental degradation fund for research, monitoring, and mediation.

[26] See Note by the Secretariat of the Third International Organotin Symposium, Monaco, 17–20 April 1990, *Use of Tributyltin Compounds in Antifouling Paints for Shipping*, for the IMO at the 30th session of the MEPC, July 1990, pp. 5–6.

[27] *Supra* Note 26, p. 5.

5. Request the International Maritime Organisation (IMO) to register and certify every vessel as to type of antifouling paint used.[28]

Many of the above suggestions were considered to be either impracticable, for example, environmental charges for use of the paints, restrictions on the amount of time spent in waterways, or not relevant to the IMO, such as a ban on the use of TBT on vessels of less than 25 m in length. Of the remainder, two measures attracted the greatest support, firstly, the regulation of the maximum release rate of TBT from antifouling paints, and secondly, to encourage member countries to introduce carefully controlled drydocking practices.[29]

The Symposium adopted the four following recommendations for submission to the IMO concerning the global regulation of TBT compounds with a view to establishing a standard regulatory policy.

1. To establish the limit for vessel hull release rates for TBT compounds used in antifouling paints in $\mu g\,TBT/cm^2/day$ from vessel hulls, including an agreement on the method of measurement.
2. To establish uniform industrial process instructions and regulations for the application, removal, and disposal of all TBT based antifouling application in boatyards or drydocks, including certification of operators for application and removal.
3. To establish public information leaflets to serve as self-regulatory information strategies for small boat owners, who have previously painted their vessels with TBT based antifouling paints.
4. To establish an IMO record system to register and certify every vessel as to type of antifouling paint used.[30]

The Symposium's proposals were reviewed at the 30th session of the IMO's Marine Environment Protection Committee (MEPC) on 12–16 November 1990, together with proposals submitted by various countries.

The MEPC, noting that scientific studies and investigations had shown that TBT compounds could pose substantial risk of toxicity and other chronic impacts to ecologically and economically important marine organisms, and that use of TBT compounds in antifouling paints for vessels was a significant source of TBT found in the marine environment, recognised that there was a need for actions to control the use of TBT compounds in antifouling paints in order to reduce or eliminate potential

---

[28] *Ibid*, 6.   [29] *Ibid*, 6.   [30] *Supra* Note 26, Annex.

adverse impacts to the marine environment.[31] On 16 November 1990 the MEPC adopted resolution MEPC 46(30) whereby it was agreed:

(a) to recommend that Governments adopt and promote effective measures within their jurisdictions to control the potential for adverse impacts to the marine environment associated with the use of TBT compounds in antifouling paints, and as an interim measure specifically consider actions as follows:

    (i) to eliminate the use of antifouling paints containing TBT compounds on non-aluminium hulled vessels of less than 25 m in length,

    (ii) to eliminate the use of antifouling paints containing TBT compounds which have an average release rate of more than $4 \, \mu g$ organotin/cm^2/day,

    (iii) to develop sound management practice guidance applicable to ship maintenance and construction facilities to eliminate the introduction of TBT compounds into the marine environment as a result of painting, paint removal, cleaning, sandblasting, or waste disposal operations, or run-off from such facilities,

    (iv) to encourage development of alternatives to antifouling paints containing TBT compounds, giving due regard to any potential environmental hazards which might be posed by such alternative formulations, and

    (v) to engage in monitoring to evaluate the effectiveness of control measures adopted and provide for sharing such data with other interested parties,

(b) and to consider appropriate ways towards the possible total prohibition in the future on the use of TBT compounds in antifouling paints for ships.[32]

8.3 Regulation of TBT

In environmental law terms, the political response to TBT and the IMO resolutions has been lightning fast. Many countries have imposed legislative regulations on the use of TBT based antifouling paints. Following resolution MEPC 46(30), the trend has been to enact partial

---

[31] See the Report of the MEPC on its 30th Session, MEPC 30/24, 8 January 1991, Annex 10.
[32] *Supra* Note 31.

bans which aim at reducing TBT in water to an acceptable level. In order to achieve this, two principal strategies have been employed.

The first approach has been to ban the use of TBT based antifoulants on small pleasure craft, which are recognised as the major source of contamination. Recreational boaters tend to use the cheaper high leach rate antifoulants, and their boats, more often than not, densely populate specific localities in areas of low tidal flushing. As stated in section 8.1, the highest TBT levels have been recorded in marinas.[33] In January 1982 France banned the application of all antifouling paints containing TBT to the hulls of most boats shorter than 25 m in length. Aluminium boats were exempted because application of the alternative antifouling agent, copper, would have resulted in severe corrosion.[34]

The second approach has been to regulate the composition rather than the use of TBT based antifouling paints. TBT based paints are mainly of two basic types: TBT free association paints, and TBT copolymer paints.[35] The TBT free association paints have the TBT physically mixed into the paint, from which it can slowly leach. This type of paint system has been the basis of most traditional antifouling paints which have been in use since the last century. It is difficult to control the rate of release of TBT from a free association paint system so as to provide a constant leaching level. Such paints often give rise to high initial concentrations of TBT in the marine environment, with an exponential decay in activity. The effective life-cycle of free association paint systems depends upon the concentration of active ingredients used and the film thickness applied; however, they usually have relatively short periods of effectiveness, 18–24 months, and when the paint fails, it often still contains a significant amount of unused TBT trapped deep within the paint film. This often finds its way into the environment as these paints must be completely scrubbed off to allow for the application of another coat.[36]

The more effective TBT copolymer paints have the TBT chemically bound within and throughout a copolymer resin system. The coating on a boat that has been painted with a TBT copolymer paint will react, by hydrolysis with sea water, resulting in the slow release of TBT which combats fouling. The remaining surface of the paint system is mechanically weak and is eroded by moving sea water, resulting in the exposure of a

[33] *Supra* Note 7, pp. 565, 566.
[34] *Supra* Note 1, p. 19.
[35] Organotin Antifouling Paints Report Released, 1988, *Catch* 15(8), 15.
[36] *Supra* Note 16, p. 4.

fresh surface of TBT polymer. This hydrolysis/erosion process is repeated until no paint is left on the surface and all of the TBT is exhausted. The hydrolysis/erosion process confers two key properties on TBT copolymer paint systems: firstly, it confers the ability to control/regulate the TBT leaching rate; secondly, this process results in smoother surfaces, that is, the paint polishes the surfaces. Copolymer paints have become the basis of the present generation of slow release TBT based antifouling paints. Copolymer paints offer up to five years of fouling free performance dependent upon specification and are much less damaging to the marine environment.[37]

In 1985 the United Kingdom restricted the sale of TBT copolymer paints with more than 7.5% TBT and effectively banned all free association paints.[38] The Control of Pollution Act 1987 subsequently banned the application of any TBT antifoulants on vessels less than 25 m in length. No exemption was granted to allow the continued use of TBT based paints on aluminium hulls or fittings, in the belief that manufacturers were close to a solution to the corrosion problem caused by copper. For vessels larger than 25 m in length, TBT based antifoulings can only be supplied in containers of at least 20 l capacity. Further, all antifoulings must be registered.[39]

These two initiatives have been widely followed. In the United States the coastal states of Virginia, Maryland, New York, Oregon, Alaska and California have all passed laws modelled after those of France and the United Kingdom, prohibiting the use of TBT based paints on most boats less than 25 m in length. However, like the French example, aluminium hulled vessels and vessels larger than 25 m in length are allowed to use the slow release TBT copolymer paints. TBT based paints applied in Virginia are limited to a maximum daily leach-rate. A total ban has been imposed in Washington State. The Federal Government itself has enacted the Organotin Antifouling Paint Control Act 1988, which has many of the same features as the various state laws: prohibition on painting vessels less than 25 m in length, and a maximum daily leach rate of $4 \mu g \, TBT/cm^2/$ day on all vessels larger than 25 m in length. Moreover, all antifoulings must be registered, and since 1 March 1990 TBT based antifouling paints can only be applied by certified applicators. The regulations in place in Canada mirror those of the United States with the exception that TBT

[37] *Ibid*, 4, 5, 6.
[38] *Supra* Note 1, p. 20.
[39] See United Kingdom submission to the IMO, *Critical Review of Current and Future Marine Antifouling Coatings*, at the 35th session of the MEPC, January 1994, p. 15.

based antifouling paints do not have to be applied by only certified applicators.[40]

In the EC and South Africa the use of TBT based paints has also, since 21 June 1991, been prohibited on all vessels less than 25 m in length, with no exemptions for aluminium hulls and fittings. For vessels larger than 25 m in length, TBT antifouling paints can only be supplied in containers of at least 20 l capacity. As well as the United Kingdom – Ireland, Holland and South Africa also require registration of all antifoulings.[41]

Throughout the rest of Europe, most Western European countries have prohibited the use of TBT based paints on vessels less than 25 m in length, for example, Norway, Sweden, and Finland. However, for vessels larger than 25 m in length there are few restrictions; only Sweden restricts registrations to coatings with leach rates of less than $4 \mu g \, TBT/cm^2/$ day. Further, since 1 January 1992, all types of antifouling coatings in Sweden now require registration. In Switzerland and Austria the use of TBT based antifouling paints is totally banned, irrespective of the length of the vessel, in fresh water lakes. This is perhaps not surprising, considering that both countries are land-locked and possess no large ships. Significantly, in Turkey and the former Yugoslavia there are as yet no restrictions on the use of TBT based antifouling paints.[42]

There are also no restrictions on the use of the TBT based antifouling paints in Brazil, China, Hong Kong, Korea, Singapore, Malaysia, Russia, Ukraine and the United Arab Emirates. This is also significant as all of these countries have major building/drydock trade involving TBT copolymer usages. However, although some nations have not introduced legislative restrictions, the TBT copolymer paints which are used conform to IMO international legislative proposals by virtue of their product specifications. TBT copolymer products are supplied to dockyards in bulk containers of at least 20 l capacity, thereby only being suitable for industrial applicators, and not small boat owners.[43]

In the Asia–Pacific region, TBT oxide usage in Japan is effectively prohibited by law, although *in situ* manufacture of TBT oxide as an intermediate is authorised. TBT copolymer production and usage in antifoulants is, however, permitted under the law, concerning the examination, regulation, and manufacture, and requiring annual production reports to Japan's Ministry of International Trade and Industry (MITI). Further, guidance instructions and a voluntary self control programme have been in operation since December 1990.[44] In Australia, similarly to

---

[40] *Supra* Note 39, p. 15.   [41] *Ibid*, 15.   [42] *Supra* Note 39, p. 15.
[43] Information provided by R. Bennett.

the regulations in place in the EC and South Africa, the use of TBT based paints is prohibited on vessels less than 25 m in length with no exemptions for aluminium hulls and fittings. However, for vessels larger than 25 m in length, TBT based coatings are restricted to those with a leach rate less than 5 µg TBT/cm^2/day.[45]

At the 35th session of the MEPC, the European Council of Chemical Manufacturers' Federations (CEFIC) submitted the results of TBT monitoring programmes carried out in four major European ports[46] and their adjacent coastlines, as well as in four coastal regions of the United States.[47] Additional results from monitoring programmes carried out in the respective countries were provided by the Environmental Agency of Japan and the Department of the Environment of the United Kingdom.

These monitoring programmes revealed that TBT water concentrations have significantly decreased since the above regulations went into effect. It appears that reducing inputs of TBT has been effective in alleviating their adverse biological effects, even in areas where impacts had been high. However, the results submitted by CEFIC also revealed that drydocks were still found to be a major source of TBT contamination.[48] This indicates that further legislation is required to regulate TBT based paint application and removal procedures at an international level, as recommended in resolution MEPC 46(30).

8.4   Regulation of TBT in New Zealand

In New Zealand, the use of TBT and its close relatives in antifouling paints have been *de facto* subject to the Pesticides Act 1979 since 18 April 1985.[49]

In 1987, due to the growing international concern for the environmental consequences resulting from the use of TBT based antifouling paints, a need was identified for Government action to address those concerns in New Zealand. In October of that year, the Ministry for the Environment of New Zealand convened a Working Party to review the use of TBT

---

[44] Information provided by R. Bennett.
[45] *Supra* Note 39, p. 16.
[46] They were: Milford Haven/United Kingdom, Bremerhaven/Germany, Rotterdam/ Netherlands and Genova/Italy. See CEFIC submission to the IMO, *Results of TBT Monitoring Studies*, at the 35th session of the MEPC, January 1994, p. 3.
[47] They were: Puget Sound/Washington, Galveston Bay/Texas, Lake Erie/Great Lakes, and Narragansett Bay/Rhode Island. See *Ibid*, 5.
[48] *Ibid*, 7.   [49] *Gazette Notice* No. 3471.

antifoulants. The Working Party comprised members representing the New Zealand Paint Manufacturers Association, the Department of Scientific and Industrial Research, the Ministry of Agriculture and Fisheries, and Department of Conservation, the Auckland Harbour Board, the New Zealand Catchment Authorities Association, the Environment and Conservation Organisations of New Zealand, the Huakina Development Trust, the New Zealand Yachting Federation, the Ministry of Works and Development, and, of course, the Ministry for the Environment. Also present were advocates for the New Zealand Fishing Industry Board and the New Zealand Federation of Commercial Fishermen.[50]

The terms of reference for the Working Party were to determine the extent to which regulations on the use and/or availability of TBT antifoulants were necessary and the ways in which such regulations could be effectively implemented.[51] The Working Party's resulting Report of the Working Party Reviewing the Use of Antifoulants Containing Organotins in New Zealand contained three recommendations, presented in the form of options, for the consideration of the Ministry for the Environment. The options were as follows.

1. An immediate total ban on the import, manufacture, sale, and use of all antifoulants containing organotins.
2. A progressive total ban, which would have involved a two-stage ban where all organotin containing antifoulants would be banned with the exception that for a two year transition period only, slow-release copolymer organotin antifoulants could be used, under controlled conditions, on vessels greater than 25 m in length.
3. A progressive partial ban, which was similar to the progressive total ban, but with exemptions for aluminium hulled vessels and outdrives, and that after a two year transition period, the continued use of copolymer organotin antifoulants be subject to review of the extent of the problem and availability of proven alternative antifoulants.[52]

8.4.1   Option one: immediate total ban

The logic behind the immediate total ban option was expressed by the Working Party as follows. First, doubt remained over the effects of TBT within the aquatic environment at undetectable levels, and detectable levels had demonstrable environmental impacts in minute quantities.

[50] *Supra* Note 18, p. 9.   [51] *Ibid*, 9.   [52] *Ibid*, 11.

Second, the antifoulant manufacturing industry had been able to respond to restrictions on organotin formulations and develop alternatives that are commercially available and are used by ocean-going vessels, and some manufacturers were then formulating antifoulants suitable for aluminium craft. Third, a total ban would have been easier and cheaper to police and monitor than a partial ban, because there would have been no exemptions. Fourth, a total ban would have been better suited to Treaty of Waitangi issues. Fifth, a total ban would have been consistent with the promotion of New Zealand as a clean, green, non-polluted nation and major food producer. This reason was viewed as being particularly important as aquaculture is a significant feature of the changing scene of primary production in New Zealand. Finally, there was strong public support for a total ban.[53]

8.4.2   Option two: progressive total ban

The logic behind the progressive total ban was that if an immediate total ban put into effect this would have meant that paint manufacturers and users would have had a limited time to adjust to the new regulations. The main difference between an immediate total ban and a progressive total ban would have been that the latter would have allowed paint manufacturers and users a two year period of time to adjust to the new regulations, during which time they could make specific provision for the disposal of their existing stock and replenish their resulting depleted stocks with commercially available alternative antifoulants.[54]

8.4.3   Option three: progressive partial ban

The logic behind this final option was that the main problem of TBT pollution is recognised as being associated with recreational boating. Thus, the problem may be alleviated if the conventional type of antifouling paints are completely replaced by copolymer formulations and the use of copolymers is restricted to craft of 25 m or greater in length.

It was argued by the Paint Manufacturers Association, the New Zealand Navy, and the advocate for the Fishing Industry Board and the Federation of Commercial Fishermen, that cautious use of TBT antifoulants is acceptable. The logic applied was that vessels greater than 25 m in length spend almost all of their time in port where leach rates from TBT copolymer paints are low, or in deepwater offshore where the impact of leaching is unlikely to be of significant concern. It was assumed that the

---

[53] *Supra* Note 18, p. 38.   [54] *Ibid*, 39.

use of TBT antifoulants on those vessels did not appear to pose a significant pollution threat to the marine environment.[55]

Of the three options, option one and option three received the greatest support. In 1989 the Pesticides (Organotin Antifouling Paints) Regulations made pursuant to s. 76(1)(m) of the Pesticides Act 1979 were enacted. These regulations imposed a partial ban on TBT antifouling paints similar to the third option recommended in the Working Party's report. The restrictions were three-fold.

1. First, the regulations limited the *types* of paints that can be sold or used. They prohibited absolutely the sale (s.2(1)) or use (s.4(1)) by any person of any organotin antifouling paint of a contact leaching or soluble matrix (including ablative) type. That is, the more harmful free association paints are completely banned.

2. Secondly, the regulations restricted paints according to their *leach rate*. They prohibited the sale (s.3(2)) and use (s.4(2)) by any person of any copolymer antifouling paint which exceeded (i) a maximum leach rate of 168 µg organotin/cm^2 over a period of 14 consecutive days, or (ii) a maximum average daily leach rate of 4.0 µg organotin/cm^2 (with a permissible positive deviation of 0.8 microgrammes). That is, low-leaching, less harmful formulations may be sold or applied.

3. Thirdly, the ban extended to the *objects* to which the paint was applied. The regulations prohibited the application of copolymer antifouling paint to any construction or device, with three exceptions: (i) the hull of an aluminium boat, or (ii) the aluminium outdrive of a boat, or (iii) the hull of any boat that exceeds 25 m in length. Therefore, the application of this paint to wharves, jetties, fishpens, nets, buoys, and many boats was banned.

Thus these regulations were by no means innovatory. They closely followed the European and American precedents. In line with the northern hemisphere experience, it appeared that the elimination of the prime source of TBT pollution (small pleasure craft) by a partial ban significantly reduced the flux of TBT to the marine environment. However, given the relatively small period of time that has passed since regulations were imposed, the observed trends cannot yet be interpreted definitively.

Notwithstanding any reduction in the level of organotin contamination

---

[55] *Supra* Note 18, p. 42.

which resulted from the partial ban, the New Zealand Ministry for the Environment (the Ministry) stated that banning TBT completely from antifouling paints was still desirable to control its effects on the marine environment.[56] This argument was based on the following considerations.

1. The well documented adverse effects on the environment resulting from the use of the highly toxic organotins.
2. Treaty of Waitangi issues. Maori revere water as a component of the natural world. Organotin contamination of waters from which food is harvested is, therefore, sacrilegious. Organotin pollution is also abhorrent to Maori because the toxins bio-concentrate and bioaccumulate within the food chain.
3. The promotion of New Zealand as a tourist destination and the growth in the importance of 'green' tourism.
4. The importance for New Zealand's primary produce export industry of our 'clean green image'. The development of the aquaculture industry is seen as a significant feature of the changing trade scene of primary production in New Zealand. The effects of organotins are particularly marked in economically important marine organisms such as oysters. Any suggestion that pollutants affect marine produce could affect our trade.
5. The concern that undetectable levels of organotins may be biologically significant. There remained doubt about the environmental damage caused because toxicological levels occur at the limits of chemical detection. This uncertainty, in conjunction with the fact that minute quantities of organotin have such demonstrable environmental impacts and that the toxin is a particularly persistent environmental contaminant, excludes room for complacency.
6. When the issue was being considered by the New Zealand Government in 1988 there was strong public support for a total ban on the sale and use of organotin antifoulants. There was no reason for believing that since then there has been a shift in public opinion.[57]

On 1 December 1990, as agreed to when the 1989 Regulations went into effect, the continued use of copolymer organotin antifoulants came up for review with respect to the extent of the problem and the availability of suitable alternative antifoulants.

[56] See the Ministry for the Environment Report, *Review of Partial Ban on Antifoulants Containing Organotin*, December 1990, p. 5. [57] *Supra* Note 56, p. 5.

The review of the partial ban consisted of two main parts. The first related to the availability of alternative antifouling products for those classes of vessels that were exempted by the ban. The second drew on the results of research undertaken at the University of Auckland relating to the effectiveness of the New Zealand Government's 1988 policy response to the organotin pollution problem. Two Ministry of Agriculture and Fisheries (MAF) reports on organotin studies undertaken since the 1989 review were also discussed.

8.4.4 Availability of suitable alternatives

In order to assess the availability of suitable alternatives to organotin copolymer antifoulants in New Zealand the Ministry sought information from the paint industry, MAF, the Department of Conservation, the Railways Corporation, and the Royal New Zealand Navy.[58]

Information was sought on whether non-organotin antifoulants had been developed for use in New Zealand conditions, including those for use on vessels with aluminium hulls or outdrives, on the comparative costs and antifouling effectiveness of non-organotin and organotin antifoulants, on antifoulants currently being used by major operators and the implications for them of a total ban on organotin antifoulants, on the results of tests of non-organotin antifoulants on MAF's aluminium-hulled fisheries research vessels, and on any other matters relevant to the review.[59]

The responses to the Ministry's request for information were received from three of the four paint companies contacted directly: Epiglass NZ Ltd (Taubmans), Altex Coatings Ltd and Hitchins-Jotun Protective Coatings. Their responses were taken to represent the antifouling paint industry in New Zealand. All of the respondents manufacture organotin-free antifoulants. This reflects that the market for organotin antifoulants was considerably reduced by the implementation of the partial ban. Generally it was indicated that the antifouling effectiveness of the organotin-free products is improving all the time. One major manufacturer, Epiglass NZ Ltd, has ceased production of antifoulants containing organotin again reflecting the size of the market for such products. The new products were also reported to be price competitive with organotin antifoulants.[60]

However, non-organotin antifoulants for vessels with aluminium hulls and outdrives were not as advanced. Although they were available, they were not as effective as the organotin types. All three paint companies indicated that development costs, rather than a technology gap, had

[58] *Supra* Note 56, p. 6.   [59] *Ibid*, 7.   [60] *Supra* Note 56, p. 8.

impeded the development of non-organotin antifoulants for application on aluminium.[61]

The Railways Corporation advised that since mid-1989 all three inter-island ferries in New Zealand had been coated with organotin-free antifoulants.[62]

The Royal New Zealand Navy announced in September 1989 that it was to convert all naval vessels to organotin-free antifoulants, as a result of investigations into the organotin levels in effluent from Calliope Dock, Devonport, Auckland, and tests on organotin-free paints. The Navy also banned the application of organotin antifoulants to all vessels in Calliope Dock as from 30 September 1989 and announced that it would operate collection and treatment facilities for organotin-coated vessels using the docks until a total ban on the use of the docks by such vessels was enforced on 31 December 1991. Only HMNZS *Canterbury* remained coated with low release rate organotin antifoulant and was to be recoated with organotin-free antifoulant in November 1991. However, the Navy did report that organotin-free antifoulants were effective but did not provide as long a life as was being achieved by the organotin types.[63]

Ministry of Transport (MOT) inquiries also showed that other shipping companies operating coastal vessels which drydock in New Zealand were using organotin-free antifoulants. This was also the case with the trans-Tasman Express Line which drydocked its ships at Calliope Dock, Devonport.[64]

However, the MOT also reported that the larger New Zealand operated trans-Tasman ships which drydock in Australia and four coastal tankers which drydock in Singapore were using antifoulants containing organotin. The MOT believed that there was substantial commercial advantage from the use of these antifoulants for such vessels. Further, as Government policy was to open up the New Zealand coast to international shipping, the MOT advised that there was concern on the part of the New Zealand shipping industry that it should not be put at a disadvantage by controls applying only to New Zealand ships when the coastal trade became more competitive.[65]

8.4.5 Monitoring the effects of the partial ban

In order to assess the effectiveness of the partial ban in reducing the input of organotin (TBT) to the marine environment, sediment cores from Westhaven Marina, Auckland, were analysed, to determine whether the

[61] *Ibid*, 8.   [62] *Ibid*, 8.   [63] *Ibid*, 8.
[64] *Supra* Note 56, p. 8.   [65] *Ibid*, 9.

sedimentary record reflected the anticipated decrease in TBT input since the 1989 Regulations went into effect, and a survey of the geochemical distribution of TBT in surface marine sediments of the Waitemata Harbour and inner Hauraki Gulf was taken, to identify TBT sources and their respective significance.[66]

In the sediments from Westhaven Marina, consistently elevated TBT concentrations were measured. Levels of TBT generally decreased markedly with depth. The presence of TBT at depths of 45 cm, corresponding to a deposition date of 18 years prior to sampling indicated, however, that the processes which break TBT down to its less toxic metabolites operate on a relatively slow time scale of years or even decades. The researchers were of the view that this evidence justified concerns about TBT as a persistent environmental contaminant.[67]

Analysis of TBT distribution in marine surface sediments from the Waitemata Harbour, Tamaki Estuary, and inner Hauraki Gulf consistently showed 'hot spots' of contamination near slipways, washdown yards and the outfall from the drydock at the Devonport naval dockyards. This demonstrated that boat hull cleaning operations were a major point source of TBT, evidence which was supported by sediment analysis overseas. Further, from observations of the behaviour of TBT-containing wastewaters it was concluded that the real contribution of hull cleaning wastewaters was probably underestimated by a sediment survey.[68]

The study also sought to test the conclusion of the Working Party that vessels over 25 m in length do not pose a significant pollution threat to NZ marine ecosystems. Using the typical dimensions of container ships visiting the Port of Auckland, TBT release rates were calculated for large vessels as compared to yachts. After taking account of the lower biocide release rates of antifouling formulation for commercial vessels, it was estimated that a typical container ship is likely to release a quantity of TBT similar to roughly 100 yachts. It is usual for eight to ten such container vessels to be moored in the Port of Auckland at any time, giving a perspective on their probable significance as a source of TBT.[69]

The data obtained in the Auckland study supported these concerns. Elevated concentrations of TBT were found in sediments collected from under the commercial harbour. The levels were roughly half those found in Westhaven marina where restricted circulation and tidal flushing favoured the buildup of contaminants. Given the contrasting hydrological regimes in the marina and commercial harbour areas, inferences about

[66] *Ibid*, 7.   [67] *Supra* Note 56, p. 9.
[68] *Ibid*, 9.   [69] *Supra* Note 56, p. 10.

relative source strengths were not possible from a direct comparison of sediment TBT concentrations. Similar regimes exist, however, between swing mooring areas and the commercial harbour. TBT concentrations in the commercial harbour were found to be consistently higher. This evidence led the researchers to conclude that their evidence did not support the assumption that large vessels are not significant contributors to TBT contamination.[70]

8.4.6   Further evidence of organotin pollution in New Zealand

Two studies by MAF Fisheries on organotin pollution were also completed since the partial ban on organotin antifoulants was implemented. First, a survey of imposex in the dogwhelk showed that the area of the NZ coastal environment affected by biologically significant levels of organotin pollution was much greater than that known at the time of the 1989 Working Party review. Imposex was shown to be widespread in the North Island and northern South Island.[71]

Second, a joint DSIR-MAF Fisheries on organotin levels in sea water showed a biologically significant input in marina sites in the partial ban.[72]

8.4.7   Outcome

Increased awareness and concern at the impact of organotin antifoulants on the environment resulted from, amongst other things, the implementation of the partial ban on their sale and use in New Zealand. This provided an incentive for the development of alternative products suitable for use in New Zealand conditions and prompted moves by major users, such as the Royal NZ Navy and The Interisland Line, to phase out the use of antifoulants containing organotin.

The voluntary move taken by the Navy in September 1989 to cease its use of organotin antifoulants and to close its docking facilities to ships coated with organotin antifoulants from 31 December 1991 reflected a radical shift in opinion from that expressed by the Ministry of Defence in November 1988. At that time, Defence was monitoring the testing of organotin-free antifoulants overseas but did not expect definitive results for three to four years. Further, it was expressed that the use of the then current organotin-free antifoulants would have resulted in increased fuel and docking costs. It was the results of investigations into organotin pollution levels from Calliope Dock and the continuing development and

[70] *Ibid*, 10.   [71] *Ibid*, 10.   [72] *Ibid*, 10.

availability of more effective alternative antifoulants which brought about the shift in Defence policy.[73]

At the time of the Working Party's investigation into the use of antifoulants containing organotin in 1987/88, the Railways Corporation submitted that a total ban on such products would have resulted in additional costs of NZ$1 million per annum, assuming that suitable alternatives were unavailable. The Corporation's decisions to cease the use of organotin antifoulants suggested that suitable alternative products are now available at competitive prices.[74]

The evidence and conclusions of the Auckland University study of the distribution and concentration of TBT in marine sediments suggested that the partial ban was having the desired effect of reducing organotin pollution in areas where the targeted vessels (pleasure craft) were predominant. However, the study reaffirmed previous concerns regarding persistence of TBT as a toxic environmental contaminant. It also highlighted the significance of slipways and washdown yards etc. and large vessels as major contributors to TBT contamination of large marine ecosystems. Previously, vessels over 25 m were considered to be insignificant in terms of organotin pollution.

The contamination emanating from washdown yards etc. could have been expected to decrease over time as the partial ban took further effect and the majority, and eventually all, of small pleasure craft (under 25 m) had been recoated with non-organotin antifoulant. In the meantime, the actual scraping-down and recoating procedure represented a major point source of organotin contamination. It was, however, one which, in management terms, could have been simply controlled, and perhaps eliminated.

Since the 1988 partial ban, the Royal NZ Navy had implemented a collection and treatment system for TBT-contaminated wastewaters from ships hydroblasted in the drydock in Devonport. The Marine Pollution Bulletin described this initiative as a precedent for both other New Zealand and overseas shipyards. Smaller slipways and washdown yards are rarely designed to contain waste materials and in the majority of cases, wastewaters flow either straight back into the water or into stormwater drains. The researchers who undertook the TBT pollution study in Auckland recommended that wastewater collection and treatment systems be mandatory for all boat hull cleaning facilities to prevent the unnecessary entry of toxic components of marine antifouling paints, including TBT

[73] *Supra* Note 56, p. 11.   [74] *Ibid*, 11.

into the marine environment. Such a move would have had the added advantage of preventing unnecessary inputs into the marine environment of other undesirable components of marine antifouling paints (e.g. copper oxide, copper thiocyanate, organic 'boosters' and the resin itself).[75]

Large vessels have been identified as significant sources of organotin contamination. Extending the ban on the sale and use of organotin antifoulants to all vessels would not, however, allow for control of antifoulants used by international shipping, nor could it prevent New Zealand ocean-going vessels from going aboard for drydock maintenance and servicing.

The Ministry for the Environment stated that, given the highly toxic nature of organotins, though, any further reduction in inputs of the contaminant into the marine environment would be of considerable environmental benefit. Further, the move away from organotin antifoulants by the Interisland Line and the Royal NZ Navy suggested that there would not be a substantial commercial disadvantage resulting from an extension of the ban on the sale and use of antifoulants containing organotin. Extending the ban would also be consistent with the promotion of NZ as a non-polluted nation and major food producer. In extending the ban NZ could advocate internationally for the phasing out of the use of TBT in the marine environment.[76]

Following the review, the New Zealand Government, in 1993, enacted the Pesticides (Organotin Antifouling Paints) Regulations 1993, whereby the use of any organotin containing antifouling paint was totally prohibited (S.4.), and whereby the 1989 Regulations were consequentially revoked (S.6).

In enacting the 1993 Regulations the New Zealand Government accepted that TBT was still having a detrimental effect on the aquatic environment at detectable levels, and that doubt remained over the effects of TBT at undetectable levels. Therefore, due to the commercial availability of other suitable antifoulants, and in keeping with the promotion of New Zealand as a 'clean and green' country, the only effective regulatory strategy for the control of TBT in New Zealand is a total ban on its import, manufacture, sale and use.

8.5   Conclusions

The regulatory trend around the world, with respect to TBT containing antifouling paints, has been to enact partial bans which aim at reducing

[75] *Supra* Note 56, p. 12.   [76] *Ibid*, 12.

TBT in water to an acceptable level. At present only Switzerland, Austria, and New Zealand have banned the use of TBT containing antifouling paints totally. As mentioned earlier, in the cases of Switzerland and Austria a total ban is perhaps not surprising, as both countries are landlocked and possess no large ships. On the other hand the New Zealand position is unique, in that it does have extensive marine coastal waters. However, New Zealand does not possess major drydock facilities capable of receiving large vessels exceeding 100 m. Therefore, antifouling paint usage for the repainting of large vessels in New Zealand does not arise. Thus, the New Zealand experience without major ship-building or major ship repair and repainting operations, cannot be compared to other countries with significant shipyard operations.

At the 35th session of the IMO's MEPC in March 1994, the Committee adopted the following actions.

1. The Committee, noting that evaluation of TBT concentrations in water below 10 µg/l is often hampered by contamination problems within the analytical procedure, decided to urge Governments to co-operate towards the improvement of the analytical monitoring procedure, including further efforts towards international validation and calibration of the analytical methods.

2. The Committee recognised that an extension of present TBT restrictions to a total ban is not justified at present on cost/benefit grounds. The Committee also acknowledged that a total ban would not be possible since, at present, alternative antifouling systems were not available.

3. The Committee recalled its decision to eliminate TBT paints with the average release rate of more than $4 \, \mu g/cm^2/day$ and reconfirmed that TBT antifouling paints other than the copolymer types should be banned.

4. The Committee agreed that an international standard method of measuring leaching rates of biocides of anti-fouling paints should be developed as a matter of urgency and instructed the Secretariat to contact ISO and OECD to enquire about the possibility of establishing an international standard.

5. The Committee noted that CEFIC related the reduced level of TBT concentration in water partly to the ban on the use of TBT paints for ships of 25 m or less, and that the United Kingdom proposed that this ban should be extended to cover ships of 50 m or less and those engaged primarily in coastal trade. Germany

also proposed that those engaged in coastal trade should not use TBT antifouling paints. The Committee decided to urge Governments to take appropriate steps to reduce the use of TBT paints on small ships and those operating in coastal waters.

6. The Committee noted that the monitoring indicates that dockyards are the main source of TBT pollution and decided to urge Governments to develop guidelines for sound dockyard practices, if they have not yet done so, and,

7. The Committee, recalling again MEPC.46(30), urged Governments to continue efforts to develop environmentally less harmful antifouling alternatives.[77]

Importantly, the Committee stated that an extension of present TBT restrictions to a total ban is not justified at present on cost/benefit grounds and that a total ban would not be possible since, at present, alternative antifouling systems were not available.[78]

This action appears to sit incongruously with the findings of the New Zealand Ministry for the Environment in their review of New Zealand's 1989 Partial Ban Regulations, which stated that there were suitable alternative antifouling paint systems available for use in New Zealand; however, it supports the conclusion made by the UK delegation in its critical review of current and future marine antifouling paints presented at the 35th session of the MEPC which stated that

> An extension of present TBT restrictions to a total ban is not justified at present on cost/benefit grounds. The economic benefits of TBT are greatest and the environmental impact is least for large vessels in global trade[79]

and it supported the conclusion of the Netherlands delegation in its presentation at the 35th session which stated that

> The development of environmentally friendly paint systems and completely new anticorrosion and antifouling systems is still in its infancy. Practicable results cannot be expected for another 10 to 20 years.[80]

Thus, it appears that research to find satisfactory antifoulant systems which are more environmentally friendly than TBT copolymers continues

[77] See the Report of the MEPC on its 35th Session MEPC 35/21, 28 March 1994, p. 32.
[78] *Supra* Note 1, p. 32.
[79] See United Kingdom submission to the IMO, *A Summary of a Critical Review of Current and Future Marine Antifouling Coatings*, at the 35th session of the MEPC, MEPC 35/17/2, 14 January 1994, p. 2.
[80] See Netherlands submission to the IMO, *The Effect of TBT on Dredging Material*, at the 35th session of the MEPC, MEPC 35/17/1, 14 January 1994, p. 4.

worldwide. Due to the long time scale required to obtain product registration (one to two years) and functional performance testing (three to five years), it appears that commercial and thus legal solutions to the TBT dilemma, will be a long way ahead.

8.6   Bibliography

Articles

Arthur, TBT: Why Scientists are still looking for a Conviction, 1987, *Water and Waste Treatment* **30**, 36.

Callow, Ship Fouling: Problems and Solutions, 1990, *Chemistry and Industry*, p. 123.

Clark, Sterritt and Lester, The Fate of Tributyltin in the Aquatic Environment, 1988, *Environ. Sci. Technol.* **22**, 600.

de Mora, Stewart and Phillips, Sources and Rate of Degradation of Tri(n-butyl)tin in Marine Sediments near Auckland, New Zealand, 1995, *Marine Pollution Bulletin* **30**, 50.

Fulford, Environmental Concern brings Stricter Controls for Coatings, 1989, *Chemistry and Industry*, p. 10.

Goldberg, TBT: An Environmental Dilemma, 1986, *Environment* **28**, 17.

Huggett, Unger, Seligman and Valkirs, The Marine Biocide Tributyltin: Assessing and Managing the Environmental Risks, 1992, *Environ. Sci. Technol.* **26**, 232.

King, Miller and de Mora, Tributyl Tin Levels for Sea Water Sediment, and Selected Marine Species in Coastal Northland and Auckland, New Zealand, 1989, *New Zealand Journal of Marine and Freshwater Research* **23**, 293.

Organotin Antifouling Paints Report Released, 1988, *Catch* **15**, 15.

Phillips, *The Fate of Tri(n-butyl)tin in the Marine Environment*, Masters Thesis, University of Auckland 1993.

Stewart and de Mora, A Review of the Degradation of Tri(n-butyl)tin in Marine Sediment, 1990, *Environ. Technol.* **11**, 565.

Stewart, de Mora, Jones and Miller, Imposex in New Zealand Neogastropods, 1992, *Marine Pollution Bulletin* **24**, 207.

Wade, Garcia-Romero and Brooks, Tributyltin Contamination in Bivalves from United States Coastal Estuaries, 1988, *Environ. Sci. Technol.* **22**, 1488.

Waldock, Thain, Waite and Milton, TBT and the Marine Environment: Recovery Following Legislation in the UK, 1991. *Wat. Sci. Technol.* **24**, 303.

Weis and Cole, Tributyltin and Public Policy, 1989, *Environ. Impact Assess. Rev.*, p. 33.

Williamson, Burton, Clarke and Fleming, Gathering Danger: The Urgent Need to Regulate Toxic Substances That Can Bioaccumulate, 1993, *Ecology Law Quarterly* **20**, 605.

Texts

Henry, *The Carriage of Dangerous Goods by Sea*, 1985, Frances Pinter (Publishers), London.

Reports

International Maritime Organisation Paper, *Critical Review of Current and Future Marine Antifouling Coatings*, January 1994, 35th Session of the Marine Environment Protection Committee.

International Maritime Organisation Paper, *Draft Report of the Marine Environment Protection Committee on its Thirty-fifth Session*, March 1994, 35th Session of the Marine Environment Protection Committee.

International Maritime Organisation Paper, *Results of Research to replace Antifouling Paints with Tributyltin*, January 1994, 35th Session of the Marine Environment Protection Committee.

International Maritime Organisation Paper, *Results of TBT Monitoring Studies*, January 1994, 35th Session of the Marine Environment Protection Committee.

International Maritime Organisation Paper, *The Effect of TBT on Dredging Material*, January 1994, 35th Session of the Marine Environment Protection Committee.

International Maritime Organisation Paper, *Use of Tributyl Tin Compounds in Antifouling Paints for Shipping*, July 1990, 30th Session of the Marine Environment Protection Committee.

NZ Ministry of the Environment, *Report of the Working Party Reviewing the Use of Antifoulants Containing Organotins in New Zealand*, 1988.

NZ Ministry for the Environment Report, *Review of Partial Ban on Antifoulants Containing Organotin*, December 1990.

9

○ ○

The efficacy of legislation in controlling tributyltin in the marine environment

Carol Stewart

9.1 Introduction

The tri(n-butyl)tin (TBT) cation, the active ingredient in some marine antifouling paints, is perhaps the most toxic substance ever deliberately introduced into natural waters (Goldberg, 1986). At seawater concentrations of only a few nanograms per litre, TBT disrupts reproduction in some molluscs, notably neogastropod whelks; and at levels of $10–100 \, \text{ng} \, \text{l}^{-1}$ is toxic to many embryonic and larval organisms (Bryan and Gibbs, 1991). Throughout the 1980s, concentrations of TBT exceeded these levels in inshore waters at sites throughout the world, and in recognition of the accelerating ecosystem damage caused by TBT contamination, several countries legislated controls on its use as an antifouling biocide.

The hazard posed by a toxic substance is a function not only of environmental concentrations and its intrinsic toxicity, but also of its persistence. The resistance of a compound to environmental degradation processes increases the potential for long-range transport and uptake by non-target organisms.

TBT can now be considered to have exceeded initial estimates of its persistence and mobility in the marine environment, as its influence in offshore waters has been noted in several recent studies. Seas around the United Kingdom are perhaps the most intensively studied. TBT has been detected in the open surface waters of the North Sea (Coghlan, 1990). Although a survey of contaminants in seawater around England and Wales (Law *et al.*, 1994) found TBT in only two out of six offshore samples, both sites were located offshore from major shipping rivers. Imposex was also reported in benthic whelks from the open North Sea (ten Hallers-Tjabbes, Kemp and Boon, 1994). Perhaps the most alarming sign of the scale of TBT contamination and its persistence in the marine food web is the presence of butyltin residues in the blubber of porpoises, dolphins and whales resident in the Indian and North Pacific Oceans (Iwata *et al.*, 1994).

The heyday of TBT antifouling paints is over: between 1988 and 1993, the global market share held by organotin-based antifoulants declined from 83% to 70%. The debate has shifted away from whether or not legislative controls are required, to what level of regulation is necessary for adequate protection of the marine environment.

The main purpose of this chapter is to evaluate the current status and efficacy, on a global scale, of management action intended to reduce the environmental damage perpetrated by the profligate use of TBT-based marine antifouling paints. The legislative history of TBT will be reviewed briefly; and the body of scientific evidence evaluating the efficacy of legislation discussed. Current knowledge of recovery of marine ecosystems will be discussed, and factors that may hinder or limit recovery identified. Other organotin compounds which may have significance as marine pollutants will be briefly discussed, and possible hazards that may arise from new biocides in a regime of declining TBT popularity will be considered. Finally, priorities for management will be briefly outlined.

9.2 Legislative history

A summary of global organotin legislation is presented in Table 9.1. In 1982, the French government, motivated by concerns about their Atlantic coast oyster culture industry, implemented the first legal restrictions on the use of TBT-based antifouling paints. The *French Temporary Ban* was immediately effective in improving spatfall, oyster condition and commercial production. Following this lead, the majority of the industrialised nations have adopted similar legislative controls on organotin compounds. These controls have adopted the underlying assumption that the problem of TBT contamination of coastal waters was very largely due to pleasure boats rather than commercial shipping, and hence tin-based coatings are prohibited on small craft under 25 m in length, with exemptions made for aluminium outdrives, hulls and structures because of corrosion problems with copper-based coatings. Vessels larger than 25 m in length are still permitted to use tin-based coatings, although a low release rate (usually less than $4\,\mu g\,TBT/cm^2/day$) is commonly specified.

More recent refinements to legislation have included the compulsory registration of antifoulants in some countries, the provision for water quality monitoring to evaluate the efficacy of the control measures, and the setting of environmental quality targets (in the United Kingdom) for

Table 9.1 *Summary of global organotin legislation*

Country	Year	Comments
France	1982	*French Temporary Ban* implemented to protect oyster industry on Atlantic coast: paints with over 3% organotin prohibited on boats smaller than 25 tons, in intensive oyster culture areas. Regulations later expanded and strengthened to prohibit all organotin-based paints on boats smaller than 25 m for the entire coastline, but with exemptions for aluminum structures.
United Kingdom	1987	*UK Retail Ban*: all antifouling products containing triorganotins banned on vessels <25 m, and on fish-farming equipment. All antifoulants registered as pesticides, sale and use must be approved by Advisory Committee on pesticides. Triorganotin paints only sold in drums of 20 l or more; must contain $<7.5\%$ total tin in copolymers or 2.5% total tin as free tin. Establishment in 1987 of environmental quality standards (EQS) of 20 ng l^{-1} TBT-Sn in freshwater and 2 ng l^{-1} in seawater.
USA	1988	Regulations on a state-by-state basis from 1986, starting in Virginia, then in California, Maryland and several other coastal states. 1988: Federal Organotin Antifouling Paint Control Act through Congress. *Vessels <25 m*: all organotin-based antifouling coatings prohibited; exemptions for aluminum structures. *Vessels >25 m*: low release rate (<4 μg TBT/cm^2/day) permitted. All antifoulants must be registered; since 1990 TBT paints can only be applied by certified operators.
Canada New Zealand Australia	1989	Along similar lines to USA federal legislation; all antifoulants must be registered as pesticides.
Japan	1990	Classification of TBTO as Class I chemical halts its importation, but its manufacture has continued. Some controls introduced: TBT-based antifoulants prohibited on vessels which drydock at one year intervals; TBT paints must not contain more than 10% TBT for copolymers and 8% for diffusion type paints.

Table 9.1 *continued*

Country	Year	Comments
European Community and South Africa	1991	*Vessels <25 m*: TBT-based antifouling coatings prohibited; no exemptions for aluminum structures. *Vessels >25 m*: TBT antifouling available only in 20 l containers. UK, Ireland, Holland and South Africa require registration of all antifoulings.
Europe (non-EC members)	various	*Vessels <25 m*: TBT-based antifouling coatings prohibited in most European non-EC countries; no restrictions in Turkey and the former Yugoslavia. In Switzerland and Austria, TBT-containing antifoulings are prohibited on vessels of all lengths in lakes. *Vessels >25 m*: few restrictions. Sweden restricts registration to low release rate (<4 µg TBT/cm²/day) coatings.

the concentrations of TBT in natural waters. In the USA, TBT paints may only be applied by certified operators, and there is a general push towards the establishment of industrial standards for the application and removal of all organotin-based antifoulants, along with compulsory certification of the operators (recommendations of Third International Organotin Symposium to IMO, 1990). Austria, Switzerland and New Zealand have imposed a total ban on the use of TBT-based paints (see Chapter 8).

However, it is also evident from Table 9.1 that lesser-developed countries do not have similar legislative restraints on organotin compounds, and in fact on a global scale the proportion of the world's coastline protected by legislation is rather small. Of particular concern are the coastal waters of the tropics, which support an enormous diversity of marine life. The Third International Organotin Symposium, held in Monaco in 1990, identified massive deficiencies in the data from tropical areas, and suggested an urgent need to fill these gaps. Certainly, the few studies that exist support these concerns, with reports of heavy TBT contamination or associated deleterious effect on marine life in Indonesia (Ellis and Pattisina, 1990), Bahrain (Hasan and Juma, 1992) and Fiji (Stewart and de Mora, 1992).

9.3   Efficacy of organotin legislation and recovery of the marine environment

Studies addressing the efficacy of organotin legislation have been summarised in Tables 9.2, 9.3 and 9.4, under the headings of coastal waters, sediments and ecosystem recovery.

9.3.1   Coastal waters

TBT has a short residence time in the water column. It exists in solution as a large univalent cation and, in seawater, forms an equilibrium mixture of neutral complexes with the anions chloride, carbonate and hydroxide. These complexes are very surface active and readily adsorbed onto suspended particulate material. Degradation processes for TBT in natural waters are primarily microbially mediated, and are thought to obey first order kinetics and be inhibited by colder ambient temperatures (Stewart and de Mora, 1990). Rates of degradation in the water column tend to be slow relative to sedimentation, with half-lives ranging from days to weeks, and thus, TBT tends to accumulate in underlying sediments.

In addition to being strongly adsorbed onto sediment, TBT is moderately lipophilic, with a log K_{ow} in seawater of about 3.7 (Laughlin et al., 1985), and hence has the potential to bioaccumulate strongly. Uptake of TBT from ambient waters has been observed in species ranging from algae to salmon to eelgrass (Laughlin and Linden, 1987; François, Short and Weber, 1989). Bioaccumulation factors ranging from 1000 up to 30 000 are reported (Bryan and Gibbs, 1991). Partitioning into the sea surface microlayer is another important removal process in natural waters (Adelman, Hinga and Pilson, 1990), increasing TBT concentrations over subsurface water by factors of between two and ten (Cleary, 1991). The surface microlayer is believed to be an important route of exposure to lipophilic contaminants for the intertidal region.

As TBT is rapidly removed from the water column, one would predict that aqueous concentrations would respond rapidly to reduced inputs, and in general this appears to be the case (Dixon, 1989; Alzieu et al., 1986; Cleary, 1991; Waite et al., 1991; Dowson et al., 1992a,b, 1993a,b, 1994; CEFIC review, 1994; Wade et al., 1991; Garcia-Romero et al., 1993; Batley et al., 1992). In England and France, commercial imperatives (the decline of the Pacific oyster fisheries) led to a rapid political response, and TBT levels soon declined sufficiently for the oyster fisheries to resume production at pre-TBT rates (Alzieu, 1991). Similar improvements were noted in the United Kingdom (Dixon, 1989): for example, an oyster

Table 9.2   *Efficacy of organotin regulations: coastal waters*

Author	Date	Trend in water quality	Comments
France – regulations introduced in 1982			
Alzieu *et al.*	1986	Improvement	Organotin levels and shell malformations monitored in Pacific oysters *Crassostrea gigas* in Arcachon Bay, Atlantic Coast, between 1982 and 1985. Spatfall has recovered, the oyster industry has resumed normal production, and tin levels have decreased five-to-tenfold. However, some shell malformations persist.
Alzieu *et al.*	1989	Improvement	Monitoring of Atlantic coast between 1980 and 1987. Both TBT levels and oyster condition responded quickly to regulations; but residual TBT contamination reported for some marinas, with shell malformations persisting. It is thought that illegal use of TBT paints is responsible.
Alzieu *et al.*	1991	No improvement	On the Mediterranean coasts of France, Spain and Italy seawater TBT concentrations have not decreased since a previous MED-POL survey in the region. Apparently the 1982 French regulations have not been effective in protecting the French Mediterranean coastline. The movement of foreign vessels, painted outside the scope of the French legislation, is thought to be the cause.

Table 9.2 *continued*

Author	Date	Trend in water quality	Comments
United Kingdom – Regulations introduced in 1987			
Dixon	1989	Improvement	Signs of rapid recovery – seawater TBT levels have decreased by at least twofold in many estuaries with high yachting activity. Oyster fisheries closed since the mid-1970s have been able to resume production.
Cleary	1991	Improvement	Organotin concentrations in seawater at 15 sites monitored. Levels began to decline in 1988, and continued to do so in 1989 in the surface microlayer and in subsurface waters. However, in 1989 TBT concentrations frequently exceed the EQT and known toxicity thresholds for many aquatic species.
Waite *et al.*	1991	Improvement	Monitoring of TBT concentrations in seawater, sediment and shellfish from 1986 to 1989 indicates a clear improvement in water quality, with levels reduced to a third to a quarter of 1986 levels. Normal oysters can now be grown in 5 out of 6 estuaries studied. However, the EQT is still frequently exceeded. No trend of decrease was apparent in sediments over this time scale.
Dowson *et al.*	1992a	Improvement	In 1990, a survey of organotin concentrations in the water column from 17 sites in 7 Suffolk and Essex estuaries was undertaken. The EQT was exceeded at 7 of these sites.

Table 9.2 *continued*

Author	Date	Trend in water quality	Comments
Dowson *et al.*	1993a,b	Improvement	Above monitoring programme continued through to 1992. Most seawater TBT concentrations had declined below detection limits. A 5-year time span seems to be necessary for TBT regulations to reduce water column concentrations effectively.
Dowson *et al.*	1994	Improvement	Butyltins monitored in 1992 at 12 freshwater sites in the Norfolk Broads known to have a history of contaminations. TBT was detected at only one site.
CEFIC review	1994	Improvement	TBT monitored in the U.K. from 1986 to 1992 by MAFF. In estuarine waters, TBT levels have decreased from a mean of around $60 \, \mathrm{ng \, l^{-1}}$ to below the detection limit of $2 \, \mathrm{ng \, l^{-1}}$. Levels in oysters have decreased approximately tenfold.

USA – regulations introduced in 1988

Author	Date	Trend in water quality	Comments
Wade *et al.*	1991	Improvement	Butyltins monitored in oysters at 59 sites in the Gulf of Mexico from 1987 to 1990. By 1990, TBT levels have decreased at 85% of sites, shown no change at 5% of sites, and increased at 10% of sites. Organotin controls have apparently reduced environmental loading of these compounds.

Table 9.2 *continued*

Author	Date	Trend in water quality	Comments
Garcia-Romero *et al.*	1993	Improvement	Butyltins monitored in oysters at 53 sites in the Gulf of Mexico from 1989 to 1991. The geometric mean concentration decreased from 1989 to 1990, but increased again in 1991, although not to 1989 levels.
Uhler *et al.*	1993	Not clear	Butyltins monitored in sentinel molluscs at 69 East Coast and 42 West Coast sites from 1988 to 1990. In general, West Coast sites showed decreasing trends, but East Coast sites showed no clear trends towards reducing levels, with more sites showing a tendency to increase than to decrease.

Australia – regulations introduced in 1989

Author	Date	Trend in water quality	Comments
Batley *et al.*	1992	Improvement	TBT concentrations and condition monitored in oysters *Saccostrea commercialis* from 1988 to 1991 near Sydney, NSW. TBT concentrations decreased below the detection limit in all cases; shell deformities decreased and oyster condition improved.

Table 9.2 *continued*

Author	Date	Trend in water quality	Comments
The Netherlands – regulations introduced in 1990			
Ritsema	1994	No improvement	Dissolved butyltins monitored in 6 Dutch marinas from 1990 to 1992. Significant decreases were not observed, and all mean TBT concentrations exceeded the Dutch EQT of $10\,\mathrm{ng}\,l^{-1}$. It was suggested that TBT-based paints had already been phased out to a large extent by the late 1980s, thus the 1990 controls may not have produced further reductions in butyltin contamination. Desorption from contaminated sediments was implicated as the source of dissolved butyltins, and higher concentrations during the summer perhaps indicated some fresh input via the illegal use of tin-based antifouling paints.
Japan – regulations introduced in 1990			
CEFIC review	1994	Improvement	The Environment Agency of Japan has monitored TBT in fish, sediments and birds since 1985; and in water and sediments since 1988. In general, water, sediment and shellfish TBT levels have decreased approximately twofold by 1992, but levels in fish tissue have not decreased. The most marked reduction was evident after 1990.

Table 9.3 *Efficacy of organotin regulations: sediment studies*

Author	Year	Country	Type of study	Comments
Valkirs et al.	1991	USA	Surface sediments	Surface sediment TBT concentrations in San Diego Bay, California, were not found to reflect recent decreases in water column values.
Wuertz et al.	1991	USA	Surface sediments	Significantly less TBT found at all 4 sites in 1990 than in 1988 in Boston Harbour.
Waite et al.	1991	UK	Surface sediments	TBT concentrations in sediments from 6 estuaries monitored between 1986 and 1989 did not reflect recent three-to-fourfold decreases in water and bivalve values. One trend was for 'hotspots' of contamination to become less evident: it is thought that the movement of contaminated sediments has caused a more even distribution of TBT in the estuaries.
Sarradin et al.	1991	France	Surface sediments	Surface sediments in some mooring areas of Arcachon Bay still exhibited high concentrations of TBT, despite 1982 regulations.
Dowson et al.	1993a	UK	Surface sediments	TBT levels in surface sediments monitored from 1990 to 1992 in 6 estuaries. High spatial variability was observed. Of 22 sites, about half showed a clear reduction; others showed erratic and fluctuating concentrations consistent with movement of contaminated sediments.
Dowson et al.	1993b	UK	Sediment cores	Vertical distributions of butyltins were measured in sediment cores collected at 8 contaminated freshwater and estuarine sites in Suffolk and Essex in 1990 and 1991. Sedimentary TBT

concentrations decreased with depth, but in undisturbed cores maximum enrichment tended to occur just below the surface, suggesting that accumulation of organotins in underlying sediments is declining in response to the 1987 UK Retail Ban.

Dowson et al.	1994	UK	Surface sediments	Butyltins were measured in surface river sediments from 12 Norfolk Broad sites. A marked decline was evident at all sites, in comparison to 1989 measurements. At over half the sites, TBT loadings in sediment had declined to undetectable levels. Localised hotspots of contamination still occur, but are limited to boatyards and marinas.
de Mora et al.	1995	New Zealand	Sediment cores	Cores collected in Auckland harbour and marinas in 1990, following 1989 regulations. An decreasing gradient in TBT concentrations with increasing depth was observed. Slight subsurface maxima were evident in marina cores, indicating declining inputs of TBT.
Quevauviller et al.	1994	France	Sediment core	An intertidal sediment core was collected in Arcachon Harbour. Butyltin levels increased from the bottom of the core to a maximum at 15 cm depth, then decreased slightly towards the surface. Maximum butyltin enrichment below the surface suggests that the 1982 controls may be taking effect, but recent sediment concentrations do not appear to be following trends in overlying waters.

Table 9.3 *(continued)*

Author	Year	Country	Type of study	Comments
Sarradin et al.	1994	France	Sediment cores	Intertidal sediment cores were collected in Arcachon Bay in 1990. Despite 1982 controls, significant quantities of TBT remain in the sediments and may be the cause of persistent oyster shell malformations. However, no relationship was apparent between TBT concentrations and depth, suggesting that hydrodynamic perturbations (storms, tidal flushing) of the upper sediment layers of this estuary are a more important influence on sediment composition than TBT inputs or degradation phenonema.

Table 9.4 *Efficacy of organotin regulations: ecosystem recovery*

Author	Year	Country	Species	Comments
Pacific oysters C. gigas				
Alzieu et al.	1986	France	C. gigas	Following 1982 regulations, a clear improvement in shell quality and condition of Arcachon Bay oysters. But in 1985, 40% of oysters still develop shell malformations.
Alzieu et al.	1989, 1991	France	C. gigas	Long-term monitoring of oyster conditions at Arcachon Bay from 1982 to 1989 * 1983 Spatfall normal * 1984 Oyster production normal * 1985–9 TBT concentration and shell malformations improve, but there is a resurgence of anomalies in 1986–7 because of insufficient compliance with regulations.
Waite et al.	1991	England	C. gigas	MAFF monitoring programme from 1986 to 1989 of oysters in 6 estuaries shows rapid recovery in oyster condition. As TBT contamination has declined, oyster shell thickness and flesh weight have improved, with the greatest rate of improvement being between 1987 and 1988. British regulations judged to be very effective: an oyster fishery on the River Blackwater, closed since the 1970s, re-opened in 1987 and was producing marketable oysters by 1988.

Table 9.4 *(continued)*

Author	Year	Country	Species	Comments
Dyrynda	1992	England	*C. gigas*	Survey of oyster condition and growth in 1989 and 1990 in Poole Harbour. Oysters transplanted to a poorly flushed site with a high boat density showed the onset of shell thickening as well as reduced growth and poor condition. However, two less impacted sites supported normal oyster growth.
Other bivalves Minchin *et al.*	1987	Ireland	*Pecten maximus*	Extensive use of TBTO as net-dip for salmonid cage netting in Mulroy Bay suspected in poor settlement and deterioration of populations of the scallop *P. maximus*, as well as the flame shell *Lima hians* and other bivalves. Following restrictions on use of TBTO in 1985, there was a good settlement of *P. maximus* in 1986; settlements of other bivalves except for *L. hians* were also recorded. Effects of organotins on bivalve reproduction and larval development not known.
Langston *et al..*, Langston and Burt	1990, 1991	England	*Scrobicularia plana*	A decade of field observations suggests an inverse correlation between levels of TBT contamination in sediments and abundance of sediment-dwelling clams *S. plana*. No recovery of recruitment in clams by late 1980s.

Batley et al.	1992	Australia	*Saccostrea commercialis*	The Sydney rock oyster S. *commercialis* respond to TBT intoxication by developing 'shell curl'. Oysters from 4 of 6 places surveyed in 1988 showed deformities; by 1991 deformities were absent at all sites. Improvements in oyster growth and condition were also noticed.
Neogastropod whelks				
Bailey and Davies	1991	Scotland	*Nucella lapillus*	TBT used on salmon pens in Loch Laxford. There was evidence that the 1987 regulations curtailing the use of TBT in mariculture was beginning to be reflected in the imposex condition of dogwhelks by 1989. The reproductive condition of adult female dogwhelks was found to be static or deteriorating, but the rate of imposex development in juveniles decreased, suggesting recruitment from 'clean' populations. The time scale for complete recovery is likely to be many years.
Gibbs et al.	1991	Atlantic coast of Europe	*N. lapillus*	Imposex known to be widespread in population of *N. lapillus* throughout the entire Atlantic coast of Europe. In Atlantic France, the longest time has now elapsed since regulations on TBT usage in 1982. A 1989 resurvey of *N. lapillus* in south-west Brittany showed that populations were still significantly impacted by imposex, with around a third of adult females sterile. Significant TBT contamination was still found in this region: a shipyard was thought to be partly responsible.

Table 9.4 (continued)

Author	Year	Country	Species	Comments
Evans et al.	1991	England	*Nucella lapillus*	There were signs of recovery in *N. lapillus* populations of the Northumbrian coast resurveyed in 1989. Although 4 of 7 populations surveyed still had an incidence of imposex of 100%, RPSI values were lower, sex ratios indicated better survivorship of adult females, and there was evidence of improved recruitment enhancing the proportion of juveniles.
Stewart et al.	1992	New Zealand	*Lepsiella scobina*	Auckland Harbour neogastropod whelks *L. scobina* were found to be severely affected by imposex, so that breeding activity had ceased in most inner harbour populations. A survey of *L. scobina* populations in 1990/1991 indicated no recovery in breeding capacity, with egg capsules absent and only a few large adult *L. scobina* in advance stages of imposex found.
Bryan et al.	1993	England	*Nassarius reticulatus*	*N. reticulatus* is a useful alternative to *Nucella lapillus* as an imposex-based TBT indicator at contaminated sites, as it is less sensitive to TBT and also is not sterilised by TBT and thus is not as prone to local population extinctions. However, its use as an indicator in a regime of rapidly declining seawater TBT levels is problematic. Imposex intensity declines very slowly because of the longevity of *N. reticulatus*, the slow reversibility of imposex in adult females, and limited recruitment. Tissue TBT burdens provide a much better indication of change.

Reference	Year	Location	Species	Notes
Evans et al.	1994	Scotland	Nucella lapillus	Imposexed populations of Nucella lapillus were resurveyed on the Isle of Cumbrae in 1988 and 1991–3. The following indices point toward good recovery: RPSI, abundance, adult sex ratio and % juveniles. At one site, whelks were absent in 1988 but are now abundant with a low and declining incidence of imposex. The origin of this 'recolonisation' is not clear, as N. lapillus do not have a dispersive larval stage.
Douglas et al.	1993	England	Nucella lapillus	'Recolonisation' of parts of the Durham coastline, northeast England, by Nucella lapillus.
ten Hallers-Tjabbes et al.	1994	North Sea	Buccinum undatum	Imposex in benthic female whelks B. undatum reported for the first time in a 1991/1992 survey of the southern North Sea. Even whelk populations sampled as far as 200 km offshore were imposexed; close to the Euro Channel shipping route 100% of female whelks were imposexed or no live individuals were captured. In 1974, whelks from the area were not affected by imposex. Imposex frequency increased with shipping traffic intensity.
Tester and Ellis (pers. comm.)	1994	Canada	Nucella lamellosa; N. emarginata; N. canaliculata; Searlesia dira	TBT regulations were introduced in Canada in 1989. Four local coastal whelk species were surveyed in 1987–9 and again in 1994. These indices point towards good recovery: RPSI, imposex frequency and vas deferens formation frequency. Sites classified as low boating activity appear to be recovering more rapidly.

ground in Essex, which was closed in the mid-1970s because of shell malformations and reduced growth, was able to re-open in 1988, after TBT controls were imposed in 1987.

A rapid initial response to TBT controls has generally been observed, but the time scale of recovery to pre-TBT levels is longer. Dowson *et al.* (1993a,b) monitored the decline of aqueous TBT contamination in six estuarine systems on the east coast of England; over a five-year period, levels diminished to below analytical detection limits ($<3 \, \text{ng} \, \text{l}^{-1}$). However, residual low-level contamination has been a problem in some regions, especially those with heavily contaminated sediments (Alzieu *et al.*, 1986, 1989; Astruc *et al.*, 1989; Ritsema, 1994).

9.3.2   Sediments

TBT is relatively transient in the water column, but tends to accumulate in underlying sediments where degradation processes are considerably slower, with half-lives measured in years or even decades (Stewart and de Mora, 1990). Thus, sedimentary concentrations should reflect trends in inputs of TBT much more slowly. The studies summarised in Table 9.3 indicate that this is generally the case.

Nearshore sediments are frequently mobile and reworked, in which case one would not expect a record of contaminant input to be preserved. For example, both Astruc *et al.* (1989) and Sarradin *et al.* (1994) observed no relationship between TBT concentrations and depth in Arcachon Bay sediment cores, suggesting that these sediments have been reworked by hydrodynamic processes such as storms and tides. A sediment core from a Fijian marina showed the same pattern (Stewart and de Mora, 1992).

However, other authors have observed a relationship between butyltin concentrations and depth, indicating that some vertical structure has been preserved in these sediment profiles. In particular, a marked and rapid decrease of TBT with increasing depth has been observed (de Mora *et al.*, 1989, 1995; Aboul Dahab, El-Sabrouti and Halim, 1990; Quevauviller *et al.*, 1994). De Mora *et al.* (1989) considered the data to be best described by first order kinetics of decay, indicative of *in situ* degradation of TBT. For the three sediment cores, a mean rate constant of $0.375 \, \text{y}^{-1}$ yielded an estimated half-life of degradation of about 1.8 years. More recent work on a wider range of sediment cores around Auckland Harbour yielded half-lives ranging from 1.3 to 4.3 years. Similarly, Waldock *et al.* (1990) reported a half-life for TBT in anaerobic sediment of almost two years. Degradation of TBT is primarily microbially mediated, and the rate is thought to be dependent on temperature (Stewart and de Mora, 1990). In a sediment core collected from Ballenas Basin in the Strait of Georgia, Western Canada (Macdonald *et al.*, 1991), a sediment chronology has been inferred from ^{210}Pb dating. Butyltins were analysed at several depths within this core (Stewart and Thompson, 1994); and TBT concentrations diminished only slightly from 2.2 to 1.9 ng g^{-1} TBT-Sn at 25 cm depth (corresponding to sediment laid down around 1972). Application of the simple first-order kinetic model would imply degradation of TBT in Ballenas Basin sediments to be a much slower process, with a half-life measurable in decades or longer. However, as the authors acknowledge, it is not known to what extent the observed butyltin distribution may be influenced by mixing or post-depositional diffusion processes.

Thus, certain coastal sediments appear to be capable of preserving a depositional history of TBT, and as time elapses after TBT controls, one would expect to see decreased TBT concentrations in surface sediments. Subsurface maxima in sedimentary TBT concentrations were indeed apparent in several studies (Dowson *et al.*, 1993b; de Mora *et al.*, 1995; Quevauviller *et al.*, 1994). Contamination is, in a sense, being buried by cleaner sediment, presumably reflecting reduced TBT inputs to surface waters.

9.3.3 Ecosystem recovery

TBT is known to be an extremely potent toxin, disrupting reproduction in some molluscan species at seawater concentrations of only a few nanograms per litre, and exerting toxic effects on numerous larval and embryonic organisms at levels of 10–100 ng l^{-1}. Bryan and Gibbs (1991) have reviewed the impact of low seawater TBT concentrations, known to exist in inshore waters worldwide in the 1980s, on non-target marine organisms. However, the real extent of the damage to marine ecosystems will almost certainly never be known. Many marine ecosystems are now entering a 'recovery phase' as loadings of TBT diminish, but dynamics and the time scale of recovery are not well understood. Studies which have assessed the effect of organotin legislation on the recovery of marine organisms from the toxic effects of TBT are summarised in Table 9.4, and several general points emerge.

Recovery depends on habitat: in general, residents of the water column will tend to recover more quickly than benthic infauna, such as clams, for the obvious reason that TBT is transient in the water column but relatively long-lived in sediment. Thus, Pacific oysters have responded rapidly to decreasing aqueous TBT contamination, with normal spatfall, an increased growth rate and reduced shell malformations in Arcachon Bay populations in the year following the implementation of TBT controls (Alzieu *et al.*, 1989, 1991). Similar improvements in the health of Sydney rock oysters *Saccostrea commercialis* were observed following imposition of controls in New South Wales, Australia (Batley *et al.*, 1992); also in England in the Pacific oyster (Waite *et al.*, 1991). On the other hand, larval recruitment of the sediment-dwelling clam *Scrobicularia plana* appears to be severely retarded by TBT contamination of sediments (Langston *et al.*, 1990; Langston and Burt, 1991). Field observations collected by these authors over a decade suggested an inverse correlation to exist between sedimentary TBT concentrations and abundance of *S. plana*, with no signs of recovery by the late 1980s.

Life cycle characteristics are also an important influence on recovery dynamics, with recruitment by larvae from 'clean' areas facilitated by a dispersive planktonic juvenile stage, such as the broadcast spawning of oysters. However, recovery might be expected to be retarded in species which have no free-swimming stage and hatch adults directly. In the case of Atlantic dogwhelk *Nucella lapillus*, its acute sensitivity to TBT, together with its direct development and limited adult mobility combined to make the species very vulnerable at the population level. Indeed, populations of *N. lapillus* are in decline, with occasional local extinctions, throughout the entire Atlantic coast of Europe (Bryan *et al.*, 1986; Gibbs *et al.*, 1991). Recoveries of afflicted populations were predicted to be very slow, and dependent on seawater TBT concentrations reaching low enough levels to permit the survival of juveniles rafted in from healthy populations.

It is perhaps surprising, then, that there are already signs of recovery in afflicted populations of *N. lapillus*, according to such indicators as abundance, the proportion of juveniles, adult sex ratio and the RPSI (Evans *et al.*, 1991, 1994). Douglas *et al.* (1995) have also described 'recolonisation' of parts of the Durham coastline by dogwhelks, at sites where they have apparently been absent for several years. On the West Coast of Canada, similar trends in recovery have been observed in four indigenous neogastropod species (Tester and Ellis, pers. comm.).

9.4   Factors which may limit or hinder recovery

9.4.1   Persistence of TBT in sediments

Sediments are a geochemical sink for TBT, and degradation processes are known to be much slower than in overlying waters. Indeed, the stability of TBT in the sedimentary reservoir controls its overall persistence in the marine environment. For instance, in Arcachon Bay where TBT restrictions have been in force for more than a decade, there is still heavy contamination of the estuarine sediments (Quevauviller *et al.*, 1994; Sarradin *et al.*, 1994). TBT stored in this reservoir may continue to exert toxicity after primary inputs of TBT to the water column have stopped, thus retarding the recovery of marine ecosystems. As well as the obvious possibility of dietary uptake by deposit-feeding infauna such as clams (Langston and Burt, 1991), TBT may be remobilised to the water column. Hydrodynamic perturbations such as storms, tidal currents, dredging operations and propeller wash may physically resuspend sediment; indeed, mobility of nearshore coastal sediments is probably the rule rather than the exception. Another process is desorption of TBT into the aqueous phase. The sorption coefficient for TBT is only moderately high (10^2–10^4 l kg^{-1}), and sorption is known to be reversible (Unger, MacIntyre and Huggett, 1988).

In the Netherlands, dissolved butyltins were monitored after the 1990 ban on tributyltin-containing antifoulants (Ritsema, 1994). Over the following three years, butyltin concentrations remained essentially constant in marine waters, with mean levels well in excess of the Dutch Environmental Quality Target (EQT) of 10 ng l^{-1}. The lack of success of these regulations was attributed primarily to the release of butyltins from the reservoir of contaminated sediment. Marinas which were dredged frequently were generally found to contain less dissolved butyltin. Presumably, dredging removes TBT-contaminated sediment and, because of sediment–water partitioning, eventually results in lower TBT concentrations in the water phase. Ritsema (1994) also observed that in some marinas, dissolved TBT concentrations peaked in summertime. Fresh inputs of TBT, from the illegal use of tin-based paints, may be occurring; this interpretation is supported by higher ratios, during these periods, of TBT to its degradation products DBT and MBT. An alternative explanation is that increased boat traffic in summer months resuspends sediment, facilitating desorption.

In Arcachon Bay, France, it has been estimated that between 100 and 1000 kg of TBT (as Sn) remains stored in the upper 30 cm of the sediment column (Sarradin *et al.*, 1994). Restrictions on TBT-based paint use imposed in 1982, and were rapidly effective in allowing the oyster industry to resume normal production. However, both low-level TBT contamination in the water and associated oyster shell malformations have persisted (Alzieu *et al.*, 1986, 1989). Suspicion is growing that the sedimentary reservoir of TBT contamination may be responsible, and a correlation was found between the proportion of shell malformations and TBT levels in the sediment (Sarradin *et al.*, 1994).

9.4.2   Poor water exchange

It is self-evident that areas such as embayments, with limited tidal flushing, will tend to retain contamination longer than more open areas with vigorous water exchange. As a worst-case example, Dyrynda (1992) transplanted Pacific oysters to Holes Bay, a land-locked basin with a high boat density adjoining Poole Harbour, Dorset. These oysters were characterised by abnormal shell thickening, subsequent to the 1987 TBT restrictions, illustrating that unfavourable hydrodynamic conditions can hinder recovery. Similarly, high concentrations of dissolved butyltins, arising from desorption from contaminated sediments, were associated with poor tidal flushing in Dutch marinas, following the 1990 ban (Ritsema, 1994). On the other hand, the recovery of coastal dogwhelk populations (Evans *et al.*, 1991, 1994) has been more rapid than predicted. This may be due to their rocky shore habitat, where large volumes of ocean water are available for dilution of seawater TBT concentrations to levels where juveniles can survive.

9.4.3   Regional-scale inconsistencies in control measures

Contaminants may be mobile; more importantly in the case of TBT, boats themselves are mobile. Thus, control measures imposed by a country may be compromised by visits of boats painted with TBT-based antifoulant in ports where no controls are imposed. A good example of this scenario is France. On the Atlantic coast, control measures imposed in 1982 rapidly reduced TBT contamination. However, on the Mediterranean coast, where the same regulations apply, coastal seawater TBT contamination is a continuing cause for concern (Alzieu, 1991), probably because of the free exchange of boats in the Mediterranean Sea. By 1991, most European Community nations had followed the lead of France in restricting TBT-based marine antifoulant use, but many other Mediterranean countries have no such controls (Gabrielides *et al.*, 1990; Aboul Dahab *et al.*, 1990). Other regions where an inconsistent approach may delay recovery are the Baltic Sea and the Caribbean Sea.

9.4.4   Poor dockyard practices

Drydocks, boat washdown yards and slipways, where coats of marine antifouling paint are applied and removed, are well known to be important point sources of TBT to the marine environment (Waldock and Thain, 1988; Stewart and de Mora, 1992; de Mora *et al.*, 1995). 'Hotspots' of contamination frequently occur in adjacent sediments (Waldock, Thain and Waite, 1987; de Mora *et al.*, 1995), and TBT-induced biological damage in the vicinity, such as Pacific oyster shell malformation, can be severe. In a survey of the inputs of TBT to the marine environment from shipping activities, Waldock and Thain (1988) concluded that 'unregulated drydocks practices clearly result in the release of large quantities of TBT'. In 1990, the Marine Environment Protection Committee of the International Maritime Organisation resolved to monitor carefully the entry of TBT into the sea via hull-cleaning operations.

To facilitate compliance with the newly established Environmental Quality Target of $2 \, ng \, l^{-1}$ TBT-Sn in seawater, a voluntary code of practice for British drydock operators was formulated in 1988. In New Zealand, the Navy recently implemented a wastewater collection and treatment system for the effluent from the large drydock at the Devonport base in Auckland. Sediment cores collected in the vicinity of the outfall from this drydock showed evidence of sporadic but appreciable discharges of TBT-laden waste waters (de Mora *et al.*, 1995).

9.4.5 Continued use of TBT by large vessels in deep-sea trade
The use of TBT as an antifouling toxin is declining rapidly, particularly in the pleasure craft sector. However, its use is still widespread within the shipping sector, which in fact accounts for 90% of sales by volume (Lloyd's Register Review, 1994). In particular, the continued use of TBT on large vessels in deep sea trade is likely to continue while TBT-based formulations maintain their performance advantage over tin-free alternatives. The Lloyd's Review did not consider an extension of the present TBT restrictions to a total ban justified on cost/benefit grounds, as in the deep sea trade sector the economic benefits of TBT are the greatest and the environmental impact considered to be the least.

TBT contamination is certainly a feature of waters and sediments in shipping ports (Maguire *et al.*, 1986; de Mora *et al.*, 1995; Gabrielides *et al.*, 1990; Higashiyama *et al.*, 1991), but very few studies to date have considered the environmental impact of TBT contamination from the shipping sector in isolation from other sources. Benthic whelks *Buccinum undatum* living in the open North Sea were recently discovered to be afflicted with imposex (ten Hallers-Tjabbes *et al.*, 1994); furthermore the frequency of imposex was found to correlate positively with shipping intensity, reaching its maximum at the entrance to the Euro Channel shipping lane.

In one particularly well monitored case, the oil tanker terminal at Sullom Voe in the Shetland Islands, it has been possible to obtain a direct measure of the environmental impact of TBT from large ships, because there are essentially no pleasure craft or drydock facilities in the area (Bailey and Davies, 1988; Davies and Bailey, 1991). Low but measurable concentrations of TBT were recorded near the tanker jetties, and dogwhelk populations are gradually being eliminated in the Voe itself. However, outside the Voe in the open waters of Yell Sound one mile to the north, imposex was much less severe, and a further four miles away, had declined to background levels. This, then, is the direct environmental impact for one busy harbour to be weighed against the economic benefits of TBT copolymer paint use (Lloyd's Register Review, 1994).

9.5   Overview of other toxic organotin compounds

Organotin compounds in general are known to be potent toxins, and this property has led to their use as the active ingredient in a wide variety of biocides. Triphenyltin and tricyclohexyltin compounds have been used in agricultural fungicides and miticides, and tributyltin, although best known as an antifouling toxin, has also been used as a slimicide in industrial cooling systems, a timber preservative and molluscicide for schisto-somiasis control. Certainly, the widespread use of TBT-based antifouling paints has had the greatest environmental significance, although localised instances of heavy TBT loadings to the marine environment have included the spillage of timber preservatives (de Mora and Phillips, 1992) and the discharge of industrial cooling waters (Gabrielides et al., 1990). In general, though, there is not the same potential for the large-scale broadcasting of TBT from these sources.

Several other organotin compounds have been detected in coastal and estuarine waters. Dibutyltin and monobutyltin are frequently associated with tributyltin and probably stem from its degradation; tetrabutyltin has also been detected. The origin of this compound may be as an impurity in the manufacture of TBT; another possibility is via a redistribution reaction of TBT in the surface microlayer (Matthias et al., 1986). Other organotins occurring environmentally include mono-, di- and trimethyltin, which probably originate from in situ biomethylation in anaerobic sediments (Dowson et al., 1992a; Maguire et al., 1986).

In general, agriculturally applied organotins such as triphenyltin (TPT) and tricyclohexyltin (TCT) do not have much significance for the marine environment. Like TBT, these compounds are extremely surface active; they tend to adsorb rapidly and completely to soils and have little tendency to leach into ground and surface waters. However, the direct introduction of TPT into the water column via its substitution for TBT as an antifouling toxin is of considerably more concern. In recognition of this possibility, some countries such as the United Kingdom, Canada, the USA and New Zealand legislated against all triorganotin compounds in marine paint formulations whereas other countries specifically restricted only TBT compounds. One such country is Japan, and reports of marine phenyltin contamination are becoming common (Hashimoto *et al.*, 1990; Higashiyama *et al.*, 1991; Shiraishi and Soma, 1992; Horiguchi *et al.*, 1994). TPT accumulation quite frequently exceeds TBT accumulation in sentinel bivalves, and its persistence in the marine environment may be even greater than that of TBT, according to the ratio of the tri-form to its di- and mono-degradation products (Higashiyama *et al.*, 1991). TPT contamination has also been reported in Mediterranean coastal waters (Alzieu *et al.*, 1991; Tolosa *et al.*, 1992).

The toxicological significance of these loadings of triphenyltin is yet to be established. The aquatic toxicity of TPT is comparable to that of TBT on the zooea of the mud crab *Rhithropanopeus harrisi* (Laughlin *et al.*, 1985) and on the ophiuroid *Ophiderma brevispina* (Walsh *et al.*, 1986). However, TPT showed no tendency to induce imposex in dogwhelks *N. lapillus*, even when injected directly. Horiguchi *et al.* (1994) noted that tissue concentrations of TPT in two species of Japanese dogwhelks correlated positively with imposex intensity, but this is probably because TPT also covaried with TBT.

9.6   Hazards of non-organotin-based antifoulants

In a regime of waning popularity of organotin-based antifouling formulations, care must be taken to avoid the substitution of TBT by other biocides which may damage the marine environment. Some studies addressing these concerns are listed in Table 9.5.

The use of copper as the active ingredient in marine antifouling formulations dates back to the use of copper sheeting on ship hulls. Sir Humphrey Davy discovered that the dissolution of copper into seawater was the mode of action, and since then copper has been incorporated in various ionic forms in marine antifouling paints. However, despite this historical and widespread use, there have been few reports of copper-induced damage to the marine environment. The accumulation of copper in

Table 9.5 *Alternative antifouling biocides and additives, and their possible hazards*

Active ingredient	Reference	Possible hazards
Copper	Lloyd's Register Review (1994)	The use of copper as an antifouling biocide is historic and ubiquitous. Elevated levels of copper in sediments and biota have been associated with high densities of pleasure boats and poor water exchange; but there are few, if any, reports of copper-induced environmental degradation. Copper speciation, and hence toxicity, is governed by pH and salinity. In freshwater lakes, particularly of low pH, copper toxicity may be of concern.
Tetracycline	Peterson *et al.* (1993)	Tetracycline, an antibiotic, has been advocated as an additive to antifouling paint. Is was found in this study to be rapidly leached and degraded in seawater, and is unlikely to be either an effective antifouling biocide or a marine environmental hazard.
Triazine	Readman *et al.* (1993)	The triazine Irgarol 1051 is a herbicidal additive in some antifouling preparations. In a survey of coastal Mediterranean seawater, 16 out of 20 samples were contaminated with Irgarol 1051. Levels were highly variable, ranging up to $1700\,\mathrm{ng\,l^{-1}}$, and were obviously related to boating activity. The environmental impact of these levels is unknown due to a lack of knowledge about sublethal effects of Irgarol 1051.
	Gough *et al.* (1994)	Irgarol 1051 was detected in all coastal, estuarine and marina (but not riverine) samples in a survey of southern England coastal waters. Highest levels were found in marinas.

Table 9.5 *continued*

Active ingredient	Reference	Possible hazards
		Sediments were also found to be contaminated with Irgarol 1051, suggesting that partitioning onto particulate matter may be an important influence on the compound's fate. In contrast, low levels of the agriculturally derived triazine herbicides simazine and atrazine, were found in river and estuarine waters only.
Naturally occurring marine chemicals	CEFIC review (1994)	Promising results are reported for some algal extracts, which are thought to act as repellents. No hazards are known for these compounds.
		Coatings containing powdered chilli, insect extract and iodine compounds were not found to be effective.

sediments and marine organisms is known (Young, Alexander and McDermott-Ehrlich, 1979; Claisse and Alzieu, 1993) but, as with elevated TBT levels, the problem is largely related to high boating intensities or poor tidal exchange.

Copper speciation is governed by pH, salinity and dissolved organic matter. In seawater the predominant forms of dissolved copper are expected to be Cu^{2+}, $CuOH^+$ and $CuCl^+$, but at pH 8.2 solubility is low and copper is precipitated and found in sediments, where it is considered to be relatively nontoxic. Organic chelates are also relatively nontoxic and there is no evidence of significant biotransformation processes, such as those known for mercury and arsenic. Copper is not lipophilic and only shows a slight tendency for bioaccumulation. It can be concluded that from an environmental impact point of view, there are definite advantages in switching from TBT-based to copper-based products. One possible exception is in freshwaters of low pH, where more toxic dissolved forms of copper may predominate.

One of the most intractable problems with copper-based formulations is that marine algae are frequently copper-tolerant, and it has been a common practice to add herbicidal boosters to inhibit the growth of algal slimes. The use of such compounds is of concern, as their environmental profiles are generally unknown. A good example is the herbicidal booster Irgarol 1051, an s-triazine closely related to the agricultural herbicides simazine and atrazine. A recent report has highlighted significant levels of Irgarol 1051 contamination in Mediterranean coastal waters (Readman et al., 1993), with seawater concentrations of up to $1700 \, \mathrm{ng} \, l^{-1}$. Coastal and estuarine waters in Southern England were also found to contain Irgarol 1051, with the intensity of contamination related to boating activity. The toxicological significance of Irgarol 1051 contamination is not known, although Environmental Quality Standard (EQS) levels have been set for the sum of both atrazine and simazine, at $2 \, \mu\mathrm{g} \, l^{-1}$ as an annual average and $10 \, \mu\mathrm{g} \, l^{-1}$ as a maximum allowable concentration.

9.7   Summary

In view of the accelerating degradation of coastal ecosystems attributed to TBT, there can be no doubt that legislative measures to restrict TBT input were necessary. The success of these restrictions has been evaluated, and it is evident that full recovery of the marine environment to pre-TBT levels is conditional upon a number of factors. Some, such as hydrodynamic conditions, are clearly beyond our control, but there is considerable scope for mitigation by improved management. The elimination of TBT entering waters from drydock and hull-washing facilities, as recommended by the IMO in 1990, is one example. However, on a global scale, the areas now in the greatest need of protection are the coastal tropics, and the most important management challenge is in implementing a ban on the use of TBT paints on coastal craft worldwide.

9.8   References

Aboul Dahab, O., El-Sabrouti, M. A. and Halim, Y. (1990) Tin compounds in sediments of Lake Maryut, Egypt. *Environmental Pollution* **63**, 329–44.

Adelman, D., Hinga, K. R. and Pilson, M. E. Q. (1990) Biogeochemistry of butyltins in an enclosed marine ecosystem. *Environmental Science and Technology* **24**, 1027–32.

Alzieu, Cl. (1991) Environmental problems caused by TBT in France: assessment, regulations, prospects. *Marine Environmental Research* **32**, 7–17.

Alzieu, Cl., Michel, P., Tolosa, I., Bacci, E., Mee, L. D. and Readman, J. W. (1991) Organotins in the Mediterranean: a continuing cause for concern. *Marine Environmental Research* **32**, 261–70.

Alzieu, C., Sanjuan, J., Deltreil, J. P. and Borel, M. (1986) Tin contamination in Arcachon Bay: effects on oyster shell anomalies. *Marine Pollution Bulletin* **17**, 494–8.

Alzieu, C., Sanjuan, J., Michel, P., Borel M. and Dreno, J.P. (1989) Monitoring and assessment of butyltins in Atlantic coastal waters. *Marine Pollution Bulletin* **20**, 22–6.

Astruc, M., Lavigne, R., Pinel, R., Leguille, F., Desauziers, V., Quevauviller, P. and Donard, O. (1989) Speciation of tin in sediments of Arcachon Bay, France. From *Proceedings of the 2nd International Symposium on Metal Speciation, Separation and Recovery*, Volume 2, Patterson, J.W. and Passino, R. (eds), Lewis Publishers, pp. 263–81.

Bailey, S. K. and Davies, I. M. (1988) Tributyltin contamination around an oil terminal in Sullom Voe (Scotland). *Environmental Pollution* **55**, 161–72.

Bailey, S. K. and Davies, I. M. (1991) Continuing impact of TBT, previously used in mariculture, on dogwhelk [*Nucella lapillus* L.] populations in a Scottish sea loch. *Marine Environmental Research* **32**, 187–99.

Batley, G. E., Scammell, M. S. and Brockbank, C. I. (1992) The impact of the banning of tributyltin-based antifouling paints on the Sydney rock oyster, *Saccostrea commercialis. The Science of the Total Environment* **122**, 301–14.

Bryan, G. W., Burt, G. R., Gibbs, P. E. and Pascoe, P. L. (1993) *Nassarius reticulatus* (Nassaridae: Gastropoda) as an indicator of tributyltin pollution before and after TBT restrictions. *Journal of the Marine Biological Association of the U.K.* **73**, 913–29.

Bryan, G. W. and Gibbs, P. E. (1991) Impact of low concentrations of tributyltin (TBT) on marine organisms: a review. *Metal Toxicology: Concepts and Applications.* Boston, Ann Arbor, pp. 323–61.

Bryan, G. W., Gibbs, P. E., Hummerston, L. G. and Burt, G. R. (1986) The decline of the gastropod *Nucella lapillus* around south-west England: evidence for the effect of tributyltin from antifouling paints. *Journal of the Marine Biological Association of the U.K.* **66**, 611–40.

CEFIC (Conseil European des Federations de l'Industrie Chimique), 1994. *Use of organotin compounds in antifouling paints.* Paper presented at 35th Session of the Marine Environment Protection Committee, IMO, London, March 7–11.

Claisse, D. and Alzieu, Cl. (1993) Copper contamination as a result of antifouling paint regulations? *Marine Pollution Bulletin* **26**, 395–7.

Cleary, J. J. (1991) Organotin in the marine surface microlayer and sub-surface waters of south-west England: relation to toxicity thresholds and the UK environmental quality standard. *Marine Environmental Research* **32**, 213–22.

Coghlan, A. (1990) Lethal paint makes for the open sea. *New Scientist* **128**, 16.

Davies, I. M. and Bailey, S. K. (1991) The impact of tributyltin from large vessels on dogwhelk (*Nucella lapillus* L.) populations around Scottish oil ports. *Marine Environmental Research* **30**, 297–322.

de Mora, S. J., King, N. G. and Miller, M. C. (1989) Tributyltin and total tin in marine sediments: profiles and the apparent rate of TBT degradation. *Environmental Technology Letters* **10**, 901–8.

de Mora, S. J. and Phillips, D. (1992) *The concentration of tri(n-butyl)tin in sediments from the Waikumete Stream and Henderson Creek following an accidental spill from a timber treatment works in West Auckland: a resurvey 6 months after the discharge.* Auckland Uniservices Report 4560, University of Auckland, Private Bag 92019, Auckland, New Zealand.

de Mora, S. J., Stewart, C. and Phillips, D. (1995) Sources and rate of degradation of tri(n-butyl)tin in marine sediments near Auckland, New Zealand. *Marine Pollution Bulletin* **30**, 50–7.

Dixon, T. (1989) Early response to TBT ban. *Marine Pollution Bulletin* **20**, 2.

Douglas, E. W., Evans, S. M., Frid, C. L. J., Hawkins, S. T., Mercer, T. S. and Scott, C. L. (1993) Assessment of imposex in the dogwhelk *Nucella lapillus (L.)* and tributyltin along the northeast coast of England. *Invertebrate Reproduction and Development* **24**, 243–8.

Dowson, P. H., Bubb, J. M. and Lester, J. N. (1992a) Organotin distribution in sediments and waters of selected East Coast estuaries in the UK. *Marine Pollution Bulletin* **24**, 492–8.

Dowson, P. H., Bubb, J. M. and Lester, J. N. (1993a) Temporal distribution of organotins in the aquatic environment: five years after the 1987 UK retail ban on TBT-based antifouling paints. *Marine Pollution Bulletin* **26**, 487–94.

Dowson, P. H., Bubb, J. M. and Lester, J. N. (1993b) Depositional profiles and relationships between organotin compounds in freshwater and estuarine sediment cores. *Environmental Monitoring and Assessment* **28**, 145–60.

Dowson, P. H., Bubb, J. M. and Lester, J. N. (1994) The effectiveness of the 1987 Retail Ban on TBT-based antifouling paints in reducing butyltin concentrations in East Anglia, UK. *Chemosphere* **28**, 905–10.

Dowson, P. H., Pershke, D., Bubb, J. M. and Lester, J. N. (1992b) Spatial distribution of organotins in sediments of lowland river catchments. *Environmental Pollution* **76**, 259–66.

Dyrynda, E. A. (1992) Incidence of abnormal shell thickening in the Pacific Oyster *Crassostrea gigas* in Poole Harbour (UK), subsequent to the 1987 TBT restrictions. *Marine Pollution Bulletin* **24**, 156–63.

Ellis, D. V. and Pattisina, L. A. (1990) Widespread neogastropod imposex: a biological indicator of global TBT contamination? *Marine Pollution Bulletin* **21**, 248–53.

Evans, S. M., Hawkins, S. T., Porter, J. and Samosir, A. M. (1994) Recovery of dogwhelk populations on the Isle of Cumbrae, Scotland, following legislation limiting the use of TBT as an antifoulant. *Marine Pollution Bulletin* **28**, 15–17.

Evans, S. M., Hutton, A., Kendall, M. A. and Samosir, A. M. (1991) Recovery in populations of dogwhelks *Nucella lapillus* (L.) suffering from imposex. *Marine Pollution Bulletin* **22**, 331–3.

François, R., Short, F. T. and Weber, J. H. (1989) Accumulation and persistence of tributyltin in eelgrass (*Zostera marina* L.) tissue. *Environmental Science and Technology* **23**, 191–6.

Gabrielides, G. P., Alzieu, Cl., Readman, J. W., Bacci, E., Aboul Dahab, O. and Salihoglu, I. (1990) MEDPOL survey of organotins in the Mediterranean. *Marine Pollution Bulletin* **21**, 233–7.

Garcia-Romero, B., Wade, T. L., Salata, G. G. and Brooks, J. M. (1993) Butyltin concentrations in oysters from the Gulf of Mexico from 1989 to 1991. *Environmental Pollution* **81**, 103–11.

Gibbs, P. E., Bryan, G. W. and Pascoe, P. L. (1991) TBT-induced imposex in the dogwhelk, *Nucella lapillus*: geographical uniformity of the response and effects. *Marine Environmental Research* **32**, 79–87.

Goldberg, E. D. (1986) TBT: an environmental dilemma. *Environment* **28**, 17–44.

Gough, M. A., Fothergill, J. and Hendrie, J. D. (1994) A survey of southern England coastal waters for the s-triazine antifouling compound IRGAROL 1051. *Marine Pollution Bulletin* **28**, 613–20.

Hasan, M. A. and Juma, H. A. (1992) Assessment of tributyltin in the marine environment of Bahrain. *Marine Pollution Bulletin* **24**, 408–10.

Hashimoto, S., Koshikawa, Y., Serizawa, Y. and Otsuki, A. (1990) Monitoring of organic tin compounds in Inland Sea by transplant of reference blue mussels. In *Proceeding of Oceans '90*, Marine Technology Society and IEEE Ocean Engineering Society, 1990, pp. 1049–53.

Higashiyama, T., Shiraishi, H., Otsuki, A. and Hashimoto, S. (1991) Concentrations of organotin compounds in blue mussels from the wharves of Tokyo Bay. *Marine Pollution Bulletin* **22,**, 585–7.

Horiguchi, T., Shiraishi, H., Shimuzu, M. and Morita, M. (1994) Imposex and organotin compounds in *Thais clavigera* and *T. bronni* in Japan. *Journal of the Marine Biological Association of the U.K.* **74**, 651–69.

Iwata, H., Tanaba, S., Miyazaki, N. and Tatsukawa, R. (1994) Detection of butyltin compound residues in the blubber of marine mammals. *Marine Pollution Bulletin* **28**, 607–12.

Langston, W. J., Bryan, G. W., Burt, G. R. and Gibbs, P. E. (1990) Assessing the impact of tin and TBT in estuaries and coastal regions. *Functional Ecology* **4**, 433–43.

Langston, W. J. and Burt, G. R. (1991) Bioavailability and effects of sediment-bound TBT in deposit-feeding clams, *Scrobicularia plana*. *Marine Environmental Research* **32**, 61–77.

Laughlin, R. B., Johannesen, R. B., French, W., Guard, H. and Brinckman, F. E. (1985) Structure–activity relationships for organotin compounds. *Environmental Toxicology and Chemistry* **4**, 343–51.

Laughlin, R. B. and Linden, O. (1987) Tributyltin – contemporary environmental issues. *Ambio* **16**, 252–6.

Law, R. J., Waldock, M. J., Allchin, C. R., Laslett, R. E. and Bailey, K. J. (1994) Contaminants in seawater around England and Wales: results from monitoring surveys, 1990–1992. *Marine Pollution Bulletin* **28**, 668–75.

Lloyd's Register Review (1994) *Use of triorganotin compounds in antifouling paints: critical review of current and future marine antifouling coatings*. MEPC 35/Inf. 27, International Maritime Organisation, London.

Macdonald, R. W., Macdonald, D. M., O'Brien, M. C. and Gobeil, C. (1991) Accumulation of heavy metals (Pb, Zn, Cu, Cd), carbon and nitrogen in sediments from the Strait of Georgia, B.C., Canada. *Marine Chemistry* **34**, 109–35.

Maguire, R. J., Tkacz, R. J., Chau, Y. K. and Wong, P. T. S. (1986) Occurrence of organotin compounds in water and sediment in Canada. *Chemosphere* **15**, 253–74.

Matthias, C. L., Bellma, J. M., Olson, G. J. and Brinkman, F. E. (1986) Comprehensive method for determination of aquatic butyltin and butylmethyltin species at ultratrace level using simultaneous hydridisation-extraction with gas chromatography-flame photometric detection. *Environmental Science and Technology* **20**, 609–15.

Minchin, D., Duggan, C. B. and King, W. (1987) Possible effects of organotins on scallop recruitment. *Marine Pollution Bulletin* **18**, 604–8.

Petersen, S. M., Batley, G. E. and Scammell, M. S. (1993) Tetracycline in antifouling paints. *Marine Pollution Bulletin* **26**, 96–100.

Quevauviller, Ph., Donard, O. F. X. and Etcheber, H. (1994) Butyltin distribution in a sediment core from Arcachon Harbour (France). *Environmental Pollution* **84**, 89–92.

Readman, J. W., Kwong, L. L. W., Grondin, D., Bartocci, J., Villeneuve, J-P. and Mee, L. D. (1993) Coastal water contamination from a triazine herbicide used in antifouling paints. *Environmental Science and Technology* **27**, 1940–2.

Ritsema, R. (1994) Dissolved butyltins in marine waters of the Netherlands three years after the ban. *Applied Organometallic Chemistry* **8**, 5–10.

Sarradin, P. M., Astruc, A., Desauziers, V., Pinel, R. and Astruc, M. (1991) Butyltin pollution in surface sediments of Arcachon Bay after ten years of restricted use of TBT-based paints. *Environmental Technology* **12**, 537–43.

Sarradin, P. M., Astruc, A., Sabrier, R. and Astruc, M. (1994) Survey of butyltin compounds in Arcachon Bay sediments. *Marine Pollution Bulletin* **28**, 621–8.

Shiraishi, H. and Soma, M. (1992) Triphenyltin compounds in mussels in Tokyo Bay after restrictions of use in Japan. *Chemosphere* **24**, 1103–9.

Stewart, C. and de Mora, S. J. (1990) A review of the degradation of tri(*n*-butyl)tin in the marine environment. *Environmental Technology* **11**, 565–70.

Stewart, C. and de Mora, S. J. (1992) Elevated tri(*n*-butyl)tin concentrations in shellfish and sediments from Suva Harbour, Fiji. *Applied Organometallic Chemistry* **6**, 507–12.

Stewart, C., de Mora, S. J., Jones, M. R. L. and Miller, M. C. (1992) Imposex in New Zealand neogastropods. *Marine Pollution Bulletin* **24**, 204–9.

Stewart, C. and Thompson, J. A. J. (1994) Extensive butyltin contamination in southwestern coastal British Columbia, Canada. *Marine Pollution Bulletin* **28**, 601–6.

ten Hallers-Tjabbes, C. C., Kemp, J. F. and Boon, J. P. (1994) Imposex in whelks (*Buccinum undatum*) from the open North Sea: relation to shipping traffic intensities. *Marine Pollution Bulletin* **28**, 311–13.

Tolosa, I., Merlini, N., de Bertrand, N., Bayona, J. M. and Albaiges, J. (1992) Occurrence and fate of tributyl- and triphenyltin compounds in Western Mediterranean coastal enclosures. *Environmental Toxicology and Chemistry* **11**, 145–55.

Uhler, A. D., Durell, G. S., Steinhauer, W. G. and Spellacy, A. M. (1993) Tributyltin levels in bivalve mollusks from the east and west coasts of the United States: results from the 1988–1990 national status and trends Mussel Watch project. *Environmental Toxicology and Chemistry* **12**, 139–53.

Unger, M. A., MacIntyre, W. G. and Huggett, R. J. (1988) Sorption behaviour of tributyltin on estuarine and freshwater sediments. *Environmental Toxicology and Chemistry* **7**, 907–15.

Valkirs, A. O., Davidson, B., Kear, L., Fransham, R. L., Grovhoug, J. G. and Seligman, P. F. (1991) Long-term monitoring of tributyltin in San Diego Bay, California. *Marine Environmental Research* **32**, 151–67.

Wade, T. L., Garcia-Romero, B. and Brooks, J. M. (1991) Oysters as biomonitors in the Gulf of Mexico. *Marine Environmental Research* **32**, 233–41.

Waite, M. E., Waldock, M. J., Thain, J. E., Smith, D. J. and Milton, S. M. (1991) Reductions in TBT concentrations in UK estuaries following legislation in 1986 and 1987. *Marine Environmental Research* **32**, 89–111.

Waldock, M. J. and Thain, J. E. (1988) Inputs of TBT to the marine environment from shipping activity in the U.K. *Environmental Technology Letters* **9**, 999–1010.

Waldock, M. J., Thain, J. E., Smith, D. and Waite, M. E. (1990) The degradation of TBT in estuarine sediments. In *Third International Organotin Symposium Proceedings*, Monaco, 17–20 April 1990, IAEA, pp. 46–8.

Waldock, M. J., Thain, J. E. and Waite, M. E. (1987) The distribution and potential toxic effects of TBT in UK estuaries during 1986. *Applied Organometallic Chemistry* **1**, 287–301.

Walsh, G. E., McLaughlin, L. L., Louie, M. K., Deans, C. H. and Lores, E. M. (1986)

Inhibition of arm regeneration by *Ophioderma brevispina* (Echinodermata, Ophiuroidea) by tributyltin oxide and triphenyltin oxide. *Ecotoxicological and Environmental Safety* 12, 95–100.

Wuertz, S., Miller, C. E., Doolittle, M. M., Brennan, J. F. and Cooney, J. J. (1991) Butyltins in estuarine sediments two years after tributyltin use was restricted. *Chemosphere* 22, 1113–20.

Young, D. R., Alexander, G. V. and McDermott-Ehrlich, D. (1979) Vessel-related contamination of Southern California harbours by copper and other metals. *Marine Pollution Bulletin* 10, 50–6.

Index

○ ○